TEST BANK FOR CAMPBELL'S

BIOLOGY

FOURTH EDITION

DAN WIVAGG, Editor
Baylor University

CONTRIBUTORS

RICHARD DUHRKOPF
Baylor University

GARY FABRIS
Red Deer College, Alberta, Canada

REBECCA PYLES
East Tennessee State University

KURT E. REDBORG
Coe College, Cedar Rapids, Iowa

RICHARD STOREY
The Colorado College, Colorado Springs

The Benjamin/Cummings Publishing Company, Inc.

Menlo Park, California ■ Reading, Massachusetts
New York ■ Don Mills, Ontario ■ Wokingham, U.K. ■ Amsterdam
Bonn ■ Paris ■ Milan ■ Madrid ■ Sydney ■ Singapore ■ Tokyo
Seoul ■ Taipei ■ Mexico City ■ San Juan, Puerto Rico

Sponsoring Editor: Johanna Schmid
Associate Editor: Kim Johnson
Editorial Assistants: Hilair Chism, Tabinda Khan
Senior Production Editor: Larry Olsen
Copyeditor: Marjorie Wexler
Proofreader: Anna Reynolds Trabucco
Cover Designer: Yvo Riezebos
Senior Manufacturing Coordinator: Merry Free Osborn
Composition: Fog Press
Cover Photo: Copyright © John Sexton

ISBN 0-8053-1947-6

2 3 4 5 6 7 8 9 10—VG—99 98 97

The Benjamin/Cummings Publishing Company, Inc.
2725 Sand Hill Road
Menlo Park, California 94025

Preface

This test bank for the new fourth edition of Neil Campbell's *Biology* has been throughly revised from the third edition. For this edition, each question from the third edition test bank was reviewed by at least three people with both biology and editorial backgrounds. Most of the third edition questions have been retained, sometimes with modifications, and many new questions have been added, usually at the end of each chapter.

The new questions were written by the following group of biology teachers:

- **Richard Duhrkopf**, Assistant Professor, Department of Biology, Baylor University
- **Gary Fabris**, Instructor, Department of Natural Science, Red Deer College
- **Rebecca Pyles**, Assistant Professor, Department of Biological Sciences, East Tennessee State University
- **Kurt Redborg**, Associate Professor, Department of Biology, Coe College
- **Richard Storey**, Professor and Chair, Biology Department, The Colorado College
- **Dan Wivagg**, Professor and Director of Undergraduate Studies, Department of Biology, Baylor University

I thank them for their timely submission of new questions and their careful editing.

The questions in this test bank are based on the following formats:

1. **Five-Choice Completion Items.** This type is the most widely used. A problem is framed in the form of an incomplete statement or a question. Broad experience in many types of test programs has shown that this type of question is consistently good for discriminating between the able and the less able student. In writing this type, we attempted to create many items that demand thought rather than mere rote-recall answers.

2. **Classification Questions.** This type of question has five alternative answers that are used to answer all the questions in the set. The answers are related concepts, principles, graphs, illustrations, numbers, or words. This type of question provides a quick means of determining the test-taker's mastery of the topics presented in the stem.

3. **Laboratory and Experimental Sets.** In this type of question, a laboratory situation, data, graphs, or diagrams are presented, followed by five choices for completing the items. These items tend to test higher-order levels of reasoning requiring application, analysis, synthesis, and evaluation.

I have retained the question classification scheme used in the third edition test bank. These classifications are:

1. **Factual Recall:** Involves the least complex mental ability. Emphasizes the process of remembering, recognizing, or recalling the appropriate material.

2. **Conceptual Understanding**: Requires the type of understanding that an individual needs in order to understand a concept and relate it to one or more of the answer choices.

3. **Application:** Requires the student to analyze the material he or she has learned and apply it to one or more of the answer choices. If an application has been explicitly presented in class, a question based on the application may become a recall question for that class.

How To Use This Test Bank

The answer and question type are indicated in the left column of each page. Many questions from the third edition have been reassigned to other classifications, but remember that the classification of a question for *your* students depends entirely on what went on in *your* class. Remember also that the classification does not indicate the level of difficulty of a question. For example, a factual recall question about the nature of an enzyme in the pentose phosphate pathway would be much more difficult than an application question about the effects of relative humidity on transpiration rates.

The test bank contains an average of 52 questions per chapter (ranging from 32 to 76 questions per chapter). The organization of the test bank parallels that of the textbook. The illustrations in the test bank have been adapted from artwork in Campbell's *Biology*, Fourth Edition.

Be aware of the time your students will need to finish your exam. Conceptual Understanding and Application questions will take a considerable amount of time when approached by a thoughtful, prepared student. You should take the test yourself (usually while preparing the answer key) and allow the students three times as much time as you required.

Questions can be influenced by changes in vocabulary and format. If you alter a question, the correct answer often changes. Please feel free to change the questions as you see fit, but if you change a question, please reexamine your answer.

Adopters of *Biology*, Fourth Edition, may photocopy the test questions without permission of the publisher.

A computerized version of this test bank is also available in Macintosh (31955), IBM 3.5" disk (31953), and IBM 5.25" disk (31954) from your Benjamin/Cummings sales representative.

Test bank users should inform Benjamin/Cummings of problems with software or specific test bank questions. We welcome comments and suggestions in anticipation of an even better test bank for the fifth edition of Campbell's *Biology*.

Acknowledgments

I especially thank Hilair Chism, Editorial Assistant, for her cheerfulness and patience during the editorial process for the test bank. Her feedback was positive and encouraging, with gentle suggestions for changes, a style we teachers should emulate. Hilair has resigned from Benjamin/Cummings and entered a graduate program in scientific illustration, and so we biology teachers can expect to see more of her work in the future.

I also thank Tabinda Khan, who carefully read many questions under the duress of tight deadlines. Her queries and comments indicate that she carefully considered each question. Many questions are better because of her input, and under her guidance all the parts have come together at the appropriate times.

Anna Reynolds Trabucco proofread the page proofs and suggested many changes that improved accuracy, format, and style. I am indebted to her for her careful work.

Thanks are also due to Kim Johnson and Don O'Neal for their encouragement and advice. I appreciate their input as well.

Dan Wivagg, Editor

Contents

Chapter 1

b
Factual Recall

1.1 What characterizes a prokaryotic cell?
    a. the presence of mitochondria
    b. the lack of a membrane-enclosed nucleus
    c. the presence of a nucleus with no DNA
    d. the lack of ribosomes
    e. having a cell wall without a cell membrane

a
Conceptual
Understanding

1.2 What is the primary reason for including a control within the design of an experiment?
    a. To demonstrate in what way the experiment was performed incorrectly.
    b. To insure that the results obtained are due to a difference in only one variable.
    c. To provide more data so that one can perform a more sophisticated statistical analysis.
    d. To test the effect of more than one variable.
    e. To accumulate more facts that can be reported to other scientists.

a
Factual Recall

1.3 Evolution is "the biological theme that ties together all the others." This is because the process of evolution
    a. All of the below are appropriate answers.
    b. explains how organisms become adapted to their environment.
    c. explains the diversity of organisms.
    d. explains why all organisms have characteristics in common.
    e. explains why distantly related organisms sometimes resemble one another.

c
Conceptual
Understanding

1.4 Which of the following is a correct statement about the scientific method?
    a. It can only be done by someone with a Ph.D. degree.
    b. Its methods are substantially different from the way people normally process information from and about their environment.
    c. It organizes evidence and helps us predict what will happen in our environment.
    d. It distinguishes between good and bad.
    e. It requires expensive laboratory equipment.

a
Factual Recall

1.5 Which of the following does NOT comprise a logical hierarchy of organization?
    a. molecules, atoms, organelles, tissues, systems
    b. molecules, cells, tissues, organ systems, populations
    c. cells, tissues, organs, organ systems, organisms
    d. organisms, populations, communities, biomes, biosphere
    e. family, order, class, phylum, kingdom

e
Factual Recall

1.6 Which of the following statements is NOT true about all living things?
    a. They are made of cells or cell products.
    b. They are the products of evolution.
    c. Their composition includes carbon, hydrogen, oxygen, and nitrogen.
    d. They undergo growth and development.
    e. They have a cell wall as an outer boundary.

a
Factual Recall

1.7 A common first step in the scientific method is
a. formulation of testable hypotheses.
b. collecting data.
c. formulation of a theory.
d. a search for relevant materials in the library.
e. conducting a controlled experiment.

b
Conceptual
Understanding

1.8 Why did the popular press give Reznick and Endler's research on guppies a lot of attention?
a. Their research was an excellent example of how to use the hypothetico-deductive method in science.
b. They were able to document evolution over a relatively short time period.
c. They were able to show that the physical environment caused variation in life histories of wild guppy populations.
d. They proved that water temperature can lead to genetic differences in populations.
e. They showed that pike-cichlids prey mainly on reproductively mature adults and, therefore, cause individual guppy populations to mature more quickly.

d
Conceptual
Understanding

1.9 Which of the following are properties of ALL life forms?
a. heritable programs in the form of DNA
b. photosynthesis
c. growth and development
d. Only a and c are correct.
e. a, b, and c are correct.

d
Factual Recall

1.10 Which of the following levels in the hierarchy of biological organization include all of the other levels in the list?
a. organ system
b. organism
c. population
d. ecosystem
e. community

e
Factual Recall

1.11 Which of the following levels in the hierarchy of biological organization includes all of the other levels in the list?
a. organelles
b. biological molecules
c. cells
d. atoms
e. tissues

c
Factual Recall

1.12 With each step upward in the hierarchy of biological order, novel properties emerge that were not present at the simpler levels of organization. These emergent properties result from
a. vital forces that arise at each level.
b. the physical and chemical phenomena that operate only in living things.
c. the arrangement and interactions between components.
d. simple summation of the individual behavior of the component parts.
e. the emergence of supernatural forces.

Questions 1.13–1.16 will use the answers below. Each answer may be used once, more than once, or not at all.

    a. *Antonie van Leeuwenhoek*
    b. *Robert Hooke*
    c. *Matthias Schleiden and Theodor Schwann*
    d. *Barbara McClintock*
    e. *Charles Darwin*

b
Factual Recall

1.13 Discovered, described, and named cells.

a
Factual Recall

1.14 Discovered single-celled organisms.

d
Factual Recall

1.15 Used observation of inheritance in corn to describe what turned out to be a novel molecular mechanism.

c
Factual Recall

1.16 Concluded that all living things consist of cells.

b
Conceptual
Understanding

1.17 The cell theory is an example of a conclusion based on
    a. vitalistic inspiration.
    b. inductive reasoning.
    c. controlled experimentation.
    d. microscopic study of all organisms.
    e. application of evolutionary theory.

d
Conceptual
Understanding

1.18 Many different interactions between organisms and their environment form an ecosystem. The dynamics of an ecosystem must include which of the following?
    a. cycling of nutrients
    b. flow of energy from sunlight to producers
    c. biological magnification of pesticides
    d. Only a and b are correct.
    e. a, b, and c are correct.

a
Factual Recall

1.19 Bacteria are members of which kingdom?
    a. Monera
    b. Protista
    c. Plantae
    d. Fungi
    e. Animalia

Questions 1.20–1.24 will use the following answers. Each answer may be used once, more than once, or not at all.
  a. Monera
  b. Protista
  c. Fungi
  d. Plantae
  e. Animalia

d
Factual Recall

1.20 Almost all are photosynthetic.

b
Factual Recall

1.21 Consist mostly of unicellular eukaryotes.

c
Factual Recall

1.22 Eukaryotic decomposers.

e
Factual Recall

1.23 Most obtain food by ingestion.

a
Factual Recall

1.24 All members of this group are prokaryotic.

b
Factual Recall

1.25 Diversity is a hallmark of life. Biologists have so far identified and named how many different species?
  a. 50,000
  b. 1.5 million
  c. 30 million
  d. 300 million
  e. 1.2 billion

d
Conceptual
Understanding

1.26 Two species belonging to the same class must also belong to the same
  a. family.
  b. genus.
  c. order.
  d. phylum.
  e. species.

c
Factual Recall

1.27 Which of these taxonomic categories is the most inclusive?
  a. genus
  b. class
  c. phylum
  d. family
  e. order

d
Factual Recall

1.28 Which kingdom consists of eukaryotes that nourish themselves mainly by decomposing organic material?
a. Monera
b. Protista
c. Plantae
d. Fungi
e. Animalia

c
Conceptual
Understanding

1.29 All of the following observations and ideas are incorporated into Darwin's concept of natural selection EXCEPT:
a. Individuals of a population vary.
b. Reproductive potential exceeds what the environment can support.
c. A change in the environment will create an appropriate, heritable adaptation during the lifetime of individuals coping with that environment.
d. Through natural selection, a population may adapt to the environment over many generations.
e. Members of a population in a particular environment are unequal in their potential for leaving offspring.

e
Conceptual
Understanding

1.30 What do a fungus, a tree, and a human have in common?
a. They are all members of the same kingdom.
b. They are all prokaryotic.
c. They are all members of the same class.
d. They all have cell walls.
e. They are all composed of cells with nuclei.

c
Application

1.31 A biologist discovers an organism new to science that has numerous nuclei enclosed by a single cell membrane. Assuming she is a good scientist, what should she do next?
a. hold a press conference to announce that the cell theory is no longer valid
b. destroy the organism so that the cell theory will not be challenged
c. determine how the organism is related to other organisms with more typical cell structure
d. assume that the organism is a mutant life form, and thus is unimportant
e. hide the organism from other scientists so as to maximize the financial benefits from her discovery

b
Application

1.32 Maria and Bill go to a new restaurant and do not like the food they are served. Their hypothesis is that if they go to the restaurant again, they will not like the food. If they continue to follow the hypothetico-deductive method, what should they do next?
a. never go near the restaurant again and tell all their friends not to try it
b. go back to the restaurant several times and order different items
c. try some nearby restaurants instead
d. call the city health inspectors and encourage them to inspect the restaurant
e. get some friends to go to the restaurant and order what Maria and Bill didn't like

Chapter 2

2.1 The atomic number of neon is 10. Therefore it
  a. has 8 electrons in the outer electron shell.
  b. is inert.
  c. has an atomic mass of 10.
  d. Only a and b are correct.
  e. a, b, and c are correct.

2.2 An atomic form of an element containing different numbers of neutrons is
  a. an isotope.
  b. an ion.
  c. a polar atom.
  d. an isomer.
  e. radioactive.

2.3 Oxygen has an atomic number of 8. Therefore, it must have
  a. 8 protons.
  b. 8 electrons.
  c. 8 neutrons.
  d. Only a and b are correct.
  e. a, b, and c are correct.

2.4 What are the chemical properties of atoms whose outer electron shells contain eight electrons?
  a. They form ionic bonds in aqueous solutions.
  b. They form covalent bonds in aqueous solutions.
  c. They are particularly stable and nonreactive.
  d. They tend to be gases.
  e. Both c and d are correct.

2.5 Each element is unique and different from other elements because of its
  a. atomic weight.
  b. atomic number.
  c. mass number.
  d. Only a and b are correct.
  e. a, b, and c are correct.

2.6 What do atoms form when they share electron pairs?
  a. elements
  b. ions
  c. aggregates
  d. isotopes
  e. molecules

Use the following choices to answer Questions 2.7–2.10. Each choice may be used once, more than once, or not at all.

    a. *nonpolar covalent molecule*
    b. *polar covalent bond*
    c. *ionic bond*
    d. *hydrogen bond*
    e. *hydrophobic interaction*

c
Factual Recall

2.7 Results from a transfer of electron(s) between atoms.

b
Factual Recall

2.8 Results from an unequal sharing of electrons between atoms.

d
Factual Recall

2.9 Best explains attraction of water molecules to each other.

a
Conceptual
Understanding

2.10 Would be least affected by the presence of water.

a
Conceptual
Understanding

2.11 Nitrogen (N) is much more electronegative than hydrogen (H). Which of the following statements is correct about ammonia (NH_3)?
    a. Each hydrogen atom has a partial positive charge.
    b. The nitrogen atom has a strong positive charge.
    c. Each hydrogen atom has a slight negative charge.
    d. The nitrogen atom has a partial positive charge.
    e. There are covalent bonds between the hydrogen atoms.

c
Conceptual
Understanding

2.12 What is the maximum number of covalent bonds an element with atomic number 15 can make with hydrogen?
    a. 1
    b. 2
    c. 3
    d. 4
    e. 5

c
Factual Recall

2.13 One difference between carbon-12 and carbon-14 is that carbon-14 has
    a. 2 more protons than carbon 12.
    b. 2 more electrons than carbon 12.
    c. 2 more neutrons than carbon 12.
    d. Only a and c are correct.
    e. a, b, and c are correct.

d
Factual Recall

2.14 When two atoms are equally electronegative, they will interact to form
    a. equal numbers of isotopes.
    b. ions.
    c. polar covalent bonds.
    d. nonpolar covalent bonds.
    e. ionic bonds.

c
Factual Recall

2.15 How do isotopes differ from each other?
  a. number of protons
  b. number of electrons
  c. number of neutrons
  d. valence electron distribution
  e. ability to form ions

d
Factual Recall

2.16 The combining properties of an atom depend on the number of
  a. valence shells in the atom.
  b. orbitals found in the atom.
  c. electrons in each orbital in the atom.
  d. electrons in the outer valence shell in the atom.
  e. hybridized orbitals in the atom.

a
Factual Recall

2.17 The atomic mass of an element can be easily approximated by adding together the number of
  a. protons and neutrons.
  b. electron orbitals in each energy level.
  c. protons and electrons.
  d. neutrons and electrons.
  e. isotopes of the atom.

c
Factual Recall

2.18 Atoms whose outer electron shells contain eight electrons tend to
  a. form ionic bonds in aqueous solutions.
  b. form covalent bonds in aqueous solutions.
  c. be particularly stable and nonreactive.
  d. be particularly unstable and very reactive.
  e. be biologically important since they are present in organic molecules.

Use the information extracted from the periodic table in Figure 2.1 to answer Questions 2.19–2.22.

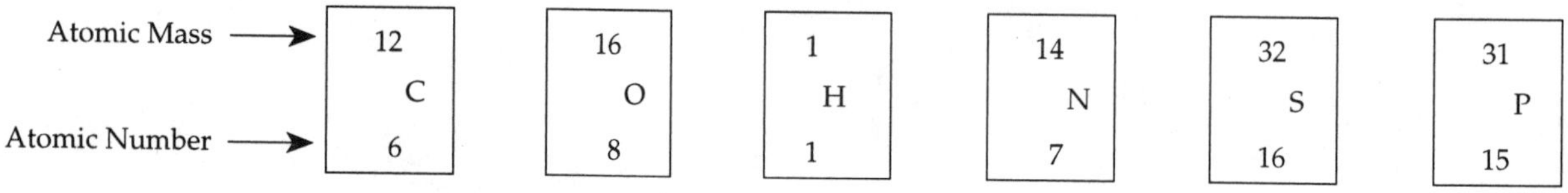

Atomic Mass ⟶

| 12 | 16 | 1 | 14 | 32 | 31 |
| C | O | H | N | S | P |

Atomic Number ⟶

| 6 | 8 | 1 | 7 | 16 | 15 |

Figure 2.1

a
Conceptual
Understanding

2.19 How many electrons does carbon have in its outermost (valence) energy level?
  a. 4
  b. 8
  c. 7
  d. 5
  e. 2

a
Conceptual
Understanding

2.20 How many neutrons does the nucleus of sulfur contain?
    a. 16
    b. 19
    c. 32
    d. 35
    e. 51

d
Conceptual
Understanding

2.21 Based on electron configuration, which of the elements would exhibit chemical behavior most like that of oxygen?
    a. C
    b. H
    c. N
    d. S
    e. P

e
Conceptual
Understanding

2.22 The atomic number of each atom is given to the left of each of the elements below. Which of the atoms has the same valence as carbon?
    a. 7 nitrogen
    b. 9 flourine
    c. 10 neon
    d. 12 magnesium
    e. 14 silicon

c
Factual Recall

2.23 Which of the following is a trace element that is essential to humans?
    a. nitrogen
    b. calcium
    c. iodine
    d. carbon
    e. oxygen

c
Conceptual
Understanding

2.24 Which of the following is a polar covalent bond?
    a. H–H
    b. C–C
    c. H–O
    d. C–H
    e. O–O

a
Factual Recall

2.25 A covalent bond is likely to be polar when
    a. one of the atoms sharing electrons is much more electronegative than the other atom.
    b. the two atoms sharing electrons are equally electronegative.
    c. the two atoms sharing electrons are of the same element.
    d. it is between two atoms that are both very strong electron acceptors.
    e. it joins a carbon atom to a hydrogen atom.

c
Conceptual
Understanding

2.26 What bond(s) is (are) easily disrupted in aqueous solutions?
    a. covalent
    b. polar covalent
    c. ionic
    d. Only a and b are correct.
    e. a, b, and c are correct.

d
Factual Recall

2.27 From its atomic number of 15, it is possible to predict that the phosphorus atom has
a. 15 neutrons.
b. 15 protons.
c. 15 electrons.
d. Only b and c are correct.
e. a, b, and c are correct.

a
Factual Recall

2.28 The ionic bond of sodium chloride is formed when
a. chlorine gains an electron from sodium.
b. sodium and chlorine share an electron pair.
c. sodium and chlorine both lose electrons from their outer energy levels.
d. sodium gains an electron from chlorine.
e. chlorine gains a proton from sodium.

c
Conceptual
Understanding

2.29 Which of the following best describes the relationship between the atoms described below?

atom 1	atom 2
6 protons	6 protons
6 neutrons	8 neutrons
6 electrons	6 electrons

a. They are isomers.
b. They are polymers.
c. They are isotopes.
d. They are ions.
e. They are both radioactive.

c
Factual Recall

2.30 How many electrons would be expected in the outer energy level of an atom with atomic number 17?
a. 2
b. 5
c. 7
d. 8
e. 17

d
Factual Recall

2.31 The atomic number of carbon is 6. ^{14}C is heavier than ^{12}C because ^{14}C has
a. six protons.
b. six neutrons.
c. eight protons.
d. eight neutrons.
e. fourteen electrons.

b
Conceptual
Understanding

2.32 Magnesium has the atomic number of 12. What kind of bonds does it form with chlorine (atomic number 17), and what is the formula for magnesium chloride?
a. covalent, MgCl
b. ionic, $MgCl_2$
c. covalent, Mg_2Cl
d. ionic, MgCl
e. ionic, ClMg

Questions 2.33–2.37 refer to Figure 2.2.

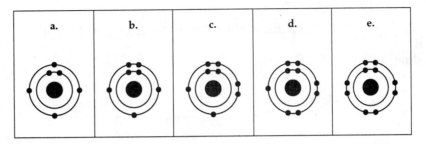

Figure 2.2

e
Factual Recall

2.33 Which of the drawings depicts the electron configuration of neon?

c
Factual Recall

2.34 Which of the drawings depicts the electron configuration of oxygen?

a
Factual Recall

2.35 Which of the drawings depicts the electron configuration of carbon?

c
Factual Recall

2.36 Which of the drawings is of an atom with the atomic number of eight?

e
Factual Recall

2.37 Which of the drawings depicts an atom which is inert?

Questions 2.38–2.42 refer to the following numbers. Each number may be used once, more than once, or not at all.
  a. 1
  b. 2
  c. 3
  d. 4
  e. 5

d
Conceptual
Understanding

2.38 The number of electrons carbon shares with oxygen molecules in a molecule of CO_2.

a
Conceptual
Understanding

2.39 The number of additional electrons needed to complete the valence shell of hydrogen.

a
Conceptual
Understanding

2.40 The valence of an atom with 7 electrons in its outer electron shell.

Chapter 2

ual
nding

2.41 The number of protons in an atom with the atomic number of 5.

ial
nding

2.42 The maximum number of electrons in the 1_s orbital.

ecall

2.43 Which four elements make up approximately 96% of living matter?
   a. carbon, hydrogen, nitrogen, oxygen
   b. carbon, sulfur, phosphorus, hydrogen
   c. oxygen, hydrogen, calcium, sodium
   d. carbon, sodium, chlorine, magnesium
   e. carbon, oxygen, sulfur, calcium

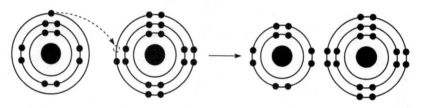

Figure 2.3

all

2.44 What results from the chemical reaction illustrated in Figure 2.3?
   a. isotopes
   b. ions
   c. a covalent bond
   d. a hydrogen bond
   e. hydrophobic interactions

ill

2.45 A covalent chemical bond is one in which
   a. electrons are removed from one atom and transferred to another atom so
       that the two atoms become oppositely charged.
   b. protons or neutrons are shared by two atoms so as to satisfy the
       requirements of both.
   c. outer shell electrons are shared by two atoms so as to satisfactorily fill the
       outer electron shells of both.
   d. outer shell electrons on one atom are transferred to the inner electron shells
       of another atom.
   e. the inner shell electrons of one atom are transferred to the outer shell of
       another atom.

d
Conceptual
Understanding

2.46 If atom $_6$X (atomic number 6) were allowed to react with hydrogen, the molecule formed would be

a.   b.   c.   d.   e.

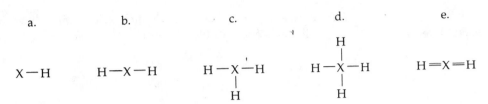

e
Conceptual
Understanding

2.47 What do the four elements most abundant in life—carbon, oxygen, hydrogen, and nitrogen—have in common?
a. They all have the same number of valence electrons.
b. Each element exists in only one isotopic form.
c. They are equal in electronegativity.
d. They are elements produced only by living cells.
e. They are all relatively light elements and can exist singly or in combination with other atoms as a gas.

a
Conceptual
Understanding

2.48 Which type of bond is important in large biological molecules?
a. All of the below are important.
b. covalent
c. hydrogen
d. ionic
e. polar covalent

a
Conceptual
Understanding

2.49 Which of the following best explains the distinction between biology and chemistry?
a. Biologists study living things while chemists study nonliving things.
b. Biology has a hierarchy of structural levels while chemistry does not.
c. Chemists study molecules while biologists do not.
d. Biological systems have emergent properties while chemical systems do not.
e. There is no clear distinction because the two sciences are parts of the same whole.

b
Factual Recall

2.50 Our best estimate is that most of the Earth's present atmosphere originated
a. from the gases present when the Earth formed.
b. from gases vented by volcanoes.
c. from gases captured by the Earth's gravity.
d. from radioactive decay and the impact of meteorites.
e. from lightning and ultraviolet radiation.

a
Conceptual
Understanding

2.51 Which of the following best describes a chemical equilibrium?
a. reactions continuing with no effect on the concentrations of reactants and products
b. concentrations of products are high
c. reactions have stopped
d. reactions stop only when all of the reactants have been converted to products
e. equal concentrations of reactants and products

Chapter 3

b
Factual Recall

3.1 Blueberries grow best in moderately acidic soil. What is an appropriate pH of a soil that is good for blueberries?
  a. 1.2
  b. 4.8
  c. 7.0
  d. 8.3
  e. 12.0

b
Factual Recall

3.2 The partial negative charge at one end of a water molecule is attracted to the partial positive charge of another water molecule. What is this attraction called?
  a. a covalent bond
  b. a hydrogen bond
  c. an ionic bond
  d. a hydration shell
  e. a hydrophobic bond

d
Factual Recall

3.3 Which bonds must be broken for water to vaporize?
  a. ionic bonds
  b. nonpolar covalent bonds
  c. polar covalent bonds
  d. hydrogen bonds
  e. Both c and d are correct.

c
Conceptual
Understanding

3.4 Which of the following liquids does not have a measurable pH?
  a. milk
  b. sea water
  c. gasoline
  d. orange juice
  e. distilled water

e
Conceptual
Understanding

3.5 Life on earth is dependent on all the properties of water as well as the abundance of water. Which property of water is probably most important for the functioning of organisms at the molecular level?
  a. cohesion and high surface tension
  b. high specific heat
  c. high heat of vaporization
  d. expansion upon freezing
  e. versatility as a solvent

b
Factual Recall

3.6 Which of the following is an example of a hydrogen bond?
  a. the bond between C and H in methane
  b. the bond between the H of one water molecule and the O of another water molecule
  c. the bond between Na and Cl in salt
  d. the bond between two hydrogen atoms
  e. the bond between Mg and Cl in $MgCl_2$

a
Factual Recall

3.7 Which of the following solutions has the greater concentration of hydrogen ions (H^+)?
   a. gastric juice at pH 2
   b. vinegar at pH 3
   c. tomatoes at pH 4
   d. black coffee at pH 5
   e. seawater at pH 8

e
Factual Recall

3.8 Which of the following solutions has the greater concentration of hydroxyl ions (OH^-)?
   a. gastric juice at pH 2
   b. vinegar at pH 3
   c. tomatoes at pH 4
   d. black coffee at pH 5
   e. seawater at pH 8

e
Factual Recall

3.9 A solution with a pH of 3 has how many more H^+ than a solution with a pH of 6?
   a. 2 times more
   b. 10 times more
   c. 100 times more
   d. 200 times more
   e. 1000 times more

b
Conceptual
Understanding

3.10 What would be the pH of a solution with a hydroxyl ion concentration (OH^-) of 10^{-10} M?
   a. 2
   b. 4
   c. 8
   d. 10
   e. 14

b
Factual Recall

3.11 Buffers are substances which help resist shifts in pH by
   a. releasing H^+ in acidic solutions.
   b. releasing H^+ in basic solutions.
   c. combining with H^+ in basic solutions.
   d. combining with OH^- in acidic solutions.
   e. releasing OH^- in basic solutions.

e
Factual Recall

3.12 Which of the following statements about water is correct?
   a. Water is more dense as a solid than it is as a liquid.
   b. Water is more dense at 100°C than it is at 37°C.
   c. Water is a good solvent for lipids.
   d. Compared to most other substances, the temperature of water rises sharply when it absorbs heat.
   e. Compared to most liquids, the evaporation of water requires a large amount of heat.

3.13 What do the following have in common with reference to water: cohesion, surface tension, specific heat?
   a. All are products of the structure of the hydrogen atom.
   b. All are produced by covalent bonding.
   c. All are properties related to hydrogen bonding.
   d. All have to do with polarity of water molecules.
   e. All are aspects of a semi-crystalline structure.

3.14 Assume that acid rain has lowered the pH of a particular lake to pH 5.0. What is the hydroxide ion concentration of this lake?
   a. 1×10^{-5} moles of hydroxide ion per liter of lake water
   b. 1×10^{-9} moles of hydroxide ion per liter of lake water
   c. 5.0 molar with regard to hydroxide ion concentration
   d. 9.0 molar with regard to hydroxide ion concentration
   e. Both b and d are correct.

3.15 The nutritional information on a cereal box shows that one serving of dry cereal has 90 "calories" (actually kilocalories). If one were to burn a serving of cereal, the amount of heat given off would be sufficient to raise the temperature of one kilogram of water how many degrees Celsius?
   a. 0.9 Celsius
   b. 9.0 Celsius
   c. 90.0 Celsius
   d. 900.0 Celsius
   e. 9000.0 Celsius

3.16 The formation of ice during colder weather helps to temper the seasonal transition to winter. This is mainly because
   a. the formation of hydrogen bonds releases heat.
   b. the formation of hydrogen bonds absorbs heat.
   c. there is less evaporative cooling of lakes.
   d. ice melts each autumn afternoon.
   e. ice is warmer than the winter air.

3.17 What bonds must be broken for water to go from a liquid to a gas?
   a. covalent
   b. polar covalent
   c. ionic
   d. hydrogen
   e. hydrophobic

18 Water's high specific heat is mainly a consequence of the
   a. small size of the water molecules.
   b. high specific heat of oxygen and hydrogen atoms.
   c. absorption and release of heat when hydrogen bonds break and form.
   d. fact that water is a poor heat conductor.
   e. inability of water to dissipate heat into dry air.

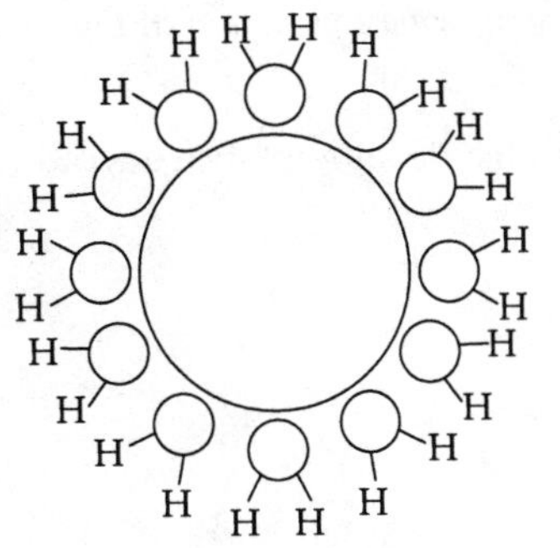

Figure 3.1

Question 3.19 is based on the diagram (Figure 3.1) of a solute molecule (large circle) surrounded by a shell of water. The small circles in the diagram depict the oxygen of the water.

a
Application

3.19 The solute molecule is most likely
   a. positively charged.
   b. negatively charged.
   c. neutral in charge.
   d. hydrophobic.
   e. polar.

e
Conceptual
Understanding

3.20 The molecular mass of glucose is 180 g. To make a one-molar solution of glucose, you should do which of the following?
   a. Dissolve 100 g of glucose in a liter of water.
   b. Dissolve 180 g of glucose in a gallon of water.
   c. Dissolve 180 g of glucose in 100 grams of water.
   d. Dissolve 180 mg (milligrams) of glucose in one liter of water.
   e. Dissolve 180 g of glucose in water, and then add more water until the total volume of the solution is one liter.

e
Factual Recall

3.21 In a lake contaminated by acid rain, fish generally die when the water is persistently below which pH?
   a. 8
   b. 7
   c. 6.5
   d. 6
   e. 5

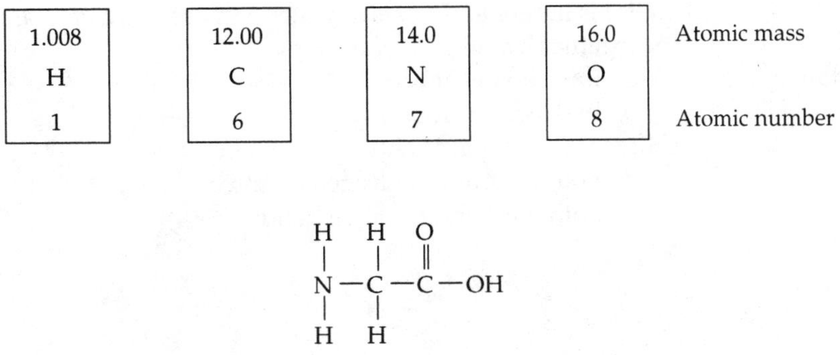

1.008	12.00	14.0	16.0	Atomic mass
H	C	N	O	
1	6	7	8	Atomic number

Figure 3.2

Questions 3.22 and 3.23 require the information shown in Figure 3.2.

c
Application

3.22 How many grams of the molecule shown above would constitute a mole of the substance?
    a. 32
    b. 40
    c. 75
    d. 114
    e. 6.02×10^{23}

c
Application

3.23 How would one make a 0.5 *M* solution of the molecule shown above?
    a. Mix 0.5 grams with enough water to yield 1 liter of solution.
    b. Mix 20 grams with enough water to yield 1 liter of solution.
    c. Dissolve 37.5 grams with enough water to yield 1 liter of solution.
    d. Dissolve 75 grams in 0.5 liter of water.
    e. Only a and d will yield a 0.5 *M* solution.

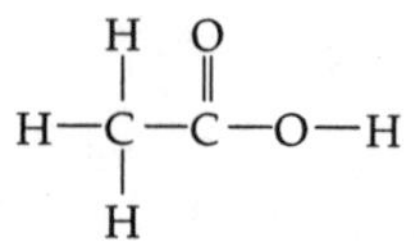

$$H-\overset{\displaystyle\overset{H}{|}}{\underset{\displaystyle\underset{H}{|}}{C}}-\overset{\displaystyle\overset{O}{\|}}{C}-O-H$$

Figure 3.3

b
Application

3.24 How many grams of the molecule in Figure 3.3 should one add to a liter of water to make a 0.2 *M* solution?
    a. 8
    b. 12
    c. 24
    d. 32
    e. 60

e
Factual Recall

3.25 Which of the following ionizes completely in solution and is therefore a strong acid?
    a. NaOH
    b. H_2CO_3
    c. CH_3COOH
    d. NH_2
    e. HCl

b
Conceptual
Understanding

3.26 Ice is lighter and floats in water because it is a crystalline structure held together by
    a. ionic bonds only.
    b. hydrogen bonds only.
    c. covalent bonds only.
    d. both ionic and hydrogen bonds.
    e. both ionic and covalent bonds.

e
Factual Recall

3.27 Assuming a temperature of 25° C, what does a pH of 7 indicate?
a. The solution is neutral.
b. The concentration of H^+ ions is 10^{-7} moles per liter.
c. There are no H^+ ions in solution.
d. There are no OH^- ions in solution.
e. Both a and b are correct.

d
Conceptual
Understanding

3.28 What does the energy to vaporize water do?
a. oxidizes the water
b. reduces (adds electrons to) the water molecules
c. decreases the number of hydrogen ions (H^+) in water
d. breaks hydrogen bonds between water molecules
e. decreases the density of water

d
Factual Recall

3.29 What determines the cohesiveness of water molecules?
a. hydrophobic interactions
b. high specific heat
c. covalent bonds
d. hydrogen bonds
e. ionic bonds

b
Factual Recall

3.30 It is correct to say that the action of buffers
a. is of relatively little significance in living systems.
b. tends to prevent great fluctuations in pH.
c. depends on the formation of a great number of hydrogen ions.
d. depends on the presence of many electron donors.
e. is to remove hydroxyl ions from organic acids.

c
Factual Recall

3.31 All of the following are true statements concerning hydrogen bonding EXCEPT:
a. In H-bonds, the hydrogen atom is also involved in a polar covalent bond.
b. H-bonds are responsible for the cohesive properties of water.
c. H-bonds are among the strongest of all chemical bonds.
d. H-bonds are rapidly formed and rapidly broken.
e. Large numbers of H-bonds confer considerable stability to a group of molecules.

c
Conceptual
Understanding

3.32 A given solution is found to contain 0.0001 mole of hydrogen ions (H^+) per liter. Which of the following best describes this solution?
a. acidic: H^+ acceptor
b. basic: H^+ acceptor
c. acidic: H^+ donor
d. basic: H^+ donor
e. neutral

b
Conceptual
Understanding

3.33 If the pH of a solution is decreased from 7 to 6, it means that the
a. concentration of H^+ has decreased to 1/10 of what it was at pH 7.
b. concentration of H^+ has increased to 10 times what it was at pH 7.
c. concentration of OH^- has increased to 10 times what it was at pH 7.
d. concentration of OH^- has increased by 1/7 of what it was.
e. solution has become more basic.

c
Conceptual
Understanding

3.34 What do cohesion, surface tension, and adhesion have in common with reference to water?
    a. All increase when temperature increases.
    b. All are produced by covalent bonding.
    c. All are properties related to hydrogen bonding.
    d. All have to do with nonpolar covalent bonds.
    e. Both a and c are correct.

e
Factual Recall

3.35 How many molecules of glycerol $C_3H_8O_3$ would be present in 1 liter of a 1 M glycerol solution?
    a. 1
    b. 14
    c. 92
    d. 1×10^{-7}
    e. 6.02×10^{25}

b
Factual Recall

3.36 Which of the following statements is true about buffer solutions? They
    a. will always have a pH of 7.
    b. tend to maintain a relatively constant pH.
    c. maintain a constant pH when bases are added to them but not when acids are added to them.
    d. cause a lowering of pH when acids are added to them.
    e. are rarely found in living systems.

Questions 3.37–3.39 refer to the following terms. Each term may be used once, more than once, or not at all.
    a. *calorie*
    b. *temperature*
    c. *heat of vaporization*
    d. *buffer*
    e. *mole*

b
Factual Recall

3.37 A measure of the average kinetic energy of the molecules in a body of matter.

e
Factual Recall

3.38 The number of grams of a substance that equals its molecular mass in daltons.

d
Factual Recall

3.39 A weak acid or base that combines reversibly with hydrogen ions.

c
Factual Recall

3.40 Which of the following is a correct definition of a kilocalorie?
    a. The amount of heat energy required to raise 1 gram of water by one degree Fahrenheit.
    b. The amount of heat energy required to raise one gram of water by ten degrees Celsius.
    c. The amount of heat energy required to raise one kilogram of water by one degree Celsius.
    d. A measure of the average kinetic energy of a pint of water.
    e. The amount of energy in one kilogram of glucose.

b
Factual Recall

3.41 At what temperature is water at its densest?
  a. 0°C
  b. 4°C
  c. 32°C
  d. 100°C
  e. 212°C

c
Factual Recall

3.42 Temperature usually increases when water condenses. Wh water is most directly responsible for this phenomenon?
  a. change in density when it condenses to form a liquid or
  b. reactions with other atmospheric compounds
  c. release of heat by formation of hydrogen bonds
  d. release of heat by breaking of hydrogen bonds
  e. high surface tension

a
Conceptual
Understanding

3.43 Water is able to form hydrogen bonds because
  a. All of the below are correct.
  b. the water molecule is shaped something like a right angl
  c. the water molecule is polar.
  d. the oxygen atom in a water molecule is weakly negative.
  e. the hydrogen atoms in a water molecule are weakly posi

d
Conceptual
Understanding

3.44 Why does ice float in liquid water?
  a. The liquid water molecules have more energy and can pi
  b. The hydrogen bonds between the molecules in ice prevei sinking.
  c. Ice always has air bubbles that keep it afloat.
  d. Hydrogen bonds keep the molecules of ice farther apart t water.
  e. The crystalline lattice of ice causes it to be denser than liq

d
Conceptual
Understanding

3.45 If liquid water molecules are hydrogen-bonded to one anoth water flows?
  a. The hydrogen bonds in liquid water are weaker than thos
  b. Hydrogen bonds form very slowly in the temperature ran
  c. In liquid water, the hydrogen bonds bend more easily tha
  d. The hydrogen bonds in liquid water are constantly breaki
  e. Only a few of the liquid water molecules actually form hy

Chapter 4

b
Conceptual
Understanding

4.1 A compound contains hydroxyl groups as its predominant functional group. Which of the following statements is true concerning this compound?
  a. It is probably a lipid.
  b. It should dissolve in water.
  c. It should dissolve in a nonpolar solvent.
  d. It won't form hydrogen bonds with water.
  e. It is hydrophobic.

b
Factual Recall

4.2 Which of the following elements is the most abundant (percent dry weight) in both humans and bacteria?
  a. oxygen
  b. carbon
  c. hydrogen
  d. nitrogen
  e. phosphorus

b
Factual Recall

4.3 What is the reason why hydrocarbons are not soluble in water?
  a. They are hydrophilic.
  b. The C–H bond is nonpolar.
  c. They do not ionize.
  d. They are large molecules.
  e. They are lighter than water.

a
Factual Recall

4.4 Which of the following is true of geometric isomers?
  a. They have variations in arrangement around a double bond.
  b. They have an asymmetric carbon that makes them mirror images.
  c. They have the same chemical properties.
  d. They have different molecular formulas.
  e. Their atoms and bonds are arranged in different sequences.

b
Conceptual
Understanding

4.5 Which property of the carbon atom gives it compatibility with a greater number of different elements than any other type of atom?
  a. Carbon has 6–8 neutrons.
  b. Carbon has a valence of 4.
  c. Carbon forms ionic bonds.
  d. Only a and c are correct.
  e. a, b, and c are correct.

c
Factual Recall

4.6 Organic chemistry is a science based on the study of
  a. functional groups.
  b. vital forces interacting with matter.
  c. carbon compounds.
  d. water and its interaction with other kinds of molecules.
  e. the properties of oxygen.

d
Conceptual
Understanding

4.7 How many electron pairs does carbon share in order to complete its valence shell?
  a. 1
  b. 2
  c. 3
  d. 4
  e. 5

c
Factual Recall

4.8 Glucose and fructose differ in
  a. the number of carbon, hydrogen, and oxygen atoms.
  b. the types of carbon, hydrogen, and oxygen atoms.
  c. the arrangement of carbon, hydrogen, and oxygen atoms.
  d. Only a and b are correct.
  e. a, b, and c are correct.

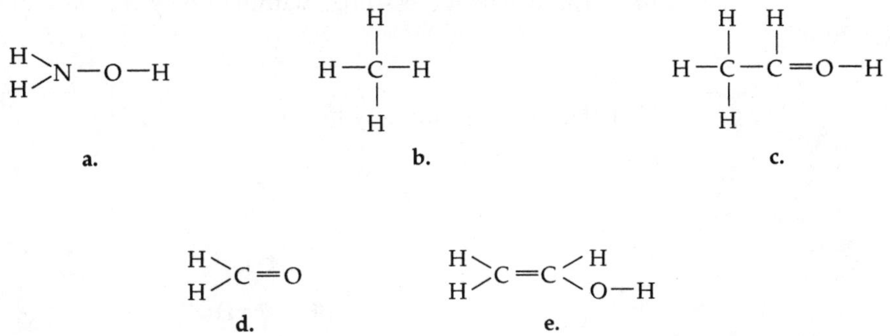

Figure 4.1

Questions 4.9 and 4.10 are based on the molecules in Figure 4.1.

c
Application

4.9 Which of the structures in Figure 4.1 is an impossible covalently bonded molecule?

d
Conceptual
Understanding

4.10 Which of the legitimate molecules in Figure 4.1 is likely to be the most soluble in water?

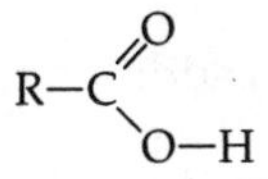

Figure 4.2

d
Factual Recall

4.11 What is the name of the functional group shown in Figure 4.2?
  a. carbonyl
  b. methyl
  c. dehydroxyl
  d. carboxyl
  e. acetyl

Questions 4.12–4.15 refer to the following list of functional groups. Each group may be used once, more than once, or not at all.

a. –OH
b. –C=O
c. –COOH
d. –NH₂
e. –SH

c
Factual Recall

4.12 Acidic, can dissociate and release H⁺.

d
Factual Recall

4.13 Basic, accepts H⁺ and becomes positively charged.

e
Factual Recall

4.14 Forms covalent cross-links within or between protein molecules.

a
Factual Recall

4.15 Polar, confers solubility in water.

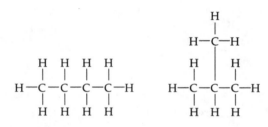

Figure 4.3

c
Factual Recall

4.16 The two molecules shown in Figure 4.3 are best described as
a. optical isomers.
b. radioactive isotopes.
c. structural isomers.
d. positive ions.
e. geometric isomers.

d
Conceptual
Understanding

4.17 Which of the following contains nitrogen in addition to carbon, oxygen, and hydrogen?
a. an alcohol such as ethanol
b. a compound of fats such as glycerol
c. a steroid such as testosterone
d. an amino acid such as glycine
e. a hydrocarbon such as benzene

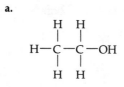

Figure 4.4

Questions 4.18–4.23 refer to the molecules shown in Figure 4.4.

d
Conceptual
Understanding

4.18 Which represents a molecule that increases the concentration of hydroge in a solution and is, therefore, an organic acid?

d
Factual Recall

4.19 Which molecule contains a carboxyl group?

b
Factual Recall

4.20 Which two of the molecules above contain a carbonyl group?
      a. a and b
      b. b and c
      c. c and d
      d. d and e
      e. a and e

c
Factual Recall

4.21 Which molecule has a carbonyl group in the form of a ketone?

a
Factual Recall

4.22 Which of the above is water soluble because it has a functional group that i alcohol?

e
Factual Recall

4.23 Which molecule contains a functional group known as an amine?

e
Factual Recall

4.24 Which is the best description of a carbonyl group?
      a. a carbon and hydrogen atom
      b. an oxygen double-bonded to a carbon and a hydroxyl group
      c. a nitrogen and a hydrogen bonded to a carbon atom
      d. a sulfur and a hydrogen bonded to a carbon atom
      e. a carbon atom joined to an oxygen atom by a double bond

d
Conceptual
Understanding

4.25 Which of the following is a FALSE statement concerning amine groups?
   a. They are basic.
   b. They are found in amino acids.
   c. They contain nitrogen.
   d. They are nonpolar.
   e. They are components of urea.

b
Application

4.26 How many structural isomers are possible for butane having the molecular formula C_4H_{10}?
   a. 1
   b. 2
   c. 4
   d. 5
   e. 8

c
Conceptual
Understanding

4.27 Which two functional groups are always found in amino acids?
   a. amine and sulfhydryl
   b. carbonyl and carboxyl
   c. carboxyl and amine
   d. alcohol and aldehyde
   e. ketone and amine

c
Factual Recall

4.28 Amino acids are acids because they possess which functional group?
   a. amino
   b. alcohol
   c. carboxyl
   d. sulfhydryl
   e. aldehyde

d
Application

4.29 A carbon skeleton is covalently bonded to both an amino group and a carboxyl group. When placed in water,
   a. it would function only as an acid because of the carboxyl group.
   b. it would function only as a base because of the amino group.
   c. it would function as neither an acid nor a base.
   d. it would function as both an acid and a base.
   e. it is impossible to determine how it would function.

c
Conceptual
Understanding

4.30 Which functional groups can act as acids?
   a. amine and sulfhydryl
   b. carbonyl and carboxyl
   c. carboxyl and phosphate
   d. alcohol and aldehyde
   e. ketone and amine

Questions 4.31–4.33 refer to the following functional groups. Each group may be used once, more than once, or not at all.
   a. –OH
   b. –C=O
   c. –COOH
   d. –NH$_2$
   e. –SH

a
Factual Recall

4.31 Which is an alcohol?

d
Factual Recall

4.32 Which is an amine?

c
Factual Recall

4.33 Which is a carboxyl group?

d
Application

4.34 A chemist wishes to make an organic molecule less acidic. Which of the following functional groups should she add to the molecule?
a. carboxyl
b. sulfhydryl
c. hydroxyl
d. amino
e. phosphate

c
Conceptual
Understanding

4.35 Which type of molecule would be most abundant in a typical cell?
a. hydrocarbon
b. protein
c. water
d. lipid
e. carbohydrate

c
Factual Recall

4.36 The concept of vitalism is based on a belief in a life force outside the jurisdiction of physical and chemical laws. According to this belief, organic compounds can arise only within living organisms. Which of the following did the most to refute the concept of vitalism?
a. Wohler's synthesis of urea
b. Berzelius's definition of organic molecules
c. Miller's experiments with ancient atmospheres
d. Rodriguez's studies of phytochemicals
e. Kolbe's synthesis of acetic acid

Chapter 5

c
Factual Recall

5.1 Which type of lipid is most important in biological membranes?
   a. fats
   b. steroids
   c. phospholipids
   d. oils
   e. triglycerides

d
Factual Recall

5.2 Which type of interaction stabilizes the alpha helix structure of proteins?
   a. hydrophobic interactions
   b. nonpolar covalent bonds
   c. ionic interactions
   d. hydrogen bonds
   e. polar covalent bonds

c
Factual Recall

5.3 Polymers of polysaccharides, fats, and proteins are all synthesized from monomers by
   a. connecting monosaccharides together.
   b. the addition of water to each monomer.
   c. the removal of water (condensation synthesis).
   d. ionic bonding of the monomers.
   e. the formation of disulfide bridges between monomers.

c
Factual Recall

5.4 Which of the following statements best summarizes structural differences between DNA and RNA?
   a. RNA is a protein while DNA is a nucleic acid.
   b. DNA is not a polymer, but RNA is.
   c. DNA contains a different sugar than RNA.
   d. RNA is a double helix, but DNA is not.
   e. DNA has different purine bases than RNA.

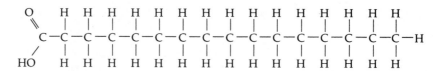

Figure 5.1

a
Factual Recall

5.5 What is the molecule illustrated in Figure 5.1?
   a. a saturated fatty acid
   b. an unsaturated fatty acid
   c. a polyunsaturated triglyceride
   d. a common component of plant oils
   e. similar in structure to a steroid

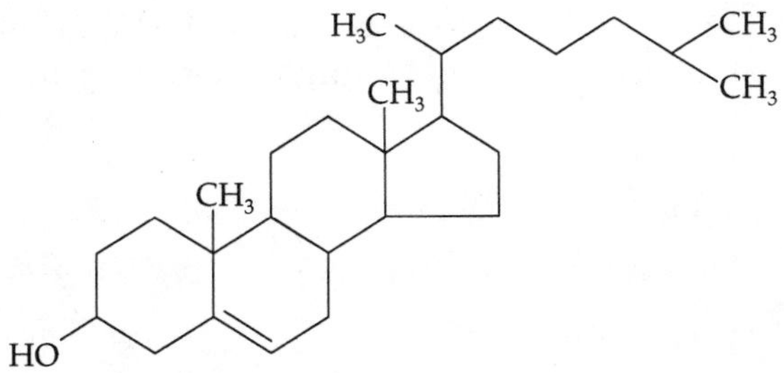

Figure 5.2

b
Factual Recall

5.6 What is the structure shown in Figure 5.2?
   a. a starch molecule
   b. a steroid
   c. a protein
   d. a cellulose molecule
   e. a nucleic acid polymer

e
Factual Recall

5.7 The formation of polymers is most precisely described as an example of
   a. catabolism.
   b. metabolism.
   c. hydrolysis.
   d. hydrophilia.
   e. anabolism.

a
Factual Recall

5.8 Which of the following is TRUE both of starch and of cellulose?
   a. They are both polymers of glucose.
   b. They are geometric isomers of each other.
   c. They can both be digested by humans.
   d. They are both used for energy storage in plants.
   e. They are both structural components of the plant cell wall.

b
Conceptual
Understanding

5.9 Which of the following is TRUE of an amino acid and starch?
   a. Both contain nitrogen.
   b. Both contain oxygen.
   c. Both are polymers.
   d. Both are hydrophobic.
   e. Both are found in proteins.

e
Factual Recall

5.10 Hydrolysis is involved in which of the following?
   a. synthesis of starch
   b. hydrogen bond formation between nucleic acids
   c. peptide bonding in proteins
   d. the hydrophylic interactions of lipids
   e. the digestion of maltose to glucose

d
Factual Recall

5.11 Carbohydrates normally function in animals as
   a. the functional units of lipids.
   b. enzymes in the regulation of metabolic processes.
   c. a component of cell membranes.
   d. energy storage molecules.
   e. sites of protein synthesis.

c

Application

5.12 If three molecules of a fatty acid that has the formula $C_{16}H_{22}O_2$ are joined to a molecule of glycerol ($C_3H_8O_3$), then the resulting molecule would have the formula
  a. $C_{48}H_{66}O_6$.
  b. $C_{48}H_{72}O_8$.
  c. $C_{51}H_{68}O_6$.
  d. $C_{51}H_{72}O_8$.
  e. $C_{51}H_{74}O_9$.

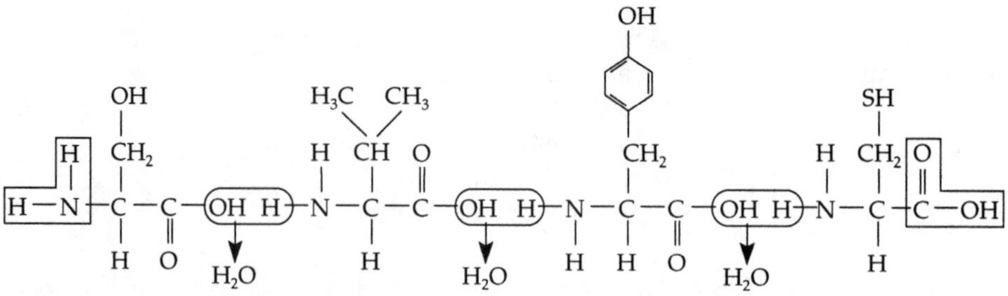

Figure 5.3

a

Factual Recall

5.13 The chemical reactions illustrated in Figure 5.3 result in the formation of
  a. peptide bonds.
  b. ionic bonds.
  c. glycosidic bonds.
  d. hydrogen bonds.
  e. an isotope.

a

Conceptual
Understanding

5.14 Large organic molecules are usually assembled by polymerization of a few kinds of simple subunits. Which of the following is an exception to the above statement?
  a. a steroid
  b. cellulose
  c. DNA
  d. an enzyme
  e. a contractile protein

c

Factual Recall

5.15 At which level of protein structure are interactions between R groups most important?
  a. primary
  b. secondary
  c. tertiary
  d. quaternary
  e. They are equally important at all levels.

b

Factual Recall

5.16 What maintains the secondary structure of a protein?
  a. peptide bonds
  b. hydrogen bonds
  c. disulfide bridges
  d. ionic bonds
  e. electrostatic charges

e
Factual Recall

5.17 Condensation synthesis reactions are used in forming which of the following compounds?
   a. triglycerides
   b. polysaccharides
   c. proteins
   d. Only a and c are correct.
   e. a, b, and c are correct.

c
Factual Recall

5.18 Which of the following is TRUE concerning saturated fatty acids?
   a. All of the below are true.
   b. They have double bonds between the carbon atoms of the fatty acids.
   c. They have a higher ratio of hydrogen to carbon than unsaturated fatty acids.
   d. They are usually liquid at room temperature.
   e. They are usually produced by plants.

a
Factual Recall

5.19 Which bonds are created during the formation of the primary structure of a protein?
   a. peptide bonds
   b. hydrogen bonds
   c. disulfide bonds
   d. Only a and c are correct.
   e. a, b, and c are correct.

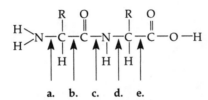

   a.   b.   c.   d.   e.

Figure 5.4

c
Factual Recall

5.20 At which bond in Figure 5.4 would water need to be added to achieve hydrolysis of the dipeptide shown, back to its component amino acids?

b
Factual Recall

5.21 The alpha helix and the beta pleated sheet are both common forms found in which level of structure of proteins?
   a. primary
   b. secondary
   c. tertiary
   d. quaternary
   e. Both a and d are correct.

a
Conceptual
Understanding

5.22 Altering which of the following levels of structural organization could alter the function of an enzyme?
   a. All of the below are correct.
   b. primary
   c. secondary
   d. tertiary
   e. quaternary

e
Factual Recall

5.23 In the double helix structure of nucleic acids, cytosine hydrogen-bonds to
    a. deoxyribose
    b. ribose
    c. adenine
    d. thymine
    e. guanine

b
Factual Recall

5.24 The structural feature that allows DNA to replicate itself is the
    a. sugar-phosphate backbone.
    b. complementary pairing of the bases.
    c. phosphodiester bonding of the helices.
    d. twisting of the molecule to form an alpha helix.
    e. three-part structure of the nucleotides.

b
Factual Recall

5.25 The major purpose of RNA is to
    a. transmit genetic information to offspring.
    b. function in the synthesis of proteins.
    c. make a copy of itself, thus insuring genetic continuity.
    d. act as a pattern to form DNA.
    e. form the genes of an organism.

a
Factual Recall

5.26 If one strand of a DNA molecule has the sequence of bases 5'-ATTGCA-3', the other strand would have the sequence
    a. 3'-TAACGT-5'.
    b. 3'-UAACGU-5'.
    c. 3'-TUUCGU-5'.
    d. 3'-TAAGCT-5'.
    e. 3'-TUUGCT-5'.

e
Factual Recall

5.27 The tertiary structure of a protein is the
    a. bonding together of several polypeptide chains by weak bonds.
    b. order in which amino acids are joined in a peptide chain
    c. bonding of two amino acids together to form a dipeptide.
    d. twisting of a peptide chain into an alpha helix.
    e. three-dimensional shape.

b
Factual Recall

5.28 What is a triacylglycerol?
    a. a protein with tertiary structure
    b. a lipid made of three fatty acids and glycerol
    c. a kind of lipid that makes up much of the plasma membrane
    d. a molecule formed from three alcohols
    e. a carbohydrate with three sugars

e
Factual Recall

5.29 The element nitrogen is always present in all of the following EXCEPT
    a. proteins.
    b. nucleic acids.
    c. amino acids.
    d. DNA.
    e. lipids.

c
Application

5.30 Which of the following would yield the most energy per gram when oxidized?
   a. starch
   b. glycogen
   c. fat
   d. protein
   e. monosaccharides

e
Factual Recall

5.31 All of the following molecules are proteins EXCEPT
   a. hemoglobin.
   b. antibodies.
   c. collagen.
   d. enzymes.
   e. cellulose.

c
Factual Recall

5.32 All of the following bases are found in DNA EXCEPT
   a. thymine.
   b. adenine.
   c. uracil.
   d. guanine.
   e. cytosine.

c
Factual Recall

5.33 All of the following molecules are carbohydrates EXCEPT
   a. lactose.
   b. cellulose.
   c. hemoglobin.
   d. glycogen.
   e. starch.

c
Factual Recall

5.34 Which of the following descriptions best fits the class of molecules known as nucleotides?
   a. a nitrogen base and a phosphate group
   b. a nitrogen base and a five-carbon sugar
   c. a nitrogen base, a phosphate group, and a five-carbon sugar
   d. a five-carbon sugar and adenine or uracil
   e. a five-carbon sugar and a purine or pyrimidine

b
Factual Recall

5.35 Polysaccharides, lipids, and proteins are similar in that they
   a. are synthesized from monomers by the process of hydrolysis.
   b. are synthesized from monomers by the process of condensation synthesis.
   c. are synthesized by peptide bonding between monomers.
   d. are decomposed into their subunits by the process of condensation synthesis.
   e. all contain nitrogen in their monomers.

b
Conceptual
Understanding

5.36 Upon chemical analysis, a particular protein was found to contain 438 amino acids. How many peptide bonds are present in this protein?
   a. 20
   b. 437
   c. 438
   d. 439
   e. 876

a
Conceptual
Understanding

5.37 What would be an expected consequence of changing one amino acid in a
particular protein?
   a. All of the below are expected.
   b. The primary structure would be changed.
   c. The tertiary structure might be changed.
   d. The biological activity of this protein might be altered.
   e. The number of amino acids present would stay the same.

d
Conceptual
Understanding

5.38 Which of the following illustrates hydrolysis?
   a. the reaction of two monosaccharides to form a disaccharide with the release
       of water
   b. the synthesis of two amino acids to form a dipeptide with the utilization of
       water
   c. the reaction of a fat to form glycerol and fatty acids with the release of water
   d. the reaction of a fat to form glycerol and fatty acids with the utilization of
       water
   e. the synthesis of a nucleotide from a phosphate, a ribose sugar, and a
       nitrogen base with the production of a molecule of water

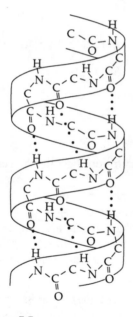

Figure 5.5

d
Factual Recall

5.39 Figure 5.5 illustrates
   a. the primary structure of a protein.
   b. beta 1-4 linkages in cellulose.
   c. the double helix of DNA.
   d. the secondary structure of a protein.
   e. the twisting of a fatty acid chain.

d
Factual Recall

5.40 What is a common feature of both starch and glycogen?
  a. Both form microfibrils that give support to connective tissue.
  b. Both contain repeated monomers of glucose and galactose.
  c. They are important structural components of plant cell walls.
  d. They are polymers of glucose.
  e. They are water-soluble disaccharides.

a
Factual Recall

5.41 The bonding of two amino acid molecules to form a larger molecule requires
  a. the release of a water molecule.
  b. the release of a carbon dioxide molecule.
  c. the addition of a nitrogen atom.
  d. the addition of a water molecule.
  e. an increase in activation energy.

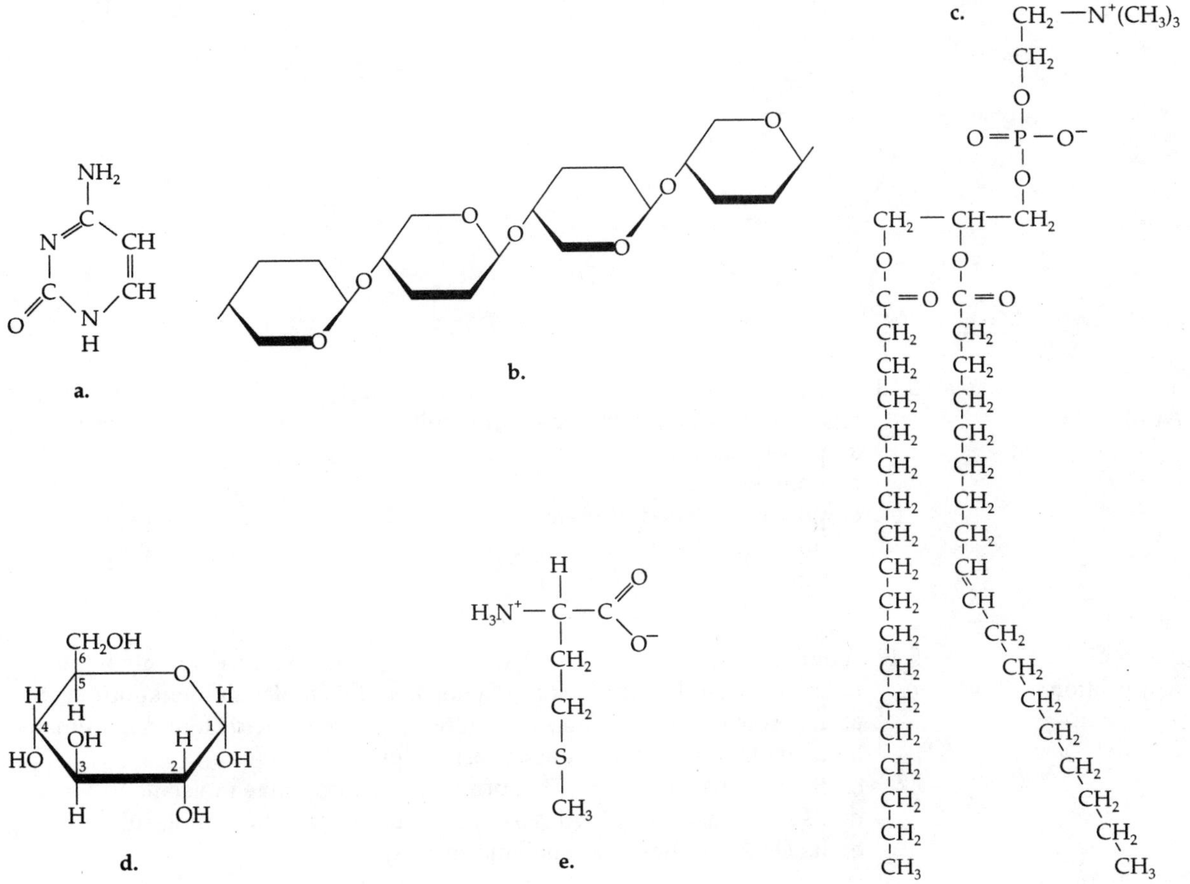

Figure 5.6

Questions 5.42–5.47 are based on the molecules illustrated in Figure 5.6.

e
Factual Recall

5.42 An amino acid.

b
Factual Recall

5.43 Makes up cellulose microfibrils.

a
Factual Recall

5.44 A component of DNA.

c
Factual Recall

5.45 A major structural component of cell membranes.

d
Factual Recall

5.46 Alpha glucose.

d
Factual Recall

5.47 A monosaccharide.

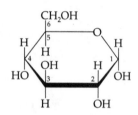

Figure 5.7

a
Application

5.48 If 100 molecules of the general type shown in Figure 5.7 were covalently joined together in sequence, the single molecule that would result would be a
a. polysaccharide.
b. polypeptide.
c. polyunsaturated lipid.
d. nucleic acid.
e. fatty acid.

b
Application

5.49 A biologist discovers a primitive organism new to science. Which of the following would be the best prediction about its molecular makeup?
a. Its proteins will contain some of the same amino acids found in humans.
b. Its membranes will contain phospholipids.
c. Its proteins will serve as the primary energy storage molecules.
d. Its monosaccharides will serve as important structural molecules.
e. Its DNA will have unusual nucleotides.

e
Conceptual
Understanding

5.50 Which of the following best explains the molecular complexity of living organisms?
a. The large number of different monomers allows the construction of many polymers.
b. Each organism has its own unique set of monomers for use in constructing polymers.
c. Condensation reactions can create many different polymers because they can use virtually any molecules in the cell.
d. While there are not many different macromolecules in cells, each one has emergent properties.
e. A small number of monomers can be assembled into large polymers with many different sequences.

a
Conceptual
Understanding

5.51 Which of the following best summarizes the relationship between condensation reactions and hydrolysis?
a. Condensation reactions assemble polymers and hydrolysis breaks them down.
b. Hydrolysis occurs during the day and condensation reactions happen at night.
c. Condensation reactions can occur only after hydrolysis.
d. Hydrolysis creates monomers and condensation reactions destroy them.
e. Condensation reactions occur in plants and hydrolysis happens in animals.

Chapter 6

c
Factual Recall

6.1 Which of the following is the most randomized form of energy?
a. light
b. electrical
c. thermal (heat)
d. mechanical
e. chemical potential energy

a
Application

6.2 When a protein forms from amino acids, the following changes apply:
a. $+ \Delta H, - \Delta S, + \Delta G$
b. $+ \Delta H, - \Delta S, - \Delta G$
c. $+ \Delta H, + \Delta S, + \Delta G$
d. $- \Delta H, - \Delta S, + \Delta G$
e. $- \Delta H, + \Delta S, + \Delta G$

b
Factual Recall

6.3 How does an enzyme catalyze a reaction?
a. by supplying the energy to speed up a reaction
b. by lowering the energy of activation of a reaction
c. by lowering the ΔG of a reaction
d. by changing the equilibrium of a spontaneous reaction
e. by increasing the amount of free energy of a reaction

b
Conceptual
Understanding

6.4 Why is ATP an important molecule in metabolism?
a. It has high-energy phosphate bonds.
b. Its phosphate bonds are easily formed and broken.
c. Its hydrolysis is endergonic.
d. It is readily obtained from an organism's environment.
e. It is extremely stable.

d
Conceptual
Understanding

6.5 The control of enzyme function is an important aspect of cell metabolism. Which of the following is LEAST likely to be a mechanism for enzyme control?
a. allosteric regulation
b. cooperativity
c. feedback inhibition
d. removing cofactors
e. reversible inhibition

a
Conceptual
Understanding

6.6 Which of the following would decrease the entropy within a system?
a. condensation reaction
b. hydrolysis
c. respiration
d. digestion
e. catabolism

e
Conceptual
Understanding

6.7 What is the change in free energy at chemical equilibrium?
   a. slightly increasing
   b. greatly increasing
   c. slightly decreasing
   d. greatly decreasing
   e. no net change

c
Conceptual
Understanding

6.8 Increasing the substrate concentration in an enzymatic reaction could overcome which of the following?
   a. denaturing of the enzyme
   b. allosteric inhibition
   c. competitive inhibition
   d. noncompetitive inhibition
   e. insufficient cofactors

a
Factual Recall

6.9 Which of the following is part of the first law of thermodynamics?
   a. Energy cannot be created or destroyed.
   b. The entropy of the universe is decreasing.
   c. The entropy of the universe is constant.
   d. Kinetic energy is stored energy that results from the specific arrangement of matter.
   e. Energy cannot be transferred or transformed.

d
Conceptual
Understanding

6.10 Whenever energy is transformed, there is always an increase
   a. in the free energy of the system.
   b. in the free energy of the universe.
   c. in the entropy of the system.
   d. in the entropy of the universe.
   e. in the enthalpy of the universe.

c
Factual Recall

6.11 In a system where temperature is uniform, free energy is
   a. the total energy of the system.
   b. the extra energy emitted by the system.
   c. the energy available to do work.
   d. equivalent to entropy.
   e. kinetic energy.

e
Application

6.12 Which of the following reactions could be coupled to the reaction ATP + $H_2O \rightarrow ADP + P_i$(-7.3 kcal)?
   a. $A + P_i \rightarrow AP$ (+9 kcal)
   b. $B + P_i \rightarrow BP$ (+8 kcal)
   c. $CP \rightarrow C + P_i$ (-4 kcal)
   d. $DP \rightarrow D + P_i$ (-10 kcal)
   e. $E + P_i \rightarrow EP$ (+5 kcal)

b
Factual Recall

6.13 Which of the following is true for exergonic reactions?
   a. The products have more free energy than the reactants.
   b. The products have less free energy than the reactants.
   c. Reactants will always be completely converted to products.
   d. A net input of energy from the surroundings is required for the reactions to proceed.
   e. The reactions upgrade the free energy in the products at the expense of energy from the surroundings.

e
Factual Recall

6.14 Which term most precisely describes the general process of breaking down large molecules into smaller ones?
   a. catalysis
   b. metabolism
   c. anabolism
   d. dehydration
   e. catabolism

c
Factual Recall

6.15 ATP generally energizes a cellular process by
   a. releasing heat upon hydrolysis.
   b. acting as a catalyst.
   c. direct chemical transfer of a phosphate group.
   d. releasing ribose electrons to drive reactions.
   e. emitting light flashes.

c
Conceptual Understanding

6.16 A solution of starch at room temperature does not spontaneously decompose rapidly to a sugar solution because
   a. the starch solution has less free energy than the sugar solution.
   b. the hydrolysis of starch to sugar is endergonic.
   c. the activation energy barrier cannot be surmounted by most of the starch molecules.
   d. starch cannot be hydrolyzed in the presence of so much water.
   e. starch hydrolysis is nonspontaneous.

c
Factual Recall

6.17 What is an organic, nonprotein component of an enzyme molecule called?
   a. an accessory enzyme
   b. an allosteric group
   c. a coenzyme
   d. a functional group
   e. an activator

e
Factual Recall

6.18 All of the following statements are representative of the second law of thermodynamics EXCEPT:
   a. Energy transfers are always accompanied by some loss.
   b. Heat energy represents lost energy to most systems.
   c. Systems tend to rearrange themselves toward greater entropy.
   d. Highly organized systems require energy for their maintenance.
   e. Every time energy changes form, there is a decrease in entropy.

a
Factual Recall

6.19 According to the second law of thermodynamics, all of the following are true
      EXCEPT that
      a. the synthesis of large molecules from small molecules is exergonic.
      b. the Earth is not a closed system.
      c. life exists at the expense of greater energy than it contains.
      d. entropy increases in a closed system.
      e. every chemical transformed represents a loss of free energy.

a
Factual Recall

6.20 A chemical reaction that has a positive ΔG is correctly described as
      a. endergonic.
      b. exergonic.
      c. enthalpic.
      d. spontaneous.
      e. exothermic.

c
Factual Recall

6.21 Which of these statements regarding enzymes is FALSE?
      a. Enzymes are proteins that function as catalysts.
      b. Enzymes display specificity for certain molecules to which they attach.
      c. Enzymes provide activation energy for the reactions they catalyze.
      d. The activity of enzymes can be regulated by factors in their immediate
          environment.
      e. An enzyme may be used many times over for a specific reaction.

d
Conceptual
Understanding

6.22 How can one increase the rate of a chemical reaction?
      a. increase the activation energy needed
      b. cool the reactants
      c. decrease the concentration of reactants
      d. add a catalyst
      e. increase entropy of reactants

b
Conceptual
Understanding

6.23 Which of the following statements regarding enzymes is TRUE? Enzymes
      a. have no effect on the rate of a reaction.
      b. increase the rate of reaction.
      c. change the direction of chemical reactions.
      d. are permanently altered by the reactions they catalyze.
      e. prevent changes in substrate concentrations from having an effect on
          reaction rates.

*Questions 6.24 and 6.25 are based on the following information. A series of enzymes
catalyze the reaction X →Y → Z → A. Product "A" binds to the enzyme that converts X
to Y at a position remote from its active site. This binding decreases the activity of the
enzyme.*

c
Application

6.24 In this example, substance X is
      a. a coenzyme.
      b. an allosteric inhibitor.
      c. a substrate.
      d. an intermediate.
      e. the product.

b
Application

6.25 In this example, substance A functions as
    a. a coenzyme.
    b. an allosteric inhibitor.
    c. the substrate.
    d. an intermediate.
    e. a competitive inhibitor.

a
Application

6.26 According to the second law of thermodynamics,
    a. the entropy of the universe is constantly increasing.
    b. for every action there is an equal and opposite reaction.
    c. every energy transfer requires activation energy from the environment.
    d. the total amount of energy in the universe is conserved or constant.
    e. energy can be transferred or transformed, but it can be neither created nor destroyed.

b
Conceptual
Understanding

6.27 Which of the following statements is TRUE concerning catabolic pathways?
    a. They combine molecules into more complex and energy-rich molecules.
    b. They are usually coupled with anabolic pathways to which they supply energy in the form of ATP.
    c. They involve endergonic reactions that break complex molecules into simpler ones.
    d. They are spontaneous and do not need enzyme catalysis.
    e. They build up complex molecules such as protein from simpler compounds.

c
Factual Recall

6.28 Which of the following statements is TRUE regarding enzyme cooperativity?
    a. A multi-enzyme complex contains all the enzymes of a metabolic pathway.
    b. A product of a pathway serves as a competitive inhibitor of an early enzyme in the pathway.
    c. A molecule bound to an allosteric site affects the active site of several subunits.
    d. Several substrate molecules can be catalyzed by the same enzyme.
    e. A substrate binds to an active site and inhibits cooperation between enzymes in a pathway.

d
Factual Recall

6.29 According to the induced fit hypothesis of enzyme function, which of the following is correct?
    a. The binding of the substrate depends on the shape of the active site.
    b. Some enzymes become denatured when activators bind to the substrate.
    c. A competitive inhibitor can outcompete the substrate for the active site.
    d. The binding of the substrate changes the shape of the enzyme slightly.
    e. The active site creates a microenvironment ideal for the reaction.

e
Application

6.30 Correct statements regarding ATP include:
    I. ATP (adenosine triphosphate) serves as a main energy shuttle inside cells.
    II. ATP drives endergonic reaction in the cell by the enzymatic transfer of the phosphate group to specific reactants.
    III. The regeneration of ATP from ADP and phosphate is an endergonic reaction.

a. I only
b. II only
c. III only
d. I and III only
e. I, II, and III

a
Factual Recall

6.31 All of the following statements regarding enzymes are true EXCEPT:
   a. Enzymes are carbohydrates that function as agents that change the rate of reaction without being consumed in the reaction.
   b. Enzymes allow molecules to react in metabolism by lowering activation energies.
   c. Each type of enzyme has a uniquely shaped active site that gives it specificity.
   d. Enzymes are very sensitive to environmental conditions that influence the weak chemical bonds responsible for their three-dimensional structure.
   e. Some enzymes change shape when regulator molecules, either activators or inhibitors, bind to specific allosteric receptor sites.

b
Factual Recall

6.32 According to the first law of thermodynamics
   a. matter can be neither created nor destroyed.
   b. energy is neither created or destroyed.
   c. all processes increase the entropy of the universe.
   d. systems rich in energy are intrinsically unstable.
   e. the universe loses energy because of friction.

b
Conceptual
Understanding

6.33 Of the following choices, which is the best analogy to the metabolic pathways in cells?
   a. a Ferris wheel that continuously turns in a circle, stopping only as people climb on and off the seats of the ride
   b. the organized street system of a large city where cars travel specific roads and visit gas stations for fuel
   c. two locomotives racing up a mountain while pulling several box cars
   d. ripples on the surface of a lake when a rock is tossed into the water
   e. a stampede of wild horses after a rifle is fired in the air

c
Factual Recall

6.34 The transfer of free energy from catabolic pathways to anabolic pathways is called
   a. feedback regulation.
   b. bioenergetics.
   c. energy coupling.
   d. entropy.
   e. cooperativity.

a
Application

6.35 Of the following, a cell without enzymes would be most like
   a. an airport without air traffic controllers to route the planes.
   b. exploring a dark cave without a flashlight to see the way.
   c. a college dormitory without a cafeteria for student dining.
   d. an automobile factory without parts to assemble into cars.
   e. a gasoline engine without fuel to burn.

d
Factual Recall

6.36 In an organism, the energy available to do work is called free energy because
 a. it can be obtained with no cost to the system.
 b. it can be spent with no cost to the universe.
 c. the organism can live free of it if necessary.
 d. it is available for work.
 e. it is equivalent to the system's total energy.

Use the information below to answer Questions 6.37 and 6.38.

A popular herbicide named glyphosate is reported to kill plants by affecting an enzyme in the pathway leading to the synthesis of several amino acids required in the assembly of plant proteins. To determine the mechanism of glyphosate action, you conduct assays of that enzyme with increasing amounts of substrate and a constant amount of glyphosate. The results are shown in Table 6.1 below.

Table 6.1. Assay of Enzyme Activity with Increasing Substrate, Constant Glyphosate

substrate concentration (mg/ml)	10	15	20	25	30
relative enzyme activity (%)	20	40	60	80	100

b
Application

6.37 These data show glyphosate most likely
 a. acts directly on one of the amino acids.
 b. is a competitive inhibitor of the enzyme-catalyzed reaction.
 c. does not affect activity at substrate concentrations below 10 mg/ml.
 d. indirectly alters the configuration of the enzyme's active site.
 e. acts as a noncompetitive inhibitor of the pathway.

a
Conceptual
Understanding

6.38 From these data, you can accurately conclude that the affected enzyme
 a. None of the below is an accurate conclusion based on the data.
 b. is an allosteric protein.
 c. alters the equilibrium of the reaction when glyphosate is added.
 d. is actually activated by glyphosate.
 e. exhibits cooperativity with increasing substrate concentrations.

b
Application

6.39 During a laboratory experiment, you discover that an enzyme-catalyzed reaction has a $\Delta G = -20$ kcal/mole. You double the amount of enzyme in the reaction and the ΔG now equals
 a. -10 kcal/mole.
 b. -20 kcal/mole.
 c. -40 kcal/mole.
 d. $+40$ kcal/mole.
 e. It is not possible to calculate the answer with the data given.

Chapter 7

e
Factual Recall

7.1 What do both mitochondria and chloroplasts have in common?
  a. ATP is produced.
  b. DNA is present.
  c. Ribosomes are present.
  d. Only b and c are correct.
  e. a, b, and c are correct.

e
Factual Recall

7.2 A cellular diameter of 40 micrometers is equivalent to
  a. 0.4 millimeter.
  b. 0.04 millimeter.
  c. 40,000 nanometers.
  d. 4,000 nanometers.
  e. Both b and c are correct.

b
Factual Recall

7.3 Grana, thylakoids and CF1 particles are all structural components found in
  a. cilia and flagella.
  b. chloroplasts.
  c. mitochondria.
  d. lysosomes.
  e. nuclei.

d
Factual Recall

7.4 Organelles that contain DNA include
  a. ribosomes.
  b. mitochondria.
  c. chloroplasts.
  d. Only b and c are correct.
  e. a, b, and c are correct.

e
Factual Recall

7.5 Which of the following would be found in an animal cell, but NOT in a bacterial cell?
  a. DNA
  b. cell wall
  c. plasma membrane
  d. ribosomes
  e. endoplasmic reticulum

b
Conceptual
Understanding

7.6 Eukaryotic cells are typically larger than prokaryotic cells because
  a. their plasma membrane has more control over the movement of materials into the cell.
  b. their internal membrane system allows compartmentalization of functions and extra surface area for nutrient exchange and placement of enzymes.
  c. their DNA is localized in the nucleus, whereas protein synthesis occurs in the cytoplasm, separating these competing functions.
  d. they have more chromosomes and a mitotic process of cell division.
  e. they have a cytoskeleton composed of microtubules and microfilaments.

b
Factual Recall

7.7 What is the innermost portion of mature plant cell walls called?
a. primary cell wall
b. secondary cell wall
c. middle lamella
d. glycocalyx
e. tonoplast

c
Conceptual
Understanding

7.8 Plasmodesmata in plant cells are similar in function to which structure in animal cells?
a. peroxisomes
b. desmosomes
c. gap junctions
d. glycocalyx
e. tight junctions

d
Factual Recall

7.9 A cell has the following molecules and structures: enzymes, DNA, ribosomes, plasma membrane, and mitochondria. It could be a cell from
a. a bacterium.
b. an animal, but not a plant.
c. a plant, but not an animal.
d. a plant or an animal.
e. any kind of organism.

a
Application

7.10 Cells can be described as having a "cytoskeleton" of internal structures that contribute to the shape, organization, and movement of the cell. All of the following are part of the "cytoskeleton" EXCEPT
a. the cell wall.
b. microtubules.
c. microfilaments.
d. intermediate filaments.
e. actin.

Refer to the following five terms to answer Questions 7.11–7.19. Choose the most appropriate term for each phrase. Each term may be used once, more than once, or not at all.
a. *lysosome*
b. *tonoplast*
c. *mitochondrion*
d. *Golgi apparatus*
e. *peroxisome*

d
Factual Recall

7.11 Secretes many polysaccharides.

a
Factual Recall

7.12 Contains hydrolytic enzymes.

a
Factual Recall

7.13 Helps to recycle the cell's organic material.

c
Factual Recall

7.14 Site of cellular respiration.

a
Factual Recall

7.15 Involved in storage diseases such as Tay-Sachs.

c
Factual Recall

7.16 Contains its own DNA and ribosomes.

e
Factual Recall

7.17 Detoxifies alcohol in the liver.

e
Factual Recall

7.18 Contains enzymes that convert fats to sugar.

c
Factual Recall

7.19 Contains cristae.

e
Factual Recall

7.20 Ions can travel directly from the cytoplasm of one animal cell to the cytoplasm of an adjacent cell through
   a. plasmodesmata.
   b. intermediate filaments.
   c. tight junctions.
   d. desmosomes.
   e. gap junctions.

b
Factual Recall

7.21 Microfilaments participate in the formation of
   a. cilia.
   b. cell cleavage furrows.
   c. flagella.
   d. mitotic spindles.
   e. basal bodies.

d
Factual Recall

7.22 An animal secretory cell and a photosynthetic leaf cell are similar in all of the following ways EXCEPT:
   a. They both have a Golgi apparatus.
   b. They both have mitochondria.
   c. They both have transport proteins for active transport of ions.
   d. They both have chloroplasts.
   e. They both have a cell membrane.

Refer to the following five terms to answer Questions 7.23–7.25. Choose the most appropriate term for each phrase. One term may be used once, more than once, or not at all.
   a. mitochondria
   b. Golgi apparatus
   c. rough endoplasmic reticulum
   d. lysosomes
   e. smooth endoplasmic reticulum

e
Factual Recall

7.23 Makes steroid hormones.

d
Factual Recall

7.24 Digests damaged organelles.

b
Factual Recall

7.25 Sorts out mixtures of substances and sends them to their proper destinations.

Refer to the following five terms to answer Questions 7.26–7.28. Choose the most appropriate term for each phrase. One term may be used once, more than once, or not at all.
  a. *centriole*
  b. *lysome*
  c. *nucleolus*
  d. *peroxisome*
  e. *ribosome*

e
Factual Recall

7.26 Found in both prokaryotic and eukaryotic cells.

a
Factual Recall

7.27 Possesses a microtubular structure similar in form to a basal body.

c
Factual Recall

7.28 Assembles ribosomal precursors.

For Questions 7.29–7.34, use the lettered answers to match the structure to its proper cell type. Choose the most inclusive category. Each answer may be used once, more than once, or not at all.
  a. *structure is a feature of all cells*
  b. *structure is found in prokaryotic cells only*
  c. *structure is found in eukaryotic cells only*
  d. *structure is found in plant cells only*
  e. *structure is found in animal cells only*

a
Application

7.29 Plasma membrane.

c
Application

7.30 Cytoskeleton.

a
Conceptual
Understanding

7.31 Ribosomes.

d
Conceptual
Understanding

7.32 Plasmodesmata.

e
Conceptual
Understanding

7.33 Tight junctions.

c
Conceptual
Understanding

7.34 Golgi bodies.

e
Application

7.35 Chloramphenicol is a drug that inhibits protein synthesis on prokaryotic ribosomes. Which of the following cells (or parts of cells) would have protein synthesis inhibited if they were grown in the presence of chloramphenicol?
   a. bacteria
   b. chloroplasts
   c. mitochondria
   d. Only a and c are correct.
   e. a, b, and c are correct.

a
Factual Recall

7.36 Which of the following is capable of converting light energy to chemical energy?
   a. chloroplasts
   b. mitochondria
   c. leucoplasts
   d. peroxisomes
   e. Golgi bodies

a
Factual Recall

7.37 Which of the following contain the 9+2 arrangement of microtubules?
   a. cilia
   b. centrioles
   c. basal bodies
   d. microfilaments
   e. nuclei

c
Factual Recall

7.38 Large numbers of ribosomes are present in cells that specialize in producing which of the following molecules?
   a. lipids
   b. starches
   c. proteins
   d. steroids
   e. glucose

a
Conceptual
Understanding

7.39 A biologist ground up some plant cells and then centrifuged the mixture. She obtained some organelles from the sediment in the test tube. The organelles took up CO_2 and gave off O_2. The organelles are most likely
   a. chloroplasts.
   b. ribosomes.
   c. nuclei.
   d. mitochondria.
   e. Golgi apparatuses.

c
Factual Recall

7.40 Which of the following pairs is mismatched?
    a. nucleolus : ribosomal RNA
    b. nucleus : DNA replication
    c. lysosome : protein synthesis
    d. cell membrane : lipid bilayer
    e. cytoskeleton : microtubules

b
Factual Recall

7.41 Which of the following relationships between cell structures and their respective functions is NOT correct?
    a. cell wall—support, protection
    b. chloroplasts—chief site of cellular respiration
    c. chromosomes—genetic control information
    d. ribosomes—site of protein synthesis
    e. mitochondria—formation of ATP

e
Factual Recall

7.42 Which of the following cell components is NOT directly involved in synthesis or secretion?
    a. ribosomes
    b. rough endoplasmic reticulum
    c. Golgi bodies
    d. smooth endoplasmic reticulum
    e. lysosome

e
Factual Recall

7.43 All of the structures listed below are associated with movement in cells or by cells EXCEPT
    a. cilia.
    b. centrioles.
    c. microtubules.
    d. flagella.
    e. microbodies.

c
Factual Recall

7.44 Which organelle is primarily involved in the synthesis of lipids?
    a. ribosomes
    b. lysosomes
    c. smooth endoplasmic reticulum
    d. mitochondria
    e. contractile vacuoles

a
Conceptual
Understanding

7.45 Of the following, which cell structure would most likely be visible with a light microscope that had been manufactured to the maximum resolving power possible?
    a. mitochondrion
    b. microtubule
    c. ribosome
    d. largest microfilament
    e. nuclear pore

b
Factual Recall

7.46 In animal cells, hydrolytic enzymes are packaged to prevent general destruction of cellular components. Which of the following organelles functions in this compartmentalization?
a. chloroplast
b. lysosome
c. central vacuole
d. peroxisome
e. glyoxysome

e
Conceptual Understanding

7.47 Which of the following does NOT contain functional ribosomes?
a. a prokaryotic cell
b. a plant mitochondrion
c. a chloroplast
d. an animal mitochondrion
e. a nucleus

c
Conceptual Understanding

7.48 When a potassium ion (K^+) passes from the soil into the vacuole of a root cell, it encounters some cellular barriers. Which of the following is the most direct path the K^+ would take through these barriers?
a. secondary cell wall → plasma membrane → thylakoid
b. primary cell wall → secondary cell wall → tonoplast
c. primary cell wall → plasma membrane → tonoplast
d. cell wall → plasma membrane → tonoplast → grana
e. cell wall → plasma membrane → grana

a
Factual Recall

7.49 Which of the following organelles is enclosed by an envelope of two membranes?
a. chloroplast
b. lysosome
c. Golgi apparatus
d. ribosome
e. central vacuole

a
Factual Recall

7.50 Which of the following is NOT a known function of the cytoskeleton?
a. to maintain a critical limit on cell size
b. to provide mechanical support to the cell
c. to maintain characteristic shape of the cell
d. to hold mitochondria and other organelles in place within the cytosol
e. to assist in cell motility by interacting with specialized motor proteins

b
Factual Recall

7.51 Cilia and flagella move
a. in opposite directions on the plasma membrane.
b. when the bending of microtubules is powered by ATP.
c. in the same direction on the plasma membrane.
d. when contraction of basal body triplets is powered by ATP.
e. when the central pair of microtubules produce ATP.

c
Factual Recall

7.52 Cell walls
   a. are unique to plant cells.
   b. are manufactured from pigments in the chloroplast.
   c. help hold plants upright, against gravity.
   d. are found inside the plasma membrane.
   e. produce cilia for locomotion.

b
Conceptual
Understanding

7.53 Which of the following is the correct sequence of plant cell wall layers, beginning with the outside and progressing inward to the plasma membrane?
   a. middle lamella, secondary wall, primary wall
   b. middle lamella, primary wall, secondary wall
   c. secondary wall, primary wall, middle lamella
   d. primary wall, middle lamella, secondary wall
   e. secondary wall, middle lamella, primary wall

e
Factual Recall

7.54 In animal cells, secreted glycoproteins
   a. digest bacteria outside the cell, in the same way that lysosomes do inside the cell.
   b. are identical to pectins between the primary walls of plant cells.
   c. are produced only on free ribosomes in the cytoplasm.
   d. provide a rigid backbone for flagella.
   e. provide support and anchorage as part of the extracellular matrix.

Chapter 8

a
Factual Recall

8.1 According to the fluid-mosaic model of cell membranes, which of the following is a true statement about membrane phospholipids?
   a. They move laterally along the plane of the membrane.
   b. They frequently flip-flop from one side of the membrane to the other.
   c. They occur in an uninterrupted bilayer, with membrane proteins restricted to the surface of the membrane.
   d. They are free to depart from the membrane and dissolve in the surrounding solution.
   e. They have hydrophilic tails in the interior of the membrane.

d
Factual Recall

8.2 What are the elevated regions (particles) seen in electron micrographs of split freeze-fractured membranes?
   a. peripheral proteins
   b. phospholipids
   c. carbohydrates
   d. integral proteins
   e. cholesterol molecules

e
Factual Recall

8.3 What are the membrane structures that function in active transport?
   a. peripheral proteins
   b. carbohydrates
   c. cholesterol
   d. cytoskeleton filaments
   e. integral proteins

b
Conceptual
Understanding

8.4 The net movement of uncharged molecules from a low concentration to a higher concentration is described by which of the following?
   a. diffusion
   b. active transport
   c. osmosis
   d. facilitated diffusion
   e. exocytosis

c
Factual Recall

8.5 What is the voltage across cell membranes called?
   a. water potential
   b. chemical gradient
   c. membrane potential
   d. osmotic potential
   e. electrochemical gradient

a
Factual Recall

8.6 All of the following cellular activities require ATP energy EXCEPT
   a. movement of O_2 into the cell.
   b. protein synthesis.
   c. Na^+ ions moving out of the cell.
   d. cytoplasmic streaming.
   e. exocytosis.

e
Factual Recall

8.7   Glycoproteins and glycolipids of animal cell membranes are most important for
   a. facilitated diffusion of molecules down their concentration gradients.
   b. active transport of molecules against their concentration gradients.
   c. maintaining the integrity of a fluid mosaic membrane.
   d. maintaining membrane fluidity at low temperatures.
   e. the ability of cells to recognize like and different cells.

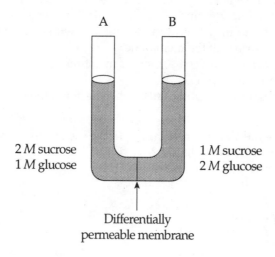

2 M sucrose
1 M glucose

1 M sucrose
2 M glucose

Differentially
permeable membrane

Figure 8.1

Use the diagram of the U-tube setup (Figure 8.1) to answer Questions 8.8 and 8.9. The solutions in the two arms of this U-tube are separated by a membrane that is permeable to water and glucose but not to sucrose. Side A is filled with a solution of 2 M sucrose and 1 M glucose. Side B is filled with 1 M sucrose and 2 M glucose.

c
Application

8.8   Initially, the solution in side A, with respect to that in side B, is
   a. hypotonic.
   b. plasmolyzed.
   c. isotonic.
   d. saturated.
   e. hypertonic.

e
Application

8.9   After the system reaches equilibrium, what changes are observed?
   a. The molarity of sucrose and glucose are equal on both sides.
   b. The molarity of glucose is higher in side A than in side B.
   c. The molarity of sucrose is increased in side A.
   d. The water level is unchanged.
   e. The water level is higher in side A than in side B.

*Suppose instead that the differentially permeable membrane in Figure 8.1 is equally permeable to **both** sucrose and glucose. Use this information to answer Questions 8.10 and 8.11.*

c
Application

8.10 Initially, the solution in side B, with respect to side A, is
    a. hypotonic.
    b. plasmolyzed.
    c. isotonic.
    d. saturated.
    e. hypertonic.

d
Application

8.11 After the system reaches equilibrium, what changes are observed?
    a. Glucose and sucrose no longer cross the membrane.
    b. The molarity of sucrose is higher on side B than on side A.
    c. The molarity of glucose has increased on both sides.
    d. The water level is unchanged.
    e. The water level is higher on side B than on side A.

c
Factual Recall

8.12 The sodium–potassium pump is called an electrogenic pump because it
    a. pumps equal quantities of Na^+ and K^+ across the membrane.
    b. pumps hydrogen ions into the cell.
    c. contributes to the membrane potential.
    d. ionizes sodium and potassium.
    e. pumps hydrogen ions into the cell.

c
Factual Recall

8.13 The movement of potassium into or out of the cell requires
    a. low cellular concentrations of sodium.
    b. high cellular concentrations of potassium.
    c. ATP as an energy source.
    d. glucose for binding and releasing ions.
    e. plant hormones embedded in the cell membrane.

d
Conceptual
Understanding

8.14 An organism with a cell wall would have difficulty doing which process?
    a. diffusion
    b. osmosis
    c. active transport
    d. phagocytosis
    e. exocytosis

e
Factual Recall

8.15 Which process accounts for the movement of solids into some animal cells?
    a. active transport
    b. facilitated diffusion
    c. diffusion
    d. osmosis
    e. phagocytosis

a
Factual Recall

8.16 All of the following are functions of membrane proteins EXCEPT
    a. enzyme synthesis.
    b. active transport.
    c. hormone reception.
    d. cell adhesion.
    e. cytoskeleton attachment.

d
Factual Recall

8.17 The membrane activity most nearly opposite to exocytosis is
   a. plasmolysis.
   b. osmosis.
   c. facilitated diffusion.
   d. phagocytosis.
   e. active transport.

c
Factual Recall

8.18 Which of the following statements is correct about diffusion?
   a. It is very rapid over long distances.
   b. It requires an expenditure of energy by the cell.
   c. It is a passive process.
   d. It occurs when molecules move from a region of lower concentration to one of higher concentration.
   e. It requires integral proteins in the cell membrane.

e
Factual Recall

8.19 Carrier molecules in the plasma membrane are required for
   a. diffusion.
   b. osmosis.
   c. facilitated diffusion only.
   d. active transport only.
   e. both facilitated diffusion and active transport.

b
Application

8.20 A cell with an internal concentration of 0.02 molar glucose is placed in a test tube containing 0.02 molar glucose solution. Assuming that glucose is not actively transported into the cell, which of the following terms describes the internal concentration of the cell relative to its environment?
   a. isotonic
   b. hypertonic
   c. hypotonic
   d. flaccid
   e. a or b, depending on the temperature

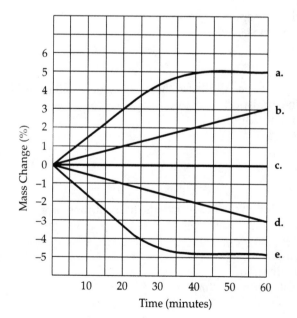

Figure 8.2

Read the following information and refer to Figure 8.2 for Questions 8.21–8.24. Five dialysis bags, impermeable to sucrose, were filled with various concentrations of sucrose and then placed in separate beakers containing an initial concentration of 0.6 M sucrose solution. At 10-minute intervals, the bags were massed (weighed) and the percent change in mass of each bag was graphed.

c
Application

8.21 Which line represents the bag that contained a solution isotonic to the 0.6 molar solution at the beginning of the experiment?

a
Application

8.22 Which line represents the bag with the highest initial concentration of sucrose?

b
Application

8.23 Which line or lines represent bags that contain a solution that is hypertonic at the end of 60 minutes?
   a. a and b
   b. b
   c. c
   d. d
   e. d and e

c
Application

8.24 What is the best explanation for the shape of line **e.** after 50 minutes?
   a. Water is no longer leaving the bag.
   b. Water is no longer entering the bag.
   c. Water is leaving and entering the bag at the same rate.
   d. Water is entering the bag at the same rate that sucrose is leaving the bag.
   e. Water is neither entering nor leaving the bag.

c
Conceptual
Understanding

8.25 One of the functions of cholesterol in animal cell membranes is to
   a. facilitate transport of ions.
   b. store energy.
   c. maintain membrane fluidity.
   d. speed diffusion.
   e. phosphorylate ADP.

a
Conceptual
Understanding

8.26 Which of the following would move through the lipid bilayer of a plasma membrane most rapidly?
   a. CO_2
   b. an amino acid
   c. glucose
   d. K^+
   e. starch

e
Application

8.27 Glucose is a six-carbon sugar that diffuses slowly through artificial phospholipid bilayers. The cells lining the small intestine, however, rapidly move large quantities of glucose from the glucose-rich food into their glucose-poor cytoplasm. Using this information, which transport mechanism is most probably functioning in the intestinal cells?
   a. simple diffusion
   b. phagocytosis
   c. active transport pumps
   d. exocytosis
   e. facilitated diffusion

b
Conceptual
Understanding

8.28 The kinds of molecules that pass through a cell membrane most easily are
    a. large and hydrophobic.
    b. small and hydrophobic.
    c. large polar molecules.
    d. ionic.
    e. monosaccharides such as glucose.

c
Factual Recall

8.29 The cotransport protein which allows two different substances to pass through
    a membrane in the same direction is
    a. usually also a uniport.
    b. usually also a biport.
    c. usually associated with a proton pump.
    d. insensitive to temperature.
    e. usually associated with an antiport.

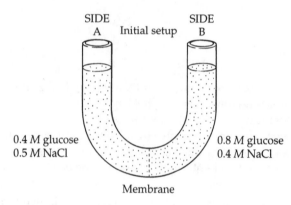

Figure 8.3

*Refer to Figure 8.3 to answer Questions 8.30 and 8.31. The solutions in the arms of a
U-tube are separated at the bottom of the tube by a selectively permeable membrane. The
membrane is permeable to sodium chloride but not to glucose. Side A is filled with a solution of 0.4 molar glucose and 0.5 molar sodium chloride (NaCl) and side B is filled with a
solution containing 0.8 molar glucose and 0.4 molar sodium chloride. Initially, the volume
in both arms is the same.*

b
Application

8.30 At the beginning of the experiment,
    a. side A is hypertonic to side B.
    b. side A is hypotonic to side B.
    c. side A is isotonic to side B.
    d. side A is hypertonic to side B with respect to glucose.
    e. side A is hypotonic to side B with respect to sodium chloride.

d
Application

8.31 If you examine side A after 3 days, you should find
    a. a decrease in the concentration of NaCl and glucose and an increase in the
       water level.
    b. a decrease in the concentration of NaCl, an increase in water level, and no
       change in the concentration of glucose.
    c. no net change in the system.
    d. a decrease in the concentration of NaCl and a decrease in the water level.
    e. no change in the concentration of NaCl and glucose and an increase in the
       water level.

Suppose instead that the selectively permeable membrane at the bottom of the tube in Figure 8.3 is permeable to both glucose and NaCl in the same way that a plasma membrane of a cell is permeable to both. It takes the system 4 hours to reach equilibrium.

e
Application

8.32 In a few minutes,
   a. the water level on side A would be higher than on side B.
   b. water would no longer move across the membrane.
   c. the glucose and NaCl would be at equilibrium, but osmosis would continue.
   d. more glucose than NaCl would have passed through the membrane.
   e. more NaCl than glucose would have passed through the membrane.

a
Application

8.33 At equilibrium, in side B you would find
   a. an increase in NaCl concentration but a decrease in glucose concentration.
   b. a decrease in the molarity of both glucose and NaCl.
   d. the water level has not changed.
   d. no net change in the system.
   e. NaCl and glucose molecules no longer crossing the membrane to side A.

e
Factual Recall

8.34 What membrane-surface molecules are thought to be most important as cells recognize each other?
   a. phospholipids
   b. integral proteins
   c. peripheral proteins
   d. cholesterol
   e. glycoproteins

b
Factual Recall

8.35 White blood cells engulf bacteria through what process?
   a. exocytosis
   b. phagocytosis
   c. pinocytosis
   d. osmosis
   e. receptor-mediated exocytosis

d
Factual Recall

8.36 Ions diffuse across membranes down their
   a. chemical gradients.
   b. concentration gradients.
   c. electrical gradients.
   d. electrochemical gradients.
   e. Both a and b are correct.

e
Conceptual
Understanding

8.37 All of the following situations are consistent with active transport EXCEPT
   a. the conversion of ATP to ADP accompanied by the movement of molecules.
   b. the rate of oxygen consumption by the cell increases when molecules move.
   c. molecules move in or out of a cell against the osmotic gradient.
   d. cells accumulate diffusible molecules in greater quantity than was found outside.
   e. the rate of movement of molecules across the cell membrane increases in an anaerobic environment.

b
Factual Recall

8.38 What does a cell use exocytosis for?
   a. to move away from danger
   b. to release substances from the cell
   c. to incorporate nutrients
   d. to pump protons
   e. to create new cells

a
Factual Recall

8.39 The presence of cholesterol in the plasma membranes of some animals
   a. enables the membrane to stay fluid more easily when cell temperature drops.
   b. enables the animal to remove hydrogen atoms from saturated phospholipids.
   c. enables the animal to add hydrogen atoms to unsaturated phospholipids.
   d. makes the membrane less flexible, so it can sustain greater pressure from within the cell.
   e. makes them more susceptible to circulatory disorders.

b
Conceptual
Understanding

8.40 Which of the following would indicate that facilitated diffusion was taking place?
   a. Substances were moving against the diffusion gradient.
   b. A substance was diffusing much faster than the physical condition indicated it should.
   c. ATP was being rapidly consumed as the substance moved.
   d. A substance was slowing as it moved down its concentration gradient.
   e. A substance was moving from a region of low concentration into a region of higher concentration of the substance.

c
Conceptual
Understanding

8.41 All of the following statements about membrane structure and function are true EXCEPT:
   a. Diffusion of gases is faster in air than across membranes.
   b. Diffusion, osmosis, and facilitated diffusion do not require any energy input from the cell.
   c. Cell recognition by animal cells is probably mediated by peripheral proteins.
   d. Voltage across the membrane depends on an unequal distribution of ions across the plasma membrane.
   e. Special membrane proteins can cotransport two solutes by coupling diffusion with active transport.

d
Application

8.42 Two similar-sized animal cells are placed in a 0.5% sucrose solution. Cell A enlarges in size for a while, then stops; cell B continues to enlarge and finally ruptures. Which of the following was true at the beginning of the experiment?
   a. Cell A was hypotonic to the solution and cell B was hypertonic.
   b. Cell A was hypertonic to the solution and cell B was hypotonic.
   c. Cell A was hypertonic to cell B.
   d. Cell B was hypertonic to cell A.
   e. Cells A and B were isotonic to each other.

a
Factual Recall

8.43 Which of the following characterizes the sodium-potassium pump?
   a. All of the below characterize the sodium-potassium pump.
   b. Sodium ions are pumped out of a cell against their gradient.
   c. Potassium ions are pumped into a cell against their gradient.
   d. Each exchange costs the cell one ATP.
   e. A carrier protein undergoes conformational change.

b
Conceptual
Understanding

8.44 Which of the following is a characteristic feature of a carrier protein in a plasma membrane?
   a. It is a peripheral membrane protein.
   b. It exhibits a specificity for a particular type of molecule.
   c. It requires energy to function.
   d. It works against diffusion.
   e. It has few, if any, hydrophobic amino acids.

b
Factual Recall

8.45 Which molecule is NOT part of the cell membrane?
   a. lipid
   b. nucleic acid
   c. protein
   d. phosphate group
   e. steroid

For Questions 8.46–8.50, match the membrane model or description with the scientist(s) who proposed the model. Each choice may be used once, more than once, or not at all.
   a. H. Davson and J. Danielli
   b. I. Langmuir
   c. C. Overton
   d. S. Singer and G. Nicolson
   e. E. Gorter and F. Grendel

c
Factual Recall

8.46 Membranes are made of lipids because substances that dissolve in lipids enter cells more rapidly than lipid-insoluble substances.

b
Factual Recall

8.47 Made artificial membranes by adding phospholipids dissolved in benzene to water.

e
Factual Recall

8.48 The first to propose that cell membranes are phospholipid bilayers.

a
Factual Recall

8.49 Membranes are a phospholipid bilayer between two layers of globular protein.

d
Factual Recall

8.50 The membrane is a mosaic of protein molecules bobbing in a fluid layer of phospholipids.

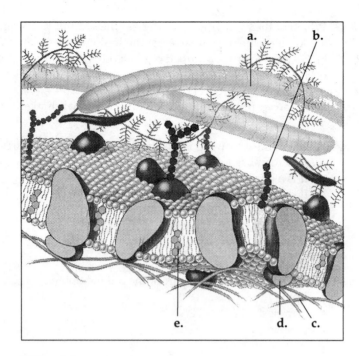

Figure 8.4

For Questions 8.51–8.55, match the labeled component of the cell membrane (Figure 8.4) with its description.

d
Factual Recall

8.51 Peripheral protein.

e
Factual Recall

8.52 Cholesterol.

a
Factual Recall

8.53 Fiber of the extracellular matrix.

c
Factual Recall

8.54 Filament of the cytoskeleton.

b
Factual Recall

8.55 Glycolipid.

b
Application

8.56 A cell lacking oligosaccharides on the external surface of its plasma membrane would likely be inefficient in
a. transporting ions against an electrochemical gradient.
b. cell–cell recognition.
c. maintaining fluidity of the phospholipid bilayer.
d. attaching to the cytoskeleton.
e. establishing the diffusion barrier to charged molecules.

Use the explanation below and data in Table 8.1 to answer Questions 8.57 and 8.58. Glucose and glucose-6-phosphate (G-6-P) do not move across artificial membranes (made of phospholipids but lacking proteins). To further your understanding of cell membrane functions, you conducted an experiment on the transport of glucose and G-6-P into living muscle and plant root cells. At the start of the experiment, the concentration of both kinds of molecules outside the cells was 1 Molar (M). The original concentration of each molecule inside the cells was 0.1 M. After 2 hours, the concentration of each molecule inside the cells was measured. The table below shows the results of this measurement.

Table 8.1 Concentrations of Glucose and G-6-P in Muscle and Root Cells after 2 Hours

	Muscle Cells	**Root Cells**
Glucose	0.5 M	0.1 M
G-6-P	0.1 M	0.1 M

b

Application

8.57 Of the following, which is the best conclusion based on the data in Table 8.1?
   a. Both glucose and G-6-P are transported into the cells.
   b. Muscle cells lack a transport system for G-6-P.
   c. Muscle cells lack a transport system for glucose.
   d. The muscle cells, but not the root cells, require glucose for metabolism.
   e. In 4 hours, glucose will reach a concentration of 1.0 M inside the muscle cells.

As a follow-up to the experiment above (Table 8.1), you add cyanide to the cells in the continuing experiment. This stops all ATP production inside the cells. Two hours later (4 hours total), you again determine the level of the test molecules inside the cells. Use the results shown in Table 8.2 to answer Question 8.58.

Table 8.2 Concentrations in the Cells 2 Hours after Treatment with Cyanide

	Muscle Cells	**Root Cells**
Glucose	1.0 M	0.1 M
G-6-P	0.1 M	0.1 M

c

Application

8.58 Of the following, which is the best conclusion based on the data in both tables?
   a. Movement of both molecules into the cells is at equilibrium after 4 hours.
   b. The addition of cyanide stimulated transport of glucose into the muscle cells.
   c. Movement of glucose into muscle cells is by passive transport.
   d. Movement of glucose into muscle cells is by active transport.
   e. Movement of G-6-P, but not glucose, is blocked by cyanide.

a
Factual Recall

8.59 Which of the following adheres to the extracellular surface of animal cell plasma membranes?
a. fibers of the extracellular matrix
b. fibers of the cytoskeleton
c. the phospholipid bilayer
d. cholesterol
e. carrier proteins

e
Application

8.60 You are conducting research on nerve cells. During an experiment, you administer an electrical stimulation to the cells. The probable result of this stimulation will be to
a. start the membrane water pump.
b. cause increased saturation of phospholipid tails.
c. result in increased membrane fluidity and asymmetry.
d. activate the active transport system.
e. open gated channels.

a
Factual Recall

8.61 You have a friend with high blood cholesterol levels and the cause is diagnosed as familial hypercholesterolemia. Of the following, the best explanation for this inherited condition is
a. defective LDL receptors on plasma membranes of her cells.
b. poor attachment of the cholesterol to the extracellular matrix of her cells.
c. a poorly formed lipid bilayer that cannot incorporate cholesterol in the membranes of her cells.
d. inhibition of the cholesterol active transport system in red blood cells.
e. a general lack of glycolipids in the blood cell membranes.

Chapter 9

9.1 Which of the following statements is true of glycolysis followed by fermentation?
  a. It produces a net gain of ATP.
  b. It produces a net gain of NADH.
  c. It is an aerobic process.
  d. It can be performed only by bacteria.
  e. It produces more energy per glucose molecule than does aerobic respiration.

9.2 The following statements compare combustion with the aerobic respiration of glucose. Which is FALSE?
  a. Combustion releases more total caloric energy from glucose than does respiration.
  b. Combustion releases energy from glucose at a more rapid rate than does respiration.
  c. Combustion releases nearly all energy as heat and light; respiration captures some of the energy in chemical bonds.
  d. Combustion uses heat to provide activation energy; respiration uses enzymes to lower activation energy.
  e. Combustion involves the direct transfer of hydrogen atoms to oxygen; respiration uses an indirect transfer of hydrogens.

9.3 Glycolysis is believed to be one of the most ancient of metabolic processes. Which statement below LEAST supports this idea?
  a. If run in reverse, glycolysis will build glucose molecules.
  b. Glycolysis neither uses nor needs O_2.
  c. Glycolysis is found in all eukaryotic cells.
  d. The enzymes of glycolysis are found in the cytosol rather than in a membrane-bound organelle.
  e. Bacteria, the most primitive of cells, make extensive use of glycolysis.

9.4 Which of the following statements about lactate fermentation is FALSE?
  a. Lactate fermentation produces ATP molecules in addition to the few produced by glycolysis.
  b. Lactate fermentation oxidizes NADH to NAD^+ to keep glycolysis functioning.
  c. Lactate fermentation takes place in vigorously exercised muscle cells.
  d. Lactate fermentation can take place under anaerobic conditions.
  e. Lactate fermentation in muscle cells often creates a need for O_2 that must be satisfied later.

9.5 Which kind of metabolic poison would most directly interfere with glycolysis?
  a. an agent that reacts with oxygen and depletes its concentration in the cell
  b. an agent that binds to pyruvate and inactivates it
  c. an agent that closely mimics the structure of glucose but is nonmetabolic
  d. an agent that reacts with NADH and oxidizes it to NAD^+
  e. an agent that inhibits the formation of acetyl coenzyme A

d
Factual Recall

9.6 All of the following statements about glycolysis are true EXCEPT:
a. Glycolysis has steps involving oxidation–reduction reactions.
b. The enzymes of glycolysis are located in the cytosol of the cell.
c. Glycolysis can operate in the complete absence of O_2.
d. The end products of glycolysis are CO_2 and H_2O.
e. Glycolysis makes ATP exclusively through substrate-level phosphorylations.

b
Conceptual
Understanding

9.7 Which of the following statements about NAD^+ is FALSE?
a. NAD^+ is reduced to NADH during both glycolysis and the Krebs cycle.
b. NAD^+ has more chemical energy than NADH.
c. NAD^+ is reduced by the action of dehydrogenases.
d. NAD^+ can receive electrons for use in oxidative phosphorylation.
e. In the absence of NAD^+, glycolysis cannot function.

b
Conceptual
Understanding

9.8 Which metabolic process is most closely associated with intracellular membranes?
a. substrate-level phosphorylation
b. oxidative phosphorylation
c. glycolysis
d. the Krebs cycle
e. ethanolic fermentation

c
Conceptual
Understanding

9.9 Pyruvate is the last product of glycolysis. Which statement below is TRUE?
a. There is more energy in 6 molecules of carbon dioxide than in 2 molecules of pyruvate.
b. There is more energy in pyruvate than in lactate.
c. There is less energy in two molecules of pyruvate that in one molecule of glucose.
d. Pyruvate is in a more oxidized state than carbon dioxide.
e. Pyruvate is in a more reduced state than glucose.

c
Factual Recall

9.10 During oxidative phosphorylation, H_2O is formed. Where do the oxygen atoms in the H_2O come from?
a. carbon dioxide
b. glucose
c. molecular oxygen
d. pyruvate
e. lactate

c
Conceptual
Understanding

9.11 What does chemiosmosis involve?
a. the diffusion of water down an electrochemical gradient that drives ATP synthesis
b. a proton gradient that drives the redox reactions of electron transport
c. a proton-motive force that drives the synthesis of ATP
d. an ATP synthase that pumps protons across the inner mitochondrial membrane
e. the uptake of NADH produced in glycolysis into the mitochondrion

e
Application

9.12 Muscle cells in oxygen deprivation convert pyruvate to _____ and in this step gain _____.
a. lactate; ATP
b. alcohol; CO_2
c. alcohol; ATP
d. ATP; NAD^+
e. lactate; NAD^+

a
Factual Recall

9.13 Phosphofructokinase is an important control enzyme. Which of the following statements concerning this enzyme is FALSE?
a. It is activated by citrate.
b. It is inhibited by ATP.
c. It is activated by ADP.
d. It is a coordinator of the processes of glycolysis and the Krebs cycle.
e. It is an allosteric enzyme.

d
Factual Recall

9.14 Which type of enzyme in cellular respiration is primarily responsible for removing electrons from organic molecules?
a. decarboxylase
b. ATP synthase
c. deaminase
d. dehydrogenase
e. phosphofructokinase

c
Factual Recall

9.15 Why is it impossible to quantify the amount of ATP derived from each glucose molecule during cellular respiration?
a. Our techniques are not good enough.
b. The mitochondria are too unstable.
c. The proton gradient is used for many purposes.
d. ATP is used up as soon as it is produced.
e. The ATP remains inside the mitochondria.

a
Conceptual
Understanding

9.16 Assume that a eukaryotic cell has abundant glucose and O_2, but needs ATP. The proton gradient in mitochondria of this cell will be generated by _____ and used primarily for _____.
a. the electron transport chain; ATP synthesis
b. the electron transport chain; substrate-level phosphorylation
c. glycolysis; production of H_2O
d. fermentation; NAD reduction
e. diffusion of protons; ATP synthesis

b
Factual Recall

9.17 Which process in eukaryotic cells will normally proceed whether O_2 is present or absent?
a. fermentation
b. glycolysis
c. Krebs cycle
d. oxidative phosphorylation
e. electron transport

d
Conceptual
Understanding

9.18 The direct energy source that drives ATP synthesis during respiratory oxidative phosphorylation is
a. oxidation of glucose to CO_2 and water.
b. the thermodynamically favorable flow of electrons from NADH to the mitochondrial electron transport carriers.
c. the final transfer of electrons to oxygen.
d. the difference in H^+ concentrations on opposite sides of the inner mitochondrial membrane.
e. thermodynamically favorable transfer of phosphate from glycolysis and Krebs cycle intermediate molecules of ADP.

c
Application

9.19 DNP is a substance that causes membranes to become more permeable to H^+. What would you expect to happen to an animal that is given an injection of this substance and kept on the same diet throughout the experiment?
a. It would become hyperactive.
b. It would have a lower body temperature.
c. It would lose weight.
d. Its metabolic rate would decrease.
e. It would produce excess amounts of ATP.

a
Application

9.20 A fatty acid is partially oxidized to form 10 molecules of acetyl CoA. Starting with these 10 molecules, how many molecules of ATP will be made directly by the Krebs cycle only?
a. 10
b. 20
c. 32
d. 320
e. 686

c
Factual Recall

9.21 The primary function of the mitochondrion is the production of ATP. To carry out this function, the mitochondrion must have all of the following EXCEPT
a. the membrane-bound electron transport chain.
b. proton pumps embedded in the inner membrane.
c. enzymes for glycolysis.
d. enzymes for the Krebs cycle.
e. mitochondrial ATP synthase.

e
Factual Recall

9.22 A major function of the mitochondrial inner membrane is the conversion of energy from electrons to the stored energy of the phosphate bond in ATP. To accomplish this function, this membrane must have all of the following features EXCEPT
a. proteins to accept electrons from NADH.
b. integral, transverse ATP synthase.
c. proton pumps embedded in the membrane.
d. the electron transport chain of proteins.
e. high permeability to protons.

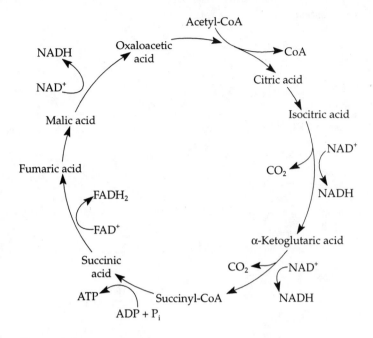

Figure 9.1

b
Conceptual
Understanding

9.23 Refer to Figure 9.1. If citric acid has six carbon atoms, how many carbon atoms does succinic acid have?
a. 1
b. 4
c. 5
d. 6
e. 12

a
Conceptual
Understanding

9.24 Refer to Figure 9.1. Starting with one acetyl CoA molecule, what is the maximum number of ATP molecules that could be made through substrate-level phosphorylation?
a. 1
b. 2
c. 11
d. 12
e. more than 20

Refer to Figure 9.2 to answer Questions 9.25–9.27. In Figure 9.2 there are some reactions of glycolysis in their proper sequence. Each reaction is lettered. Use these letters to answer the questions.

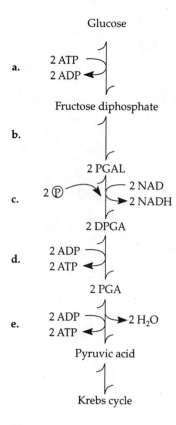

Figure 9.2

b
Conceptual
Understanding

9.25 Which reaction shows a split of one molecule into two smaller molecules?

c
Conceptual
Understanding

9.26 In which reaction is an inorganic phosphate added to the reactant?

e
Conceptual
Understanding

9.27 In which reaction is a net gain of ATP finally realized from glycolysis?

e
Conceptual
Understanding

9.28 Fermentation is not as energy productive as respiration because
   a. it does not take place in a specialized membrane-bound organelle.
   b. pyruvate is more reduced than CO_2; it still contains much of the energy from glucose.
   c. it takes place within the mitochondria of cells.
   d. it is the pathway common to fermentation and respiration.
   e. NAD^+ is regenerated by alcohol or lactate production, without the high-energy electrons passing through the electron transport chain.

b
Factual Recall

9.29 The oxygen consumed during cellular respiration is directly involved in
   a. glycolysis.
   b. accepting electrons at the end of the electron transport chain.
   c. the citric acid cycle.
   d. the oxidation of pyruvate to acetyl CoA.
   e. the phosphorylation of ADP.

a
Factual Recall

9.30 The ATP made during fermentation is generated by which of the following?
   a. substrate-level phosphorylation
   b. electron transport
   c. photophosphorylation
   d. chemiosmosis
   e. oxidation of NADH

d
Conceptual
Understanding

9.31 All of the following substances are produced in a muscle cell under anaerobic conditions EXCEPT
   a. ATP.
   b. pyruvate.
   c. lactate.
   d. acetyl CoA.
   e. NADH.

c
Factual Recall

9.32 In addition to ATP, what are the end products of glycolysis?
   a. CO_2 and H_2O
   b. CO_2 and ethyl alcohol
   c. NADH and pyruvate
   d. CO_2 and NADH
   e. H_2O and ethyl alcohol

c
Factual Recall

9.33 The Krebs cycle produces which of the following molecules that then transfer energy to the electron transport system?
   a. ATP and CO_2
   b. CO_2 and FAD
   c. $FADH_2$ and NADH
   d. NADH and ATP
   e. NADH, $FADH_2$, and ATP

b
Conceptual
Understanding

9.34 Catabolism of proteins, lipids, and carbohydrates can result in a 2-carbon molecule that enters the Krebs cycle. What is the molecule?
a. glucose
b. acetyl acid
c. a fatty acid
d. an amino acid
e. pyruvate

d
Factual Recall

9.35 In chemiosmotic phosphorylation, what is the most direct source of energy that is used to convert ADP + P_i to ATP?
a. energy released as electrons flow through the electron transport system
b. energy released from substrate-level phosphorylation
c. energy released from ATP synthase pumping hydrogen ions against their concentration gradient
d. energy released from diffusion of protons through ATP synthase
e. No external source of energy is required because the reaction is exergonic.

c
Application

9.36 Suppose a yeast cell uses 10 moles of glucose for energy production. No oxygen is available. What will be the maximum net yield of ATP in moles?
a. 12
b. 15
c. 20
d. 30
e. 36

a
Factual Recall

9.37 How many carbon atoms can each acetyl CoA feed into the Krebs cycle?
a. 2
b. 4
c. 6
d. 8
e. 10

Questions 9.38–9.41 are based on the stages of glucose oxidation listed below.
a. stage I: glycolysis
b. stage II: oxidation of pyruvate to acetyl CoA
c. stage III: Krebs cycle
d. stage IV: oxidative phosphorylation (chemiosmosis)

d
Factual Recall

9.38 Which one of the stages produces the most ATP when glucose is completely oxidized to carbon dioxide and water?

a
Factual Recall

9.39 Which one of the stages normally occurs whether or not oxygen is present?

a
Factual Recall

9.40 Which one of the stages occurs in the cytosol of the cell?

b
Factual Recall

9.41 Carbon dioxide is released during which stage(s)?
    a. stage III only
    b. stages II and III
    c. stages III and IV
    d. stages I, II, and III
    e. stages II, III, and IV

d
Factual Recall

9.42 Which of the following intermediary metabolites enters the Krebs cycle and is formed, in part, by the removal of CO_2 from a molecule of pyruvate?
    a. lactate
    b. glyceraldehyde phosphate
    c. oxaloacetic acid
    d. acetyl CoA
    e. citric acid

d
Factual Recall

9.43 When hydrogen ions are pumped from the mitochondrial matrix, across the inner membrane, and into the intermembrane space, the result is
    a. the formation of ATP.
    b. the reduction of NAD^+.
    c. the restoration of the Na^+–K^+ balance across the membrane.
    d. the creation of a proton gradient.
    e. the lowering of pH in the mitochondrial matrix.

e
Factual Recall

9.44 Carbon skeletons to be broken down during cellular respiration can be obtained from
    a. polysaccharides.
    b. proteins.
    c. lipids.
    d. Only a and b are correct.
    e. a, b, and c are correct.

d
Factual Recall

9.45 Where is ATP synthase located in the mitochondrion?
    a. ribosomes
    b. cytochrome system
    c. outer membrane
    d. inner membrane
    e. matrix

c
Application

9.46 Each time a molecule of glucose is completely oxidized, how many oxygen (O_2) molecules are required?
    a. 1
    b. 2
    c. 6
    d. 12
    e. None—you use up the CO_2 molecules instead.

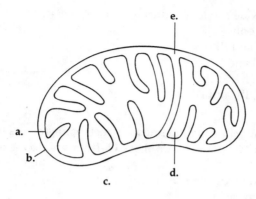

Figure 9.3

Refer to Figure 9.3 to answer Questions 9.47–9.49.

e
Factual Recall

9.47 The electron transport chain energy is used to pump H⁺ ions into which location?

c
Factual Recall

9.48 Glycolysis takes place in which location?

a
Factual Recall

9.49 Where are the proteins of the electron transport chain located?

a
Application

9.50 In order for a muscle cell to keep working, it must
    a. regenerate ATP at a very fast rate.
    b. receive sufficient oxygen for glycolysis.
    c. store sufficient ATP as a raw material for oxidative phosphorylation.
    d. rapidly replace the enzymes of respiration.
    e. effectively transport NADH into the mitochondria.

d
Factual Recall

9.51 Cellular respiration harvests the most chemical energy by
    a. substrate-level phosphorylation.
    b. forming lactate from pyruvate.
    c. converting oxygen to ATP.
    d. transferring electrons from organic molecules to oxygen.
    e. generating carbon dioxide and oxygen in the electron transport chain.

b
Application

9.52 During cellular respiration, electrons travel downhill from
    a. food → Krebs cycle → ATP → NAD⁺.
    b. food → NADH → electron transport chain → oxygen.
    c. glucose → ATP → oxygen.
    d. glucose → ATP → electron transport chain → NADH.
    e. food → glycolysis → Krebs cycle → NADH → ATP.

e
Factual Recall

9.53 During aerobic respiration, which of the following directly donates electrons to the electron transport chain at the lowest energy level?
   a. NAD^+
   b. NADH
   c. ATP
   d. $ADP + P_i$
   e. $FADH_2$

e
Application

9.54 Inside an active mitochondrion, most electrons follow which pathway?
   a. glycolysis → NADH → oxidative phosphorylation → ATP → oxygen
   b. Krebs cycle → $FADH_2$ → electron transport chain → ATP
   c. electron transport chain → Krebs cycle → ATP
   d. pyruvate → Krebs cycle → ATP → NADH → oxygen
   e. Krebs cycle → NADH → electron transport chain → oxygen

c
Factual Recall

9.55 You eat a cheeseburger and a fresh salad. Which of the following molecules in your food is NOT normally oxidized in aerobic respiration to generate ATP?
   a. sucrose
   b. lipids
   c. nucleic acids
   d. proteins
   e. amino acids

a
Factual Recall

9.56 Which of the following is not a "food" molecule accepted by glycolysis for catabolism?
   a. fatty acids
   b. sucrose
   c. glucose
   d. glycerol
   e. starch

a
Application

9.57 You have a friend who lost 15 pounds of fat on a diet. Where did the fat go (how was it lost)?
   a. It was released as CO_2 and H_2O.
   b. Chemical energy was converted to heat and then released.
   c. It was converted to ATP, which weighs much less than fat.
   d. It was broken down to amino acids and eliminated from the body.
   e. It was converted to urine and eliminated from the body.

d
Application

9.58 The complete aerobic respiration of sucrose, a disaccharide of glucose and fructose, or of maltose, a disaccharide of two glucose molecules, would release _____ molecules of CO_2?
   a. 2
   b. 3
   c. 6
   d. 12
   e. None; disaccharides are not food molecules for aerobic respiration.

c
Application

9.59 Which of the following is NOT true concerning the cellular compartmentation of the steps of respiration or fermentation?
    a. Acetyl CoA is produced only in the mitochondria.
    b. Lactate is produced only in the cytosol.
    c. NADH is produced only in the mitochondria.
    d. $FADH_2$ is produced only in the mitochondria.
    e. ATP is produced in the cytosol and the mitochondria.

a
Application

9.60 A young relative of yours has never had much energy. He goes to a doctor for help and is sent to the hospital for some tests. There they discover his mitochondria can use only fatty acids and amino acids for respiration, and his cells produce more lactate than normal. Of the following, which is the best explanation of his condition?
    a. His mitochondria lack the transport protein that moves pyruvate across the outer mitochondrial membrane.
    b. His cells cannot move NADH from glycolysis into the mitochondria.
    c. His cells contain something that inhibits oxygen use in his mitochondria.
    d. His cells lack the enzyme in glycolysis that forms pyruvate.
    e. His cells have a defective electron transport chain, so glucose goes to lactate instead of to acetyl CoA.

e
Application

9.61 An organism is discovered that consumes a considerable amount of sugar, yet does not gain much weight when denied air. Curiously, the consumption of sugar increases as air is removed from the organism's environment, but the organism seem to thrive even in the absence of air. When returned to normal air, the organism does fine. Which of the following best describes the organism?
    a. It must use a molecule other than oxygen to accept electrons from the electron transport chain.
    b. It is a normal eukaryotic organism.
    c. The organism obviously lacks the Krebs cycle and electron transport chain.
    d. It is an anaerobic organism.
    e. It is a facultative anaerobe.

c
Application

9.62 Which of the following is NOT a function of glycolysis?
    a. production of ATP
    b. production of NADH
    c. production of $FADH_2$
    d. formation of pyruvate
    e. splitting the cabon skeletons of simple sugars

e
Application

9.63 Which of the following is NOT a function of the Krebs cycle?
    a. production of ATP
    b. production or NADH
    c. production of $FADH_2$
    d. release of carbon dioxide
    e. splitting the carbon skeletons of simple sugars

Chapter 10

10.1 Plants that fix CO_2 into organic acids at night when the stoma are open and carry out the Calvin cycle during the day when the stoma are closed are called
   a. C_3 plants.
   b. C_4 plants.
   c. CAM plants.
   d. Only a and b are correct.
   e. a, b, and c are correct.

10.2 Photorespiration lowers the efficiency of photosynthesis by removing which of the following from the Calvin cycle?
   a. carbon dioxide molecules
   b. glyceraldehyde phosphate molecules
   c. ATP molecules
   d. ribulose bisphosphate molecules
   e. RuBP carboxylase molecules

10.3 Assume a thylakoid is somehow punctured so that the interior of the thylakoid is no longer separated from the stroma. This damage will have the most direct effect on which of the following processes?
   a. the splitting of water
   b. the absorption of light energy by chlorophyll
   c. the flow of electrons from photosystem II to photosystem I
   d. the synthesis of ATP
   e. the reduction of $NADP^+$

10.4 Why are C_4 plants able to photosynthesize with no apparent photorespiration?
   a. They do not carry out the Calvin cycle.
   b. They use a more efficient enzyme to initially fix CO_2.
   c. They are adapted to cold, wet climates.
   d. They conserve water more efficiently.
   e. They exclude oxygen from their tissues.

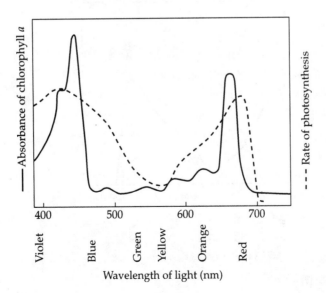

Figure 10.1

d
Conceptual
Understanding

10.5 Figure 10.1 shows the absorption spectrum for chlorophyll *a* and the action spectrum for photosynthesis. Why are they different?
   a. Green and yellow wavelengths inhibit the absorption of red and blue wavelengths.
   b. Bright sunlight destroys photosynthetic pigments.
   c. The two lines are probably the result of inaccurate measurements.
   d. Other pigments absorb light in addition to chlorophyll *a*.
   e. Anaerobic bacteria probably interfered with light absorption.

e
Conceptual
Understanding

10.6 If Figure 10.1 showed the absorbance of the total photosynthetic pigments extracted from the plant (instead of pure chlorophyll *a*), the absorption spectrum curve would
   a. still be as different from the action spectrum as the curve shown in Figure 10.1.
   b. be missing the peak between the violet and blue wavelengths.
   c. be missing the peak between the orange and red wavelengths.
   d. gain two new sharp peaks after the red wavelength.
   e. better match the action spectrum.

d
Factual Recall

10.7 Which of the following statements about the light reactions of photosynthesis is FALSE?
   a. The splitting of water molecules provides a source of electrons.
   b. Chlorophyll (and other pigments) absorb light energy which excites electrons.
   c. An electron transport chain is used to create a proton gradient.
   d. The proton gradient is used to reduce NADP.
   e. Some electrons are recycled and some are not.

c
Conceptual
Understanding

10.8 Which of the following are products of the Calvin cycle and are utilized in the light reactions of photosynthesis?
   a. CO_2 and glucose
   b. H_2O and O_2
   c. ADP, P_i, and $NADP^+$
   d. electrons and H^+
   e. Both c and d are correct.

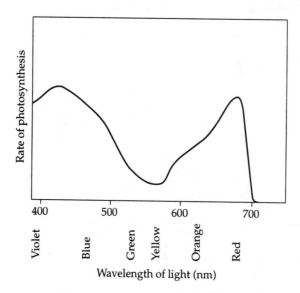

Figure 10.2

b
Application

10.9 From the photosynthetic action spectrum in Figure 10.2, we can correctly conclude that
a. chlorophyll absorbs more green than red light.
b. plants can use all colors of visible light for photosynthesis.
c. every color of light is equally good for photosynthesis.
d. light in the green range produces the most efficient photosynthesis.
e. there are two photosystems in leaves: one for red, one for blue.

b
Conceptual
Understanding

10.10 The primary function of the light reactions of photosynthesis is
a. to produce energy-rich glucose from carbon dioxide and water.
b. to produce energy-rich ATP and NADPH.
c. to produce NADPH used in respiration.
d. to convert light energy to the chemical energy of PGAL.
e. to use ATP to make glucose.

d
Factual Recall

10.11 The reactions of the Calvin cycle require all of the following molecules EXCEPT
a. CO_2.
b. ATP.
c. RuBP.
d. glucose.
e. NADPH.

c
Conceptual
Understanding

10.12 All of the following statements are true EXCEPT:
a. Thylakoid membranes contain the photosynthetic pigments.
b. The O_2 released during photosynthesis comes from water.
c. Glyceraldehyde phosphate is produced only in the light reactions of photosynthesis.
d. The light reactions of photosynthesis provide the energy for the Calvin cycle.
e. When chlorophyll is reduced, it gains electrons.

a
Conceptual
Understanding

10.13 You have just discovered a new flower species that has a unique photosynthetic pigment. The leaves of this plant appear to be reddish yellow. What wavelengths of visible light are not being absorbed by this pigment?
a. red and yellow
b. blue and violet
c. green and yellow
d. blue, green, and red
e. green, blue, and violet

c
Factual Recall

10.14 All of the events listed below occur in the energy-capturing light reactions of photosynthesis EXCEPT:
a. Oxygen is produced.
b. $NADP^+$ is reduced to NADPH.
c. Carbon dioxide is incorporated into PGA.
d. ADP is phosphorylated to yield ATP.
e. Light is absorbed.

a
Factual Recall

10.15 The chemiosomotic process in chloroplasts involves the
    a. establishment of a proton gradient.
    b. diffusion of electrons through the thylakoid membrane.
    c. oxidation of water to produce ATP energy.
    d. movement of water by osmosis into the thylakoid space from the stroma.
    e. reduction of carbon dioxide to glucose by NADPH and ATP.

c
Conceptual
Understanding

10.16 Which of the following enzymes is probably the most abundant protein in the world?
    a. PEP carboxylase
    b. hexokinase
    c. RuBP carboxylase/oxygenase
    d. aldolase
    e. pyruvate kinase

a
Factual Recall

10.17 In C_4 photosynthesis, carbon fixation takes place in the _____ cells, and then is transferred as malic or aspartic acid to ____ cells where carbon dioxide is released for entry into the Calvin cycle.
    a. mesophyll; bundle sheath
    b. stomatal; mesophyll
    c. bundle sheath; epidermal
    d. epidermal; mesophyll
    e. stomatal; epidermal

d
Factual Recall

10.18 Because bundle sheath cells are relatively protected from atmospheric oxygen, the level of ____ is held to a minimum in C_4 plants.
    a. glycolysis
    b. photosynthesis
    c. oxidative phosphorylation
    d. photorespiration
    e. decarboxylation of malic acid

c
Conceptual
Understanding

10.19 Which of the following events in the functioning of photosystem II is FALSE?
    a. Light energy excites electrons in an antenna pigment in a photosynthetic unit.
    b. The excitation is passed along to a molecule of P680 chlorophyll in the photosynthetic unit.
    c. The P680 chlorophyll donates a pair of protons to NADPH, which is thus converted to $NADP^+$.
    d. The electron vacancies in P680 are filled by electrons derived from water.
    e. The spitting of water yields molecular oxygen as a by-product.

d
Factual Recall

10.20 All of the following compounds are required (i.e., are necessary constituents for chemical reactions) at some stage of green plant photosynthesis, EXCEPT
    a. adenosine triphosphate.
    b. NADP.
    c. water.
    d. oxygen.
    e. carbon dioxide.

c
Factual Recall

10.21 When a chlorophyll molecule in photosystem I traps light, it loses an electron. In noncyclic electron flow, this electron is replaced
  a. from one of the antenna pigments.
  b. from the other end of photosystem I.
  c. by a donation from photosystem II.
  d. by a donation from an unexcited chlorophyll molecule.
  e. from one of the hydrogen atoms in NADP.

c
Application

10.22 On which of the following features do most plant and animal cells differ?
  a. active transport mechanisms
  b. mitochondrial function
  c. primary energy source
  d. transcription
  e. structure of nucleus

d
Conceptual
Understanding

10.23 Which of the following is NOT TRUE of ribulose bisphosphate carboxylase?
  a. It is a protein.
  b. It speeds up a chemical reaction.
  c. It lowers the energy of activation.
  d. It catalyzes a phosphorylation reaction.
  e. It has an affinity for both O_2 and CO_2.

a
Factual Recall

10.24 Where does the Calvin cycle of photosynthesis take place?
  a. stroma of the chloroplast
  b. thylakoid membrane
  c. cytoplasm surrounding the chloroplast
  d. chlorophyll molecule
  e. outer membrane of the chloroplast

a
Factual Recall

10.25 CAM plants can keep stomates closed in daytime, thus reducing loss of water. They can do this because they can
  a. fix CO_2 into organic acids during the night.
  b. fix CO_2 into sugars in the bundle-sheath cells.
  c. fix CO_2 into pyruvic acid in the mesophyll cells.
  d. use the enzyme phosphofructokinase, which outcompetes rubisco for CO_2.
  e. use photosystems I and II at night.

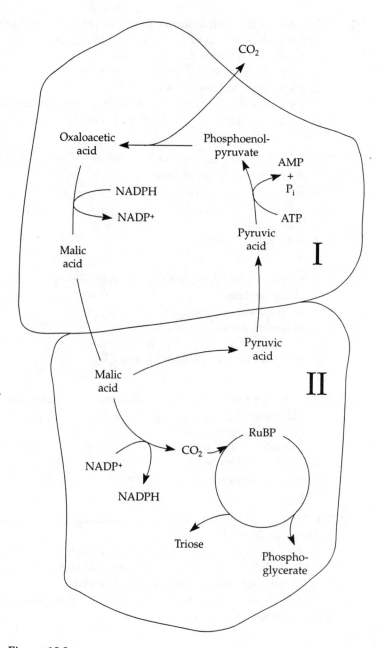

Figure 10.3

a
Factual Recall

10.26 Which of the following statements is TRUE concerning the diagram in Figure 10.3?
   a. This diagram represents some of the events of C_4 photosynthesis.
   b. Cell I is a bundle-sheath cell.
   c. The light reactions take place only in cell I.
   d. This diagram represents the type of photosynthesis found in most aquatic plants.
   e. These kinds of cells are usually found in conifers that are in hot, dry climates.

b
Conceptual
Understanding

10.27 Referring to Figure 10.3, oxygen would inhibit the CO_2 fixation reactions in
a. cell I only.
b. cell II only.
c. neither cell I nor cell II.
d. both cell I and cell II.
e. cell I during the night and cell II during the day.

b
Conceptual
Understanding

10.28 All of the following statements are correct regarding the Calvin cycle of
photosynthesis EXCEPT:
a. The energy source utilized is the ATP and NADPH obtained through the
light reaction.
b. These reactions begin soon after sundown and end before sunrise.
c. The 5-carbon sugar RuBP is constantly being regenerated.
d. One of the end products is glyceraldehyde phosphate.
e. The pathway used is the Calvin cycle.

d
Factual Recall

10.29 Of the following, the color of light least effective in driving photosynthesis is
a. blue.
b. red.
c. orange.
d. green.
e. yellow.

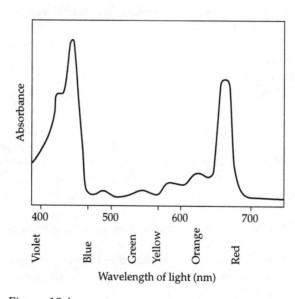

Figure 10.4

d
Conceptual
Understanding

10.30 Examine this absorption spectrum for a plant pigment (Figure 10.4). The color
of this pigment is most likely some shade of
a. blue.
b. violet.
c. red.
d. green.
e. blue and red mixed.

a

Factual Recall

10.31 Cyclic electron flow in the chloroplast produces
   a. ATP.
   b. NADPH.
   c. glucose.
   d. Only a and b are correct.
   e. a, b, and c are correct.

a

Factual Recall

10.32 In plant cells, ATP is made in response to light. An electron transport chain is involved. This electron transport chain is found in the
   a. thylakoid membranes of chloroplasts.
   b. stroma of chloroplasts.
   c. inner membrane of mitochondria.
   d. matrix of mitochondria.
   e. cytoplasm.

e

Conceptual
Understanding

10.33 If photosynthesizing green algae are provided with CO_2 synthesized with heavy oxygen (^{18}O), later analysis will show that all but one of the following compounds produced by the algae contain the ^{18}O label. That one exception is
   a. PGA.
   b. PGAL.
   c. glucose.
   d. RuBP.
   e. O_2.

c

Factual Recall

10.34 The process of noncyclic photophosphorylation uses light energy to synthesize
   a. ADP and ATP.
   b. ATP and P700.
   c. ATP and NADPH.
   d. ADP and NADP.
   e. P700 and P680.

Use the following information to answer Questions 10.35–10.37.
Thomas Engelmann illuminated a filament of algae with light that passed through a prism, thus exposing different segments of algae to different wavelengths of light. He added aerobic bacteria and then noted in which areas the bacteria congregated. He noted that the largest groups were found in the areas illuminated by the red and blue light.

c

Factual Recall

10.35 What did he conclude about the congregation of bacteria in the red and blue areas?
   a. Bacteria released excess carbon dioxide in these areas.
   b. Bacteria congregated in these areas due to an increase in the temperature of the red and blue light.
   c. Bacteria congregated in these areas because these areas had the most oxygen being released.
   d. Bacteria are attracted to red and blue light and thus these wavelengths are more reactive than other wavelengths.
   e. Bacteria congregated in these areas due to an increase in the temperature caused by an increase in photosynthesis.

d
Factual Recall

10.36 The purpose of this experiment was to determine
- a. the relationship between heterotrophic and autotrophic organisms.
- b. the relationship between wavelengths of light and the rate of aerobic respiration.
- c. the relationship between wavelengths of light and the amount of heat released.
- d. the relationship between wavelengths of light and the rate of photo-synthesis.
- e. the relationship between the concentration of carbon dioxide and the rate of photosynthesis.

b
Application

10.37 If you ran the same experiment without passing light through a prism, what would you predict?
- a. There would be no difference in results.
- b. The bacteria would be relatively evenly distributed along the algal filaments.
- c. The number of bacteria present would decrease due to an increase in the carbon dioxide concentration.
- d. The number of bacteria present would increase due to an increase in the carbon dioxide concentration.
- e. The number of bacteria would decrease due to a decrease in the temperature of the water.

a
Factual Recall

10.38 Members of the Crassulaceae (CAM plants) differ from C_4 plants in that they
- a. incorporate carbon dioxide into organic acids at night.
- b. incorporate carbon dioxide into a three-carbon compound.
- c. incorporate carbon dioxide into a four-carbon compound.
- d. do not use rubisco as an enzyme.
- e. use phosphoenolypyruvic acid as a source of carbon dioxide.

b
Conceptual
Understanding

10.39 Which one of the following statements BEST describes the relationship between photosynthesis and respiration?
- a. Respiration is the exact reversal of the biochemical pathways of photo-synthesis.
- b. Photosynthesis stores energy in complex organic molecules and respiration releases it.
- c. Photosynthesis occurs only in plants and respiration occurs only in animals.
- d. ATP molecules are produced in photosynthesis and used up in respiration.
- e. Respiration is anabolic and photosynthesis is catabolic.

d
Conceptual
Understanding

10.40 In a plant cell, where is ATP synthase located?
- a. thylakoid membrane
- b. plasma membrane
- c. inner mitochondrial membrane
- d. a and c
- e. a, b, and c

Refer to the following choices to answer Questions 10.41–10.47. Each choice may be used once, more than once, or not at all. Indicate whether the following events occur during
> a. *photosynthesis*
> b. *respiration*
> c. *both photosynthesis and respiration*
> d. *neither photosynthesis nor respiration*

c
Conceptual
Understanding

10.41 Synthesis of ATP by the chemiosmotic mechanism.

a
Conceptual
Understanding

10.42 Oxidation of water.

a
Factual Recall

10.43 Reduction of $NADP^+$.

a
Factual Recall

10.44 CO_2 fixation.

b
Factual Recall

10.45 Electron flow along a cytochrome chain.

b
Factual Recall

10.46 Oxidative phosphorylation.

c
Conceptual
Understanding

10.47 Generation of proton gradients across membranes.

a
Application

10.48 Which of the following is NOT a useful function of the light reactions?
> a. releasing oxygen for photorespiration
> b. splitting water
> c. synthesis of NADPH
> d. harvesting photons of light
> e. converting light energy to chemical energy

d
Application

10.49 Which of the following is NOT directly associated with photosystem II?
> a. splitting water
> b. release of oxygen
> c. harvesting light energy by chlorophyll
> d. photophosphorylation
> e. P680

d
Conceptual
Understanding

10.50 Which of the following is NOT directly associated with photosystem I?
   a. harvesting light energy by chlorophyll
   b. receiving electrons from plastocyanin
   c. P700
   d. splitting water
   e. passing electrons to ferredoxin

e
Factual Recall

10.51 What are the products of the light reactions that are subsequently used by the Calvin cycle?
   a. oxygen and carbon dioxide
   b. carbon dioxide and RuBP
   c. water and carbon
   d. electrons and photons
   e. ATP and NADPH

e
Conceptual
Understanding

10.52 In green plants, the primary function of the Calvin cycle is to
   a. use ATP to release carbon dioxide.
   b. use NADPH to release carbon dioxide.
   c. split water and release oxygen.
   d. transport RuBP out of the chloroplast.
   e. construct simple sugars from carbon dioxide.

c
Conceptual
Understanding

10.53 As a research scientist, you measure the amount of ATP and NADPH consumed by the Calvin cycle in 1 hour. You find 30,000 molecules of ATP consumed, but only 20,000 molecules of NADPH. Where did the extra ATP molecules come from?
   a. photosystem II
   b. photosystem I
   c. cyclic electron flow
   d. noncyclic electron flow
   e. chlorophyll

a
Application

10.54 You are a research scientist studying photosynthesis. In an experiment performed during the day, you provide a new plant, just discovered in South America, with radioactive carbon (^{14}C) dioxide as a metabolic tracer. The ^{14}C is incorporated first into oxaloacetic acid. The plant is best characterized as
   a. a C_4 plant.
   b. a C_3 plant.
   c. a CAM plant.
   d. a heterotroph.
   e. a chemoautotroph.

a
Conceptual
Understanding

10.55 Which of the following statements best represents the relationships between the light reactions and the Calvin cycle?
  a. The light reactions provide ATP and NADPH to the Calvin cycle, and the cycle returns ADP, P_i, and $NADP^+$ to the light reactions.
  b. The light reactions provide ATP and NADPH to the carbon fixation step of the Calvin cycle, and the cycle provides water and electrons to the light reactions.
  c. The light reactions supply the Calvin cycle with CO_2 to produce sugars, and the Calvin cycle supplies the light reactions with sugars to produce ATP.
  d. The light reactions provide the Calvin cycle with oxygen for electron flow, and the Calvin cycle provides the light reactions with water to split.
  e. There is no relationship between the light reactions and the Calvin cycle.

b
Factual Recall

10.56 In the thylakoid membranes, what is the main role of the antenna pigment molecules?
  a. to split water and release oxygen to the reaction center chlorophyll
  b. to harvest photons and transfer light energy to the reaction center chlorophyll
  c. to synthesize ATP from ADP and P_i
  d. to pass electrons to ferredoxin and then NADPH
  e. to concentrate photons inside the stroma

c
Conceptual
Understanding

10.57 In mitochondria, chemiosmosis translocates protons from the matrix into the intermembrane space, whereas in chloroplasts, chemiosmosis translocates protons from
  a. the stroma to the chlorophyll.
  b. the matrix to the stroma.
  c. the stroma into the thylakoid compartment.
  d. the intermembrane space to the matrix.
  e. the light reactions to the Calvin cycle.

a
Conceptual
Understanding

10.58 Biologists refer to autotrophs as the producers in an ecosystem because
  a. they make the organic material that becomes the ecosystem's food stores.
  b. they release the oxygen produced in photosynthesis.
  c. they produce the consumers.
  d. they evolved first in the line of production.
  e. they evolved from consumers.

For Questions 10.59–10.68, compare the light reactions with the Calvin cycle of photosynthesis in plants. Use the following key:
  a. light reactions alone
  b. the Calvin cycle alone
  c. both the light reactions and the Calvin cycle
  d. neither the light reactions nor the Calvin cycle
  e. occurs in the chloroplast but is not part of photosynthesis

a
Factual Recall

10.59 Produces molecular oxygen (O_2).

b
Factual Recall

10.60 Forms a proton gradient.

b
Factual Recall

10.61 Requires ATP.

a
Factual Recall

10.62 Requires ADP.

d
Factual Recall

10.63 Produces NADH.

a
Factual Recall

10.64 Produces NADPH.

b
Factual Recall

10.65 Produces triose sugars.

c
Factual Recall

10.66 Is inactive in the dark.

b
Factual Recall

10.67 Requires CO_2.

d
Factual Recall

10.68 Requires glucose.

a
Application

10.69 The three substrates (normal reactants) for the enzyme RuBP carboxylase
(rubisco) are
   a. CO_2, O_2, and RuBP.
   b. CO_2, glucose, and RuBP.
   c. RuBP, ATP, and NADPH.
   d. triose-P, glucose, and CO_2.
   e. RuBP, CO_2, and ATP.

Questions 10.70–10.72 refer to the following situation:
*A colleague from Asia brings you two plants that seem to have potential as food crops. You
conduct a series of experiments on the two plants, named Blue Ribbon (BR) and Underdog
(UD). The data in Figure 10.5 are from one experiment in which you measured the rate of
photosynthesis (PHS; rate in mmol of CO_2 fixed per hour) in both plants under experimental conditions of increasing levels of oxygen in air.*

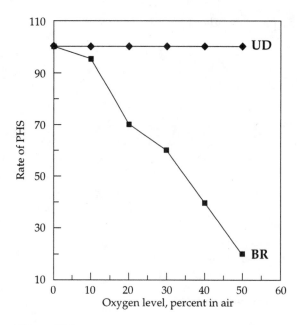

Figure 10.5

c
Conceptual
Understanding

10.70 From these data, what is the best conclusion about the mechanism of
photosynthesis in the two plants?
a.   UD is a CAM plant.
b.   UD is a C_3 plant and BR is a C_4 plant.
c.   UD is a C_4 plant and BR is a C_3 plant.
d.   Only UD has the light reactions of photosynthesis.
e.   Only BR has the Calvin cycle of photosynthesis.

b
Conceptual
Understanding

10.71 Photorespiration
a.   is not a factor in BR.
b.   explains the decrease in photosynthetic rate of BR.
c.   is most active at zero oxygen levels in both plants.
d.   keeps the level of photosynthesis high in UD.
e.   cannot be judged from these data on photosynthesis.

c
Conceptual
Understanding

10.72 If the level of CO_2 had been increased to 5% in air, the results of this
experiment most likely would have
a.   remained the same for both BR and UD.
b.   shown a higher photosynthetic rate curve for UD, but not BR.
c.   shown a higher photosynthetic rate curve for BR, but not UD.
d.   shown a higher photosynthetic rate curve for both BR and UD.
e.   decreased the rate of photosynthesis in both plants.

Use the data shown in Figure 10.6 to answer Questions 10.73–10.75.
Changes in leaf pH for three different plants grown in the same greenhouse were determined over a 24-hour period. All conditions were properly controlled. Use the key below for
your answers.

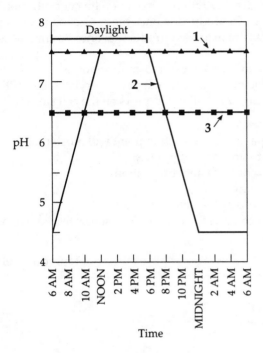

Figure 10.6

KEY:
a. a C₃ plant
b. a C₄ plant
c. a CAM plant
d. both C₃ and C₄ plants
e. both C₄ and CAM plants

a
Conceptual
Understanding

10.73 Curve 1 shows the expected response of _____.

c
Conceptual
Understanding

10.74 Curve 2 shows the expected response of _____.

b
Conceptual
Understanding

10.75 Curve 3 shows the expected response of _____.

Chapter 11

d
Factual Recall

11.1 What causes the rhythmic change in cyclin concentration in the cell cycle?
    a. an increase in production once the restriction point is passed
    b. the cascade of increased production once its enzyme is phosphorylated by cdc2
    c. the changing ratio of cytoplasm to genome
    d. its destruction by an enzyme phosphorylated by MPF
    e. the binding of PDGF to receptors on the cell surface

b
Factual Recall

11.2 A cell that passes the restriction point will most likely
    a. move into prophase of mitosis.
    b. undergo chromosome duplication.
    c. stop dividing.
    d. show a drop in MPF concentration.
    e. have just completed cytokinesis to separate into two new cells.

b
Factual Recall

11.3 Which of the following organisms does not reproduce cells by mitosis and cytokinesis?
    a. cow
    b. bacterium
    c. mushroom
    d. cockroach
    e. banana tree

b
Factual Recall

11.4 Cytokinesis usually, but not always, follows mitosis. If cells undergo mitosis and not cytokinesis, this would result in
    a. a cell with a single large nucleus.
    b. cell structures with two or more nuclei.
    c. cells with abnormally small nuclei.
    d. feedback responses that prevent mitosis.
    e. death of the cell line.

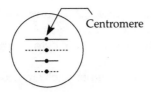

Figure 11.1

Use the following information to answer Questions 11.5 and 11.6.
Figure 11.1 shows a diploid cell with four chromosomes. There are two types of chromosomes, one long and the other short. One haploid set is symbolized by unbroken lines, while the other haploid set is represented by dotted lines. At this time, the chromosomes have not yet replicated and each chromosome has one chromatid. Now, choose the correct chromosomal conditions for the following stages:

b
Factual Recall

11.5 at metaphase of mitosis

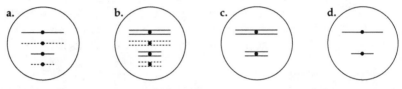

a
Factual Recall

11.6 a possible daughter cell of mitosis

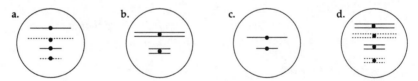

b
Conceptual
Understanding

11.7 If there are 12 chromosomes in an animal cell in the G_1 stage of the cell cycle, what is the diploid number of chromosomes for this organism?
  a. 6
  b. 12
  c. 24
  d. 36
  e. 48

c
Factual Recall

11.8 Regarding mitosis and cytokinesis, one difference between higher plants and animals is that in plants
  a. the spindles contain microfibrils in addition to microtubules, whereas animal spindles do not contain microfibrils.
  b. sister chromatids are identical, but they differ from one another in animals.
  c. a cell plate begins to form at telophase, whereas in animals a cleavage furrow is initiated at that stage.
  d. chromosomes become attached to the spindle at prophase, whereas in animals chromosomes do not become attached until anaphase.
  e. spindle poles contain centrioles, whereas spindle poles in animals do not.

d
Conceptual
Understanding

11.9 How do the daughter cells at the end of mitosis and cytokinesis compare with their parent cell when it was in G_1 of the cell cycle?
   a. The daughter cells have half the amount of cytoplasm and half the amount of DNA.
   b. The daughter cells have half the number of chromosomes and half the amount of DNA.
   c. The daughter cells have the same number of chromosomes and half the amount of DNA.
   d. The daughter cells have the same number of chromosomes and the same amount of DNA.
   e. The daughter cells may have new combinations of genes due to crossing over.

c
Factual Recall

11.10 Cells that have stopped dividing and are differentiating are
   a. cancer cells.
   b. in the G_2 phase of the cell cycle.
   c. in the G_1 phase of the cell cycle.
   d. in the S phase of the cell cycle.
   e. in the M phase of the cell cycle.

c
Application

11.11 Measurements of the amount of DNA per nucleus were taken on a large number of cells from a growing fungus. The measured DNA levels ranged from 3 to 6 picograms per nucleus. One nucleus had 5 picograms of DNA. What stage of the cell cycle was this nucleus in?
   a. G_0
   b. G_1
   c. S
   d. G_2
   e. M

Questions 11.12–11.16 consist of five phrases or sentences related to the control of cell division. For each one, select the term from below that is most closely related to it. Each term may be used once, more than once, or not at all.
   a. PDGF
   b. MPF
   c. protein kinase
   d. cyclin
   e. cdc2

a
Factual Recall

11.12 Released by platelets in the vicinity of an injury.

c
Factual Recall

11.13 A family of enzymes that catalyzes the transfer of a phosphate group from ATP to target proteins.

a
Factual Recall

11.14 Fibroblasts have receptors for this substance on their plasma membranes.

d
Factual Recall

11.15 A substance synthesized through the cell cycle that accumulates during early interphase and associates with another protein to form active enzymes.

d
Factual Recall

11.16 Triggers the activation of numerous proteins that facilitate mitosis.

Questions 11.17–11.21 consist of five phrases or sentences concerned with the cell cycle. For each phrase or sentence, select the answer letter from below that is most closely related to it. Each answer may be used once, more than once, or not at all.
  a. G_0
  b. G_1
  c. S
  d. G_2
  e. M

b
Factual Recall

11.17 The "restriction point" occurs here.

a
Factual Recall

11.18 Nerve and muscle cells are in this phase.

e
Factual Recall

11.19 the shortest part of the cycle

c
Factual Recall

11.20 Chromosomes are duplicated during this phase.

e
Factual Recall

11.21 Cyclin is destroyed toward the end of this phase.

e
Factual Recall

11.22 The formation of a cell plate is beginning across the middle of a cell and nuclei are reforming at opposite ends of the cell. What kind of a cell is this?
  a. an animal cell in metaphase
  b. an animal cell in telophase
  c. an animal cell undergoing cytokinesis
  d. a plant cell in metaphase
  e. a plant cell undergoing cytokinesis

Questions 11.23–11.26 refer to the following terms. Each term may be used once, more than once, or not at all.
  a. telophase
  b. anaphase
  c. prometaphase
  d. metaphase
  e. prophase

c
Factual Recall

11.23 Two centrosomes are arranged at opposite poles of the cell.

e
Factual Recall

11.24 Centrioles begin to move apart in animal cells.

e
Factual Recall

11.25 This is the longest of the mitotic stages.

b
Factual Recall

11.26 Centromeres uncouple, sister chromatids are separated, and the two new chromosomes move to opposite poles of the cell.

a
Factual Recall

11.27 The centromere is a region in which
   a. chromatids are attached to one another.
   b. metaphase chromosomes become aligned.
   c. chromosomes are grouped during telophase.
   d. the nucleus is located prior to mitosis.
   e. new spindle microtubules form.

d
Application

11.28 If cells in the process of dividing are subjected to colchicine, a drug that interferes with the functioning of the spindle apparatus, at which stage will mitosis be arrested?
   a. anaphase
   b. prophase
   c. telophase
   d. metaphase
   e. interphase

b
Factual Recall

11.29 All of the following are characteristic of telophase of mitosis EXCEPT:
   a. Cytokinesis begins.
   b. Each chromosome is made of two chromatids.
   c. The nuclear envelope reappears.
   d. Chromosomes begin to uncoil.
   e. Astral microtubules disappear.

d
Conceptual
Understanding

11.30 A cell containing 92 chromatids at the start of mitosis would, at its completion, produce cells containing how many chromosomes?
   a. 12
   b. 16
   c. 23
   d. 46
   e. 92

e
Application

11.31 During which phase of mitosis are chromosomes composed of two chromatids found?
   a. from interphase through anaphase
   b. from G_1 of interphase through metaphase
   c. from metaphase through telophase
   d. from anaphase through telophase
   e. from G_2 of interphase through metaphase

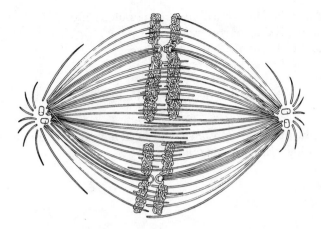

Figure 11.2

e
Factual Recall

11.32 If the cell whose nuclear material is shown in Figure 11.2 continues toward completion of mitosis, which of the following events would occur next?
   a. cell membrane synthesis
   b. spindle fiber formation
   c. nuclear envelope breakdown
   d. chromatid synthesis
   e. centromere uncoupling

b
Application

11.33 If there are 20 centromeres in a cell, how many chromosomes are there?
   a. 10
   b. 20
   c. 30
   d. 40
   e. 80

d
Factual Recall

11.34 All of the following occur during the latter stages of mitotic prophase EXCEPT:
   a. The centrioles move apart.
   b. The nucleolus disintegrates.
   c. The nuclear envelope disappears.
   d. Chromosomes are duplicated.
   e. The spindle is organized.

b
Factual Recall

11.35 If there are twelve centromeres in a cell in G_1 of the cell cycle, what is the diploid number of chromosomes?
   a. 6
   b. 12
   c. 24
   d. 36
   e. 48

d
Conceptual
Understanding

11.36 If the haploid number for a species is 3, each dividing diploid cell will have how many chromatids at metaphase?
   a. 3
   b. 6
   c. 9
   d. 12
   e. 18

d
Factual Recall

11.37 All of the following occur during mitosis EXCEPT
    a. the coiling of chromosomes.
    b. the division of centromeres.
    c. the formation of a spindle.
    d. the synthesis of DNA.
    e. the degradation of the nuclear envelope.

d
Conceptual
Understanding

11.38 How many chromosomes will a cell have during mitotic anaphase if the diploid chromosome number is 4?
    a. 1
    b. 2
    c. 4
    d. 8
    e. 16

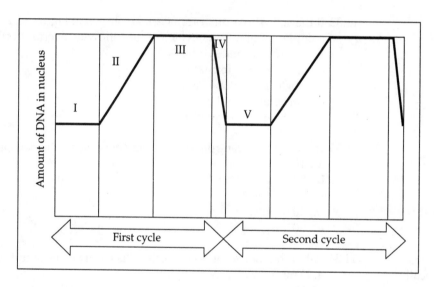

Figure 11.3

Questions 11.39–11.41 are based on Figure 11.3.

d
Application

11.39 In Figure 11.3, at which stage are centromeres uncoupling and chromatids separating?
    a. I
    b. II
    c. III
    d. IV
    e. V

c
Application

11.40 MPF reaches its threshold concentration at the end of this stage.
    a. I
    b. II
    c. III
    d. IV
    e. V

a
Application

11.41 The restriction point occurs at the end of this phase.
  a. I and V
  b. II and IV
  c. III only
  d. IV only
  e. V only

e
Application

11.42 Suppose a plant cell had a mutation that prevented the Golgi apparatus from functioning. Which of the following processes would NOT occur?
  a. cellular respiration
  b. photosynthesis
  c. mitosis
  d. DNA replication
  e. cell wall formation

Use the data in Table 11.1 to answer Questions 11.43–11.45.
The data below were obtained from a study of the length of time spent in each phase of the cell cycle by cells of three eukaryotic organisms, designated beta, delta, and gamma.

Table 11.1   Minutes Spent in Cell Cycle Phases

Cell Type	G_1	S	G_2	M
Beta	18	24	12	16
Delta	100	0	0	0
Gamma	18	48	14	20

a
Conceptual
Understanding

11.43 Of the following, the best conclusion concerning the difference between the S phases for beta and gamma is that
  a. gamma contains more chromosomes than beta.
  b. beta and gamma contain the same number of chromosomes.
  c. beta contains more chromosomes than gamma.
  d. gamma contains 48 times more DNA and RNA than beta.
  e. beta is a plant cell and gamma is an animal cell.

d
Conceptual
Understanding

11.44 The best conclusion concerning delta is that the cells
  a. contain no DNA.
  b. contain no RNA.
  c. contain only one chromosome that is very short.
  d. are actually in the G_0 phase.
  e. divide in the G_1 phase.

b
Conceptual
Understanding

11.45 The S phase was measured by
  a. counting the number of cells.
  b. determining the start and stop of increased DNA in the cells.
  c. synthesis versus breakdown of S protein.
  d. synthesis of the S chromosome.
  e. stopping G_1.

e
Factual Recall

11.46 Enzymes that control the activities of other proteins are called
    a. ATPases.
    b. microtubules.
    c. kinetochores.
    d. chromatin.
    e. protein kinases.

e
Factual Recall

11.47 Proteins that are involved in the regulation of the cell cycle, and that show fluctuations in concentration during the cell cycle, are called
    a. ATPases.
    b. kinetochores.
    c. centrioles.
    d. proton pumps.
    e. cyclins.

You analyzed for the level of a particular protein (PP) in skin cells grown in tissue culture. The results are shown in Table 11.2. Use these data to answer Question 11.48.

Table 11.2 Levels of PP in Skin Cells Over a 24-hour Period

Time (hr)	4	6	12	18	24
PP level (mg/g cells)	19	3	16	8	20

c
Application

11.48 From these data, PP is probably
    a. associated with the chromosomes.
    b. a microtubule protein in the kinetochore family.
    c. a regulatory protein in the protein kinase family.
    d. not involved in any important cell function.
    e. a membrane transport protein whose level depends on molecules in the growth medium.

a
Factual Recall

11.49 The MPF protein complex turns itself off by
    a. activating an enzyme that destroys cyclin.
    b. activating an enzyme that stimulates cyclin.
    c. binding to chromatin.
    d. exiting the cell.
    e. None of these is true; MPF is always active.

a
Factual Recall

11.50 Recent research has indicated that cancer cells
    a. transform normal cells by altering genes involved in the control of mitosis.
    b. always develop into a tumor.
    c. contain more than the normal number of chromosomes.
    d. are unable to complete the cell cycle after the S phase.
    e. enter and exit the G_0 phase three times before they divide.

b
Conceptual
Understanding

11.51 Colchicine is a drug that binds to the protein that forms microtubules, thereby preventing microtubules from forming. Colchicine has been used to study mitosis because it stops the process. Most likely this is due to
    a. prevention of sister chromatid formation.
    b. prevention of kinetochore formation.
    c. inhibition of DNA synthesis.
    d. alteration of centriole structure.
    e. prevention of cell-plate formation.

c
Factual Recall

11.52 Which of the following is NOT true of the bacterial chromosome? It
    a. contains a single, circular DNA molecule.
    b. is associated with proteins.
    c. floats freely inside the bacterial cell.
    d. is highly folded within the cell.
    e. has genes that control binary fission.

a
Conceptual
Understanding

11.53 A cell with 20 chromosomes (diploid number) goes through the cell cycle. The number of chromosomes after doubling the DNA in the S phase is
    a. 20.
    b. 40.
    c. 10.
    d. called the S number of chromosomes.
    e. called haploid.

a
Conceptual
Understanding

11.54 Taxol is an anticancer drug extracted from the Pacific yew tree. In animal cells, taxol disrupts microtubule formation by binding to microtubules and accelerating their assembly from the protein precursor, tubulin. Surprisingly, this stops mitosis. Specifically, taxol must affect
    a. the fibers of the mitotic spindle.
    b. anaphase.
    c. formation of the centrioles.
    d. chromatid assembly.
    e. the S phase of the cell cycle.

Chapter 12

12.1 When does a synaptonemal complex form?
   a. during prophase I of meiosis
   b. during fertilization or fusion of gametes
   c. during metaphase II of meiosis
   d. during prophase of mitosis
   e. during metaphase of mitosis

12.2 The phases of meiosis that cause the most variation in the four resulting daughter cells are
   a. prophase I and telophase II.
   b. prophase II and anaphase II.
   c. metaphase I and telophase II.
   d. anaphase I and prophase II.
   e. prophase I and anaphase I.

12.3 If the liver cells of an animal have 24 chromosomes, the sperm cells would have how many chromosomes?
   a. 12
   b. 24
   c. 48
   d. twice the diploid number
   e. half the haploid number

12.4 How does the sexual life cycle increase the genetic variation in a species?
   a. by producing gametes with different combinations of parental chromosomes
   b. by allowing the combination of chromosomes from two different individuals
   c. by allowing recombination of alleles on a chromosome
   d. Both a and b are correct.
   e. a, b, and c are correct.

12.5 Which of the following events occurs during prophase I of meiosis?
   a. reduction in chromosome number
   b. segregation of alleles of unlinked genes
   c. synapsis and crossing over
   d. duplication of chromatids
   e. segregation of alleles of linked genes

12.6 Which of the following is true of a species that has a chromosome number of $2n = 16$?
   a. The species is diploid with 32 chromosomes.
   b. The species has 16 different types of chromosomes.
   c. There are 16 homologous pairs.
   d. During the S phase of the cell cycle there will be 32 separate chromosomes.
   e. A gamete from this species has 8 chromosomes.

b
Factual Recall

12.7 At which stage of mitosis are chromosomes photographed in the preparation of a karyotype?
  a. prophase
  b. metaphase
  c. anaphase
  d. telophase
  e. Both c and d are correct.

b
Factual Recall

12.8 Any genetic differences in a clone are due to which process?
  a. independent assortment
  b. mutation
  c. crossing over
  d. recombination
  e. synapsis

b
Factual Recall

12.9 Referring to a plant sexual life cycle, choose the pair of terms in which the first term describes the process that leads directly to the formation of gametes, while the second term describes restoration of the diploid chromosome number from the haploid state.
  a. meiosis; fertilization
  b. gametophyte mitosis; fertilization
  c. meiosis; mitosis
  d. gametophyte meiosis; fertilization
  e. sporophyte mitosis; fertilization

b
Conceptual
Understanding

12.10 Which of the following is FALSE in comparing prophase I of meiosis and prophase of mitosis?
  a. The chromosomes condense in both.
  b. Tetrads form in both.
  c. The nuclear envelope disassembles in both.
  d. A spindle forms in both.
  e. Each chromosome has two chromatids in both.

d
Conceptual
Understanding

12.11 In a given organism, how do cells at the completion of meiosis compare with cells that are just about to begin meiosis?
  a. They have twice the amount of cytoplasm and half the amount of DNA.
  b. They have half the number of chromosomes and half the amount of DNA.
  c. They have the same number of chromosomes and half the amount of DNA.
  d. They have half the number of chromosomes and one-fourth the amount of DNA.
  e. They have half the amount of cytoplasm and twice the amount of DNA.

e
Factual Recall

12.12 What is a karyotype?
  a. the phenotype of an individual
  b. the genotype of an individual
  c. a unique combination of chromosomes found in a gamete
  d. the kind of nucleus a cell has
  e. a method of organizing the homologous chromosomes of a cell in relation to their number, size, and type

I

II

III

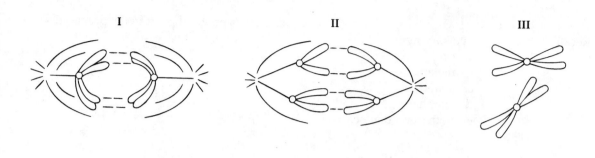

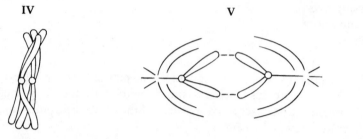

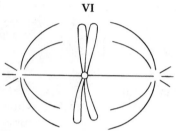

IV

V

VI

Figure 12.1

Refer to the drawings in Figure 12.1 of a single pair of homologous chromosomes as they might appear during various stages of either mitosis or meiosis, and answer Questions 12.13 and 12.14.

d
Conceptual
Understanding

12.13 Which diagram represents prophase I of meiosis?
    a. I
    b. II
    c. III
    d. IV
    e. V

d
Conceptual
Understanding

12.14 At the completion of which stage will the chromosomes have the least amount of DNA per nucleus?
    a. II
    b. III
    c. IV
    d. V
    e. VI

d
Factual Recall

12.15 Which of the following is the term for a human cell that contains 22 pairs of autosomes and two X chromosomes?
    a. an unfertilized egg cell
    b. a sperm cell
    c. a male somatic cell
    d. a female somatic cell
    e. Both a and d are correct.

e
Factual Recall

12.16 For a species with a haploid number of 23 chromosomes, how many different combinations of maternal and paternal chromosomes are possible for the gametes?
   a. 23
   b. 46
   c. 460
   d. 920
   e. more than 8 million

a
Factual Recall

12.17 Crossing over occurs during which phase of meiosis?
   a. prophase I
   b. anaphase I
   c. telophase I
   d. prophase II
   e. metaphase II

Use the following key to answer Questions 12.18–12.23. Each answer may be used once, more than once, or not at all.
   a. *The statement is true for mitosis only.*
   b. *The statement is true for meiosis I only.*
   c. *The statement is true for meiosis II only.*
   d. *The statement is true for mitosis and meiosis I.*
   e. *The statement is true for mitosis and meiosis II.*

b
Conceptual
Understanding

12.18 Homologous chromosomes synapse and crossing over occurs.

a
Conceptual
Understanding

12.19 This occurs when a cell divides to form two cells that are genetically identical.

e
Conceptual
Understanding

12.20 Centromeres uncouple and chromatids are separated from each other.

b
Conceptual
Understanding

12.21 Independent assortment of chromosomes occurs.

b
Conceptual
Understanding

12.22 The events during this process cause the majority of genetic recombinations.

d
Conceptual
Understanding

12.23 The process(es) is (are) preceded by a copying (replication) of the DNA.

For Questions 12.24–12.27, match the key event of meiosis with the stages listed below.

I.   *Prophase I*   VI.   *Prophase II*
II. *Metaphase I*   VII. *Metaphase II*
III. *Anaphase* I   VIII. *Anaphase II*
IV. *Telophase I*   IX.   *Telophase II*
V.   *Interkinesis*

b
Factual Recall

12.24 Tetrads of chromosomes are aligned at the center of the cell; independent assortment soon follows.
   a. I
   b. II
   c. III
   d. VI
   e. VIII

a
Factual Recall

12.25 Synapsis of homologous pairs occurs; crossing over may occur.
   a. I
   b. II
   c. IV
   d. VI
   e. VII

b
Factual Recall

12.26 Nuclear envelopes may form; no replication of chromosomes takes place.
   a. III
   b. V
   c. VI
   d. VII
   e. VIII

e
Factual Recall

12.27 Centromeres of sister chromatids uncouple and chromatids separate.
   a. II
   b. III
   c. VI
   d. VII
   e. VIII

e
Factual Recall

12.28 All of the following are functions of meiosis in plants EXCEPT
   a. production of spores.
   b. reduction of chromosome number by half.
   c. independent assortment of chromosomes.
   d. crossing over and recombination of homologous chromosomes.
   e. production of identical daughter cells.

Questions 12.29 and 12.30 refer to some essential steps in meiosis described below.
   1. *formation of four new nuclei, each with half the chromosomes present in the parental nucleus*
   2. *alignment of tetrads at the metaphase plate*
   3. *separation of sister chromatids*
   4. *separation of the homologs; no uncoupling of the centromere*
   5. *synapsis; chromosomes moving to the middle of the cell in pairs*

e
Conceptual
Understanding

12.29 From the descriptions above, the order that most logically illustrates a sequence of meiosis is which of the following?
   a. 1–2–3–4–5
   b. 5–4–2–1–3
   c. 5–3–2–4–1
   d. 4–5–2–1–3
   e. 5–2–4–3–1

a
Conceptual
Understanding

12.30 From the descriptions above, which play roles in generating genetic diversity?
   a. All of the below generate diversity.
   b. 1, 3, and 5
   c. 2, 4, and 5
   d. 2, 3, and 4
   e. 1, 3, and 4

d
Factual Recall

12.31 After telophase I of meiosis, what is the chromosomal makeup of each daughter cell?
   a. diploid, and the chromosomes are composed of single chromatid chromosomes
   b. diploid, and the chromosomes are composed of two chromatids
   c. haploid, and the chromosomes are composed of single chromatid chromosomes
   d. haploid, and the chromosomes are composed of two chromatids
   e. tetraploid, and the chromosomes are composed of two chromatids

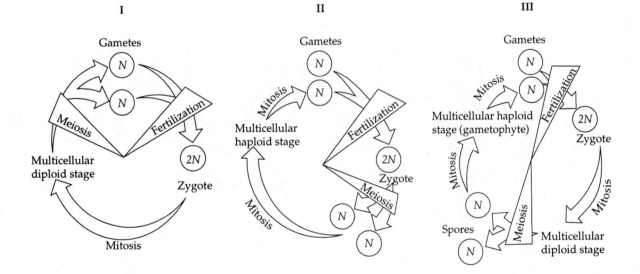

Figure 12.2

Refer to the life cycles illustrated in Figure 12.2 to answer Questions 12.32–12.35.

a
Factual Recall

12.32 Which of the life cycles is typical for animals?
    a. I only
    b. II only
    c. III only
    d. I and II
    e. I and III

c
Factual Recall

12.33 Which of the life cycles is typical for plants and some algae?
    a. I only
    b. II only
    c. III only
    d. I and II
    e. I and III

b
Factual Recall

12.34 Which of the life cycles is typical for many fungi and some protists?
    a. I only
    b. II only
    c. III only
    d. I and II
    e. I and III

e
Conceptual
Understanding

12.35 Which life cycle would generate the greatest genetic diversity and why?
    a. I—because haploid forms are less important.
    b. II—because haploid forms are more important.
    c. III—because it has the best balance between haploid and diploid forms.
    d. III—because meiosis and fertilization are more equally spaced.
    e. None—they would all generate equivalent genetic diversity.

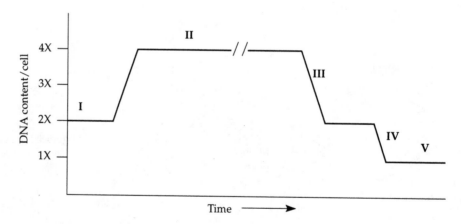

Figure 12.3

Refer to the diagram of a meiotic process in Figure 12.3 to answer Questions 12.36–12.39.

b
Conceptual
Understanding

12.36 Which number represents G$_2$?
  a. I
  b. II
  c. III
  d. IV
  e. V

e
Conceptual
Understanding

12.37 Which number represents the DNA content of a sperm cell?
  a. I
  b. II
  c. III
  d. IV
  e. V

c
Conceptual
Understanding

12.38 Which number represents the separation of homologous chromosomes?
  a. I
  b. II
  c. III
  d. IV
  e. V

c
Conceptual
Understanding

12.39 Where would you place crossing over in this diagram?
  a. I
  b. between I and II
  c. II
  d. III
  e. IV

Questions 12.40 and 12.41 are based on the representation of the cell in Figure 12.4.

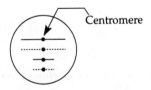

Centromere

Figure 12.4

Figure 12.4 shows a diploid cell with four chromosomes. There are two types of chromosomes, one long and the other short. One haploid set is symbolized by unbroken lines, while the other haploid set is represented by dotted lines. At this time, the chromosomes have not yet replicated. Now, choose the correct chromosomal conditions for the following stages:

b
Application

12.40 a possible daughter cell of meiosis I

a.

b.

c.

d.

c
Application

12.41 a possible daughter cell of meiosis II

a.

b.

c.

d.

e
Conceptual
Understanding

12.42 You are genetically unique. This is a result of
   a. sexual reproduction.
   b. genetic recombination.
   c. mutation.
   d. both a and c.
   e. all of the above.

c
Conceptual
Understanding

12.43 Eukaryotic sexual life cycles show tremendous variation. Of the following
   elements, which do all sexual life cycles have in common?
   I.   *copulation*
   II. *meiosis*
   III. *fertilization*
   IV. *gametes*
   V.   *spores*

   a. I, IV, and V
   b. I, II, and IV
   c. II, III, and IV
   d. II, IV, and V
   e. all of the above

d
Conceptual
Understanding

12.44 The word *homologous* literally means *same location*. How does this relate to
   homologous chromosomes?
   a. All of the below are correct.
   b. The bands resulting from staining are found in the same location.
   c. The chromosomes have the same genes in the same location.
   d. The chromosomes always move to the same location in the cell during
      division.
   e. Both a and b are correct.

e
Conceptual
Understanding

12.45 A cell has a diploid chromosome number of $2n = 4$. We will designate these four as chromosomes A, B, C, and D. If meiosis occurred WITHOUT the formation of homologous pairs, and the chromosomes were distributed randomly among the resulting cells, how many gametes could be formed?
a. 2 (AB and CD)
b. 3 (AB, BC, and CD)
c. 4 (AB, AB, BC, and CD)
d. 5 (AB, AC, AD, BC, and CD)
e. 6 (AB, AC, AD, BC, BD, and CD)

c
Application

12.46 A cell has a diploid chromosome number of $2n = 4$. We will designate these four as chromosomes A, B, C, and D. If meiosis occurred WITH the formation of homologous pairs AC and BD, and the chromosomes were then distributed randomly between the resulting cells, how many gametes could be formed?
a. 2 (AB and CD)
b. 3 (AB, BC, and CD)
c. 4 (AB, AD, BC, and CD)
d. 5 (AB, AC, AD, BC, and CD)
e. 6 (AB, AC, AD, BC, BD, and CD)

Chapter 13

13.1 A couple who are both carriers for the gene for cystic fibrosis have two children who have cystic fibrosis. What is the probability that their next child will have cystic fibrosis?
   a. 0%
   b. 25%
   c. 50%
   d. 75%
   e. 100%

13.2 A couple who are both carriers for the gene for cystic fibrosis have two children who have cystic fibrosis. What is the probability that their next child will be phenotypically normal?
   a. 0%
   b. 25%
   c. 50%
   d. 75%
   e. 100%

13.3 What is a genetic cross called between an individual of unknown genotype and a homozygous recessive?
   a. a self-cross
   b. a testcross
   c. a hybrid cross
   d. an F_1 cross
   e. a dihybrid cross

13.4 In crossing a homozygous recessive with a heterozygote, what is the chance of getting a homozygous recessive phenotype in the F_1 generation?
   a. zero
   b. 25%
   c. 50%
   d. 75%
   e. 100%

13.5 In snapdragons, heterozygotes have pink flowers, whereas the two homozygotes have red flowers or white flowers. When plants with red flowers are crossed with plants with white flowers, what proportion of the offspring will have pink flowers?
   a. zero
   b. 25%
   c. 50%
   d. 75%
   e. 100%

d
Application

13.6 Black fur in mice (B) is dominant to brown fur (b). Short tails (T) is dominant to long tails (t). What proportion of the progeny of the cross $BbTt \times BBtt$ will have black fur and long tails?
    a. 1/16
    b. 3/16
    c. 6/16
    d. 8/16
    e. 9/16

Use the following information to answer Questions 13.7–13.9. Feather color in budgies is determined by two different genes that affect the pigmentation of the outer feather and its core. Y_B_ is green; yyB_ is blue; Y_bb is yellow; and yybb is white.

e
Application

13.7 A green budgie is crossed with a blue budgie. Which of the following results is NOT possible?
    a. all green offspring
    b. all blue offspring
    c. all white offspring
    d. all yellow offspring
    e. All of the above are possible, but with different probabilities.

c
Conceptual
Understanding

13.8 Two blue budgies were crossed. Over the years, they produced 22 offspring, 5 of which were white. What are the most likely genotypes for the two blue budgies?
    a. *yyBB* and *yyBB*
    b. *yyBB* and *yyBb*
    c. *yyBb* and *yyBb*
    d. *yyBB* and *yybb*
    e. *yyBb* and *yybb*

e
Conceptual
Understanding

13.9 The inheritance of color in budgies is an example of what genetic phenomenon?
    a. pleiotropy
    b. penetrance
    c. polygenic inheritance
    d. dominance
    e. epistasis

b
Factual Recall

13.10 What was the most significant conclusion that Gregor Mendel drew from his research?
    a. There is considerable genetic variation in garden peas.
    b. Traits are inherited in discrete units, one from each parent.
    c. Dominant genes occur more frequently than recessive ones.
    d. Genes are composed of DNA.
    e. An organism that is homozygous for many recessive traits is at a disadvantage.

d
Application

13.11 A couple has three children, all of whom have brown eyes and blond hair. Both parents are homozygous for brown eyes (*BB*), but one is a blond (*rr*) and the other is a redhead (*Rr*). What is the probability that the next child will be a brown-eyed redhead?
   a. 1/16
   b. 1/8
   c. 1/4
   d. 1/2
   e. 1

Use the following information to answer Questions 13.12–13.14. A woman and her husband both show the normal phenotype for pigmentation, but both had one parent who was an albino. Albinism is an autosomal recessive trait.

b
Application

13.12 What is the probability that their first child will be an albino?
   a. 0%
   b. 25%
   c. 50%
   d. 75%
   e. 100%

b
Application

13.13 If their first two children have normal pigmentation, what is the chance that their third child will be an albino?
   a. 0%
   b. 25%
   c. 50%
   d. 75%
   e. 100%

c
Application

13.14 What is the chance that their fourth child will have a homozygous genotype?
   a. 0%
   b. 25%
   c. 50%
   d. 75%
   e. 100%

b
Factual Recall

13.15 The advantage of chorionic villi sampling over amniocentesis is that CVS
   a. detects chromosomal abnormalities not detected by amniocentesis.
   b. can give results much earlier in the pregnancy than amniocentesis.
   c. is less likely to cause miscarriage than amniocentesis.
   d. can detect chemical abnormalities not detected by amniocentesis.
   e. results are more reliable than amniocentesis results.

Use the following information to answer Questions 13.16–13.18. A woman who belongs to blood group A and is Rh positive has a daughter who is O positive and a son who is B negative. Rh positive is a simple dominant over Rh negative.

d
Application

13.16 Which of the following is a possible genotype for the son?
    a. $I^B I^B rr$
    b. $I^B I^B RR$
    c. $I^B i Rr$
    d. $I^B i rr$
    e. $I^B I^B Rr$

d
Application

13.17 Which of the following is a possible genotype for the mother?
    a. $I^A I^A RR$
    b. $I^A I^A Rr$
    c. $I^A i rr$
    d. $I^A i Rr$
    e. $I^A i RR$

c
Application

13.18 Which of the following is a possible phenotype for the father?
    a. A negative
    b. O negative
    c. B positive
    d. A positive
    e. O positive

c
Application

13.19 Given the parents $AABBCc \times AabbCc$, assume simple dominance and independent assortment. What proportion of the progeny will be expected to phenotypically resemble the first parent?
    a. 1/4
    b. 1/8
    c. 3/4
    d. 3/8
    e. 1

d
Conceptual
Understanding

13.20 The fact that all seven of the garden pea traits studied by Mendel obeyed the principle of independent assortment means that the
    a. haploid number of garden peas is 7.
    b. diploid number of garden peas is 7.
    c. seven pairs of alleles determining these traits are on the same pair of homologous chromosomes.
    d. seven pairs of alleles determining these traits behave as if they are on different chromosomes.
    e. formation of gametes in plants is by mitosis only.

a
Conceptual
Understanding

13.21 A sexually reproducing animal has two unlinked genes, one for head shape and one for tail length. Its genotype is $HhTt$. Which of the following genotypes is possible in a gamete from this organism?
    a. HT
    b. Hh
    c. $HhTt$
    d. T
    e. tt

e
Conceptual
Understanding

13.22 Which of the following is an example of polygenic inheritance?
    a. pink flowers in snapdragons
    b. the ABO blood groups in humans
    c. sex-linkage in humans
    d. white and purple color in sweet peas
    e. skin pigmentation in humans

b
Application

13.23 Two true-breeding stocks of garden peas are crossed. One parent had red, axial flowers, and the other had white, terminal flowers; all F_1 individuals had red, axial flowers. If 1,000 F_2 offspring resulted from the cross, how many of them would you expect to have red, terminal flowers? (Assume independent assortment.)
    a. 65
    b. 190
    c. 250
    d. 565
    e. 750

Questions 13.24–13.26 refer to the following terms. Each term may be used once, more than once, or not at all.
    a. incomplete dominance
    b. multiple alleles
    c. pleiotropy
    d. epistasis
    e. penetrance

c
Conceptual
Understanding

13.24 The ability of a single gene to have multiple phenotype effects.

b
Conceptual
Understanding

13.25 One example is the ABO blood group system.

a
Conceptual
Understanding

13.26 The phenotype of the heterozygote differs from the phenotypes of both homozygotes.

Use the following information to answer Questions 13.27–13.29. Albinism (lack of skin pigmentation) is caused by a recessive autosomal allele. A man and woman, both normally pigmented, have an albino child together.

b
Conceptual
Understanding

13.27 For this trait, what is the genotype of the albino child?
    a. homozygous dominant
    b. homozygous recessive
    c. heterozygous
    d. hemizygous
    e. unknown, because not enough information is provided

b
Application

13.28 The couple decides to have a second child. What is the probability that this child will be albino?
a. zero
b. 1/4
c. 1/2
d. 3/4
e. 1

d
Application

13.29 The mother is now pregnant for a third time, and her doctor tells her she is carrying fraternal twins. What is the probability that both children will have normal pigmentation?
a. 3/4
b. 1/4
c. 1/16
d. 9/16
e. 16/16

d
Conceptual
Understanding

13.30 A 9:3:3:1 phenotypic ratio is characteristic of the
a. F_1 generation of a monohybrid cross.
b. F_2 generation of a monohybrid cross.
c. F_1 generation of a dihybrid cross.
d. F_2 generation of a dihybrid cross.
e. F_2 generation of a trihybrid cross.

c
Conceptual
Understanding

13.31 A 1:2:1 phenotypic ratio in the F_2 generation of a monohybrid cross is a sign of
a. complete dominance.
b. multiple alleles.
c. intermediate inheritance.
d. polygenic inheritance.
e. pleiotropy.

c
Application

13.32 A 9 purple to 7 white phenotype in sweet peas in the F_2 generation most likely is due to
a. linkage.
b. trisomy 21.
c. epistasis.
d. crossing over.
e. pleiotropy.

b
Application

13.33 How many unique gametes could be produced through independent assortment by an individual with the genotype *AaBbCCDdEE*?
a. 4
b. 8
c. 16
d. 32
e. 1/64

a
Application

13.34 In cattle, roan coat color (mixed red and white hairs) occurs in the hetero-zygous (Rr) offspring of red (RR) and white (rr) homozygotes. When two roan cattle are crossed, the phenotypes of the progeny are found to be in the ratio of 1 red:2 roan:1 white. Which of the following crosses could produce the highest percentage of roan cattle?
a. red × white
b. roan × roan
c. white × roan
d. red × roan
e. All of the above crosses would give the same percentage of roan.

e
Application

13.35 Roan color in cattle is the result of absence of dominance between red and white color genes. How would one produce a herd of pure-breeding roan-colored cattle?
a. cross roan with roan
b. cross red with white
c. cross roan with red
d. cross roan with white
e. It cannot be done.

c
Application

13.36 Three babies were recently mixed up in a hospital. After consideration of the data below, which of the following represents the correct baby/parent combinations?

Couple #	I	II	III
Blood groups	A and A	A and B	B and O
Baby #	1	2	3
Blood groups	B	O	AB

a. I-3, II-1, III-2
b. I-1, II-3, III-2
c. I-2, II-3, III-1
d. I-2, II-1, III-3
e. I-3, II-2, III-1

d
Conceptual
Understanding

13.37 Tallness (T) is dominant to dwarfness (t), while red flower color is due to gene (R) and white is its allele (r). The heterozygous condition results in pink (Rr) flower color. A dwarf red snapdragon is crossed with a plant homozygous for tallness and white flowers. What are the genotype and phenotype of the F_1 individuals?
a. ttRr—dwarf and pink
b. ttrr—dwarf and white
c. TtRr—tall and red
d. TtRr—tall and pink
e. TTRR—tall and red

d
Application

13.38 The probability that four coins will come up heads when flipped simultaneously is
    a. 1/4 (0.25).
    b. 1/2 (0.5).
    c. 1/8 (0.125).
    d. 1/16 (0.062).
    e. 1/64 (0.016).

Questions 13.39–13.42 will use the following answers. Each answer may be used once, more than once, or not at all.
    a. Huntington's disease
    b. Tay-Sachs disease
    c. phenylketonuria
    d. hemophilia
    e. sickle-cell anemia

c
Factual Recall

13.39 Effects of this recessive single gene can be completely overcome by regulating the diet of the affected individual.

a
Factual Recall

13.40 This is caused by a dominant single gene defect and generally does not appear until the individual is 30–40 years of age.

b
Factual Recall

13.41 Individuals with this disorder are unable to metabolize gangliosides, which affects proper brain development. Affected individuals die in early infancy.

e
Factual Recall

13.42 Substitution of the "wrong" amino acid in the hemoglobin protein results in this disorder.

Use the following information to answer Questions 10.43–10.45. A man is brought to court in a paternity case. He has blood type B, Rh positive. The mother has blood group B, Rh negative.

e
Application

13.43 Which blood group of the child will exclude the man from possible paternity?
    a. AB, Rh negative
    b. B, Rh negative
    c. O, Rh negative
    d. B, Rh positive
    e. None of these choices will exclude the man from possible paternity.

c
Application

13.44 The child's blood type is A, Rh negative. What can you say about the man's chances of being the father?
    a. He is the father.
    b. He might be the father.
    c. He is not the father.
    d. He might not be the father.
    e. There is not enough information to make a decision.

b
Application

13.45 The child's blood type is A, Rh negative. The woman's male secretary has a blood type of A, Rh positive. What can you say about his chances of being the father?
a. He is the father.
b. He might be the father.
c. He is not the father.
d. He is almost certainly the father.
e. There is not enough information to make a decision.

a
Conceptual
Understanding

13.46 In a dihybrid cross, the expected proportion of offspring showing both recessive traits is
a. 1/16.
b. 3/16.
c. 9/16.
d. 1/4.
e. 1/32.

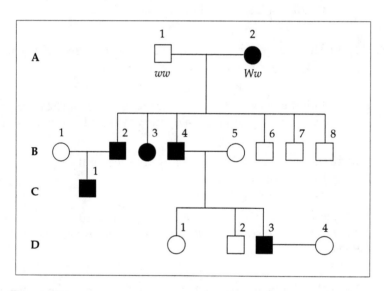

Figure 13.1

Questions 13.47 and 13.48 refer to the pedigree chart in Figure 13.1 for a family, some of whose members exhibit the recessive trait, wooly hair. Affected individuals are indicated by an open square or circle.

c
Conceptual
Understanding

13.47 What is the genotype of individual B-5?
a. WW only
b. Ww only
c. ww only
d. WW or ww
e. ww or Ww

b
Conceptual
Understanding

13.48 What is the genotype of individual D-3?
   a. *WW* only
   b. *Ww* only
   c. *ww* only
   d. *WW* or *ww*
   e. *ww* or *Ww*

b
Conceptual
Understanding

13.49 In a cross between parents who both exhibit the dominant curly- and dark-haired traits, one child has straight, light-colored hair. What is the hair genotype of the parents?
   a. *CCdd* × *ccDd*
   b. *CcDd* × *CcDd*
   c. *CcDd* × *ccdd*
   d. *Ccdd* × *CcDd*
   e. *CCDD* × *CcDd*

e
Application

13.50 In a cross *AaBbCc* × *AaBbCc*, what is the probability of producing the genotype *AABBCC*?
   a. 1/4
   b. 1/8
   c. 1/16
   d. 1/32
   e. 1/64

b
Conceptual
Understanding

13.51 Recipes are instructions on how to prepare food, just like genes are instructions the cell uses to make proteins. Using this analogy, a cookbook is analogous to
   a. a nucleus.
   b. a chromosome.
   c. a pair of homologous chromosomes.
   d. a gamete.
   e. a zygote.

d
Conceptual
Understanding

13.52 Remember how you used to make up teams in gym class? The class would line up and count off by twos. All the students who were "ones" would be on one team and the "twos" would be the other team. This is analogous to Mendel's concept of
   a. genes.
   b. ratios.
   c. dominance.
   d. segregation.
   e. a dihybrid cross.

a
Conceptual
Understanding

13.53 You are a choreographer who has been assigned the job of transforming the concepts of genetics into a dance. In order to depict independent assortment, you should have couples come together and move to the center of the floor, and then
   a. they separate, with boys and girls going randomly to two different sides of the floor.
   b. they separate, with boys going to one side of the floor and girls to the other.
   c. half the couples go to one side of the floor and half to the other side.
   d. a second group of couples enters from the side and combines with the first.
   e. they exchange partners with a neighboring couple.

e
Conceptual
Understanding

13.54 Blueprints contain specific instructions for constructing a building. The room you are sitting in now was constructed by using the instructions on a set of blueprints. The relationship between blueprint and building is analogous to the relationship between
a. chromosomes and genes.
b. genes and alleles.
c. dominant and recessive alleles.
d. segregation and independent assortment.
e. genotype and phenotype.

Chapter 14

a
Conceptual
Understanding

14.1 In birds, sex is determined by a ZW chromosome scheme. Males are ZZ and females are ZW. A lethal recessive allele that causes death of the embryo occurs on the Z chromosome in pigeons. What would be the sex ratio in the offspring of a cross between a male heterozygous for the lethal allele and a normal female?
   a. 2:1 male to female
   b. 1:2 male to female
   c. 1:1 male to female
   d. 4:3 male to female
   e. 3:1 male to female

a
Factual Recall

14.2 Which of the following statements is TRUE regarding genomic imprinting?
   a. It explains cases where the gender of the parent from whom an allele is inherited affects the expression of that allele.
   b. It is greatest in females because of the larger maternal contribution of cytoplasm.
   c. It may explain the transmission of Duchenne's muscular dystrophy.
   d. It explains sex-linked inheritance in which the sex of the parent carrying the mutant allele determines whether male or female offspring will be affected.
   e. It is found in X-inactivation in human females during early embryonic development.

b
Conceptual
Understanding

14.3 If inheritance of a human trait is sex-linked (on the X chromosome) and recessive, any of the following could result EXCEPT that
   a. expression of the trait might "skip" a generation.
   b. the trait could be more common in females than males.
   c. all females might become homozygous for the trait.
   d. the gene for the trait might mutate to a dominant allele.
   e. females could be a mosaic of two cell types.

e
Application

14.4 The following is a list of chromosomal alterations. Which one of these would automatically cause two of the others?
   a. deletion
   b. duplication
   c. inversion
   d. reciprocal translocation
   e. nonreciprocal translocation

e
Factual Recall

14.5 The finding that defective genes behave differently in offspring depending on whether they belong to the maternal or paternal chromosome is implicated in which of the following?
   a. Prader-Willi syndrome
   b. fragile X syndrome
   c. Angelman syndrome
   d. Only a and c are correct.
   e. a, b, and c are correct.

a
Factual Recall

14.6 Fragile *X* syndrome is more common in males than in females. One explanation for this is
    a. genomic imprinting by the mother.
    b. sex-linked inheritance.
    c. uniparental disomy.
    d. Only a and c are correct.
    e. a, b, and c are correct.

d
Factual Recall

14.7 What do all human males inherit from their mother?
    a. mitochondrial DNA
    b. *X* chromosome
    c. male-pattern baldness trait
    d. Only a and b are correct.
    e. a, b, and c are correct.

a
Conceptual
Understanding

14.8 A mammalian zygote with which of the following chromosomal abnormalities will NEVER develop into a viable embryo?
    a. *YO*
    b. *XO*
    c. *XXX*
    d. *XXY*
    e. *XXXY*

a
Conceptual
Understanding

14.9 All of the following statements are true about chromosomal inversions EXCEPT:
    a. They do not alter phenotype.
    b. They involve breakage of a chromosome.
    c. They do not change the normal balance of genes.
    d. They change the order of the genes on the chromosome.
    e. They involve the rearrangement of the genes.

a
Conceptual
Understanding

14.10 What does independent assortment refer to?
    a. the separation of alleles in anaphase I
    b. the random arrangement of chromosomal tetrads at metaphase I
    c. the separation of chromatids at anaphase II
    d. the random arrangement of gene loci on a chromosome
    e. the fact that any pair of chromatids in a tetrad can cross over

c
Factual Recall

14.11 A Barr body is normally found in the nucleus of which kind of human cell?
    a. unfertilized egg cells only
    b. sperm cells only
    c. somatic cells of a female only
    d. somatic cells of a male only
    e. both male and female somatic cells

d
Factual Recall

14.12 The particular position of a gene on a chromosome is known as a(n)
  a. allele.
  b. tetrad.
  c. chiasma.
  d. locus.
  e. map distance.

e
Conceptual
Understanding

14.13 The frequency of crossing over between any two linked genes is
  a. more likely if they are recessive.
  b. difficult to predict.
  c. determined by their relative dominance.
  d. the same as if they were not linked.
  e. proportional to the distance between them.

c
Application

14.14 A recessive allele on the X chromosome is responsible for red-green color blindness in humans. A woman with normal vision whose father is color-blind marries a color-blind male. What is the probability that this couple's son will be color-blind?
  a. 0%
  b. 25%
  c. 50%
  d. 75%
  e. 100%

a
Conceptual
Understanding

14.15 A man who carries an X-linked allele will pass it on to
  a. all of his daughters.
  b. half of his daughters.
  c. all of his sons.
  d. half of his sons.
  e. all of his children.

a
Factual Recall

14.16 Which of the following is a sex-influenced trait?
  a. male-pattern baldness
  b. white eyes in fruit flies
  c. hemophilia
  d. color blindness
  e. Turner syndrome

d
Application

14.17 In cats, black color is caused by an X-linked allele; the other allele at this locus causes orange color. The heterozygote is tortoise-shell. What kinds of offspring would you expect from the cross of a black female and an orange male?
  a. tortoise-shell female; tortoise-shell male
  b. black female; orange male
  c. orange female; orange male
  d. tortoise-shell female; black male
  e. orange female; black male

Refer to the following information to answer Questions 14.18–14.20. An achondroplastic dwarf man with normal vision marries a color-blind woman of normal height. The man's father was 6 feet tall, and both the woman's parents were of average height. Achondroplastic dwarfism is autosomal dominant, and red-green color blindness is X-linked recessive.

b
Application

14.18 How many of their female children might be expected to be color-blind dwarfs?
    a. all
    b. none
    c. half
    d. one out of four
    e. three out of four

c
Application

14.19 How many of their male children would be color-blind and normal height?
    a. all
    b. none
    c. half
    d. one out of four
    e. three out of four

e
Application

14.20 They have a daughter who is a dwarf with normal color vision. What is the probability that she is heterozygous for both genes?
    a. 0
    b. 0.25
    c. 0.50
    d. 0.75
    e. 1.00

c
Conceptual
Understanding

14.21 If a human interphase nucleus contained three Barr bodies, it can be assumed that the person
    a. is a female.
    b. is a male.
    c. has 4 X chromosomes.
    d. has Turner syndrome.
    e. has Down syndrome.

c
Application

14.22 Barring in chickens is due to a sex-linked dominant gene (B). The sex of chicks at hatching is difficult to determine, but barred chicks can be distinguished from non-barred at that time. To use this trait so that at hatching all chicks of one sex are differently colored from those of the opposite sex, what cross would you make?
    a. barred males × barred females
    b. barred males × non-barred females
    c. non-barred males × barred females
    d. non-barred males × non-barred females
    e. None of the above crosses will produce differently barred chicks.

b
Conceptual
Understanding

14.23 Inheritance of two different traits at two different loci can be complicated by all of the following EXCEPT
   a. the presence of the two loci on one chromosome.
   b. independent assortment of chromosomes.
   c. epistatic interactions between the two loci.
   d. environmental modification of phenotype.
   e. parental imprinting of genes.

c
Factual Recall

14.24 Which of these syndromes afflicts males only?
   a. Turner syndrome
   b. Down syndrome
   c. Duchenne's muscular dystrophy
   d. Patau syndrome
   e. Edwards syndrome

The pedigree chart in Figure 14.1 indicates the inheritance of color blindness (sex-linked).

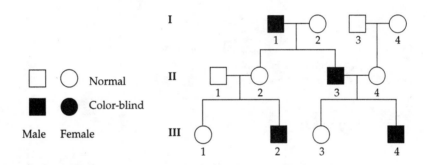

Figure 14.1

Use this answer key for Questions 14.25–14.28:
C = normal   c = color-blind   Y = Y chromosome
   a. CC
   b. Cc
   c. cc
   d. CY
   e. cY

e
Application

14.25 What is the genotype of individual III-2?

b
Application

14.26 What is the genotype of individual I-2?

c
Application

14.27 In the pedigree in Figure 14.1, what is the probability that individual III-1 is heterozygous?
   a. 0.25
   b. 0.33
   c. 0.50
   d. 0.66
   e. 0.75

a
Application

14.28 In the pedigree in Figure 14.1, what is the probability that individual III-3 is homozygous?
   a. 0
   b. 0.25
   c. 0.5
   d. 0.75
   e. 1.0

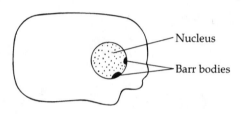

Figure 14.2

d
Conceptual
Understanding

14.29 Figure 14.2 represents the stained nucleus from a cheek epithelial cell of an individual whose genotype could be
   a. XX.
   b. XY.
   c. XYY.
   d. XXX.
   e. XXY.

d
Factual Recall

14.30 There is good evidence for linkage when
   a. two genes occur together in the same gamete.
   b. a gene is associated with a specific phenotype.
   c. two genes work together to control a specific characteristic.
   d. genes do not segregate independently during meiosis.
   e. two characteristics are caused by a single gene.

c
Factual Recall

14.31 A human individual is phenotypically female but her interphase somatic nuclei do not show the presence of sex chromatin (Barr bodies). Which of the following statements concerning her is probably true?
   a. She has Klinefelter syndrome.
   b. She has an extra X chromosome.
   c. She has Turner syndrome.
   d. She has the normal number of sex chromosomes.
   e. She has two Y chromosomes.

a
Application

14.32 If a pair of homologous chromosomes fails to separate during anaphase of meiosis I, what will be the chromosome number (*n*) of the four resulting gametes?
   a. *n*+1; *n*+1; *n*–1; *n*–1
   b. *n*+1; *n*–1; *n*; *n*
   c. *n*+1; *n*–1; *n*–1; *n*–1
   d. *n*+1; *n*+1; *n*; *n*
   e. *n*–1; *n*–1; *n*; *n*

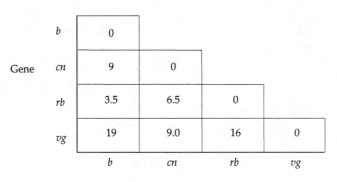

b = black body
cn = cinnabar eyes
rb = reduced bristles
vg = vestigial wings

The numbers in the boxes are the recombination frequencies between the genes (in percent).

Figure 14.3

Refer to Figure 14.3 to answer Questions 14.33 and 14.34.

d
Application

14.33 In a series of mapping experiments, the recombination frequencies for four different linked genes of *Drosophila* were determined as shown above. What is the order of these genes on a chromosome map?
   a. *rb–cn–vg–b*
   b. *vg–b–rb–cn*
   c. *cn–rb–b–vg*
   d. *b–rb–cn–vg*
   e. *vg–cn–b–rb*

e
Application

14.34 Which of the following two genes are closest on a genetic map of *Drosophila*?
   a. *b* and *vg*
   b. *vg* and *cn*
   c. *rb* and *cn*
   d. *cn* and *b*
   e. *b* and *rb*

Questions 14.35–14.38 refer to the data below and to Figures 14.4 and 14.5.

CROSS I. Purebred lines of wild-type fruit flies (gray body and normal wings) are mated to flies with black bodies and vestigial wings.

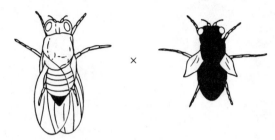

Figure 14.4

F_1 offspring all have a normal phenotype.

Figure 14.5

CROSS II. F_1 flies are crossed with flies recessive for both traits (a testcross).

Resulting Offspring	Normal	Percent
Gray body; normal wings	965	42
Black body; vestigial wings	944	41
Black body; normal wings	206	9
Gray body; vestigal wings	185	8

KEY:
a. *CROSS I results give evidence supporting the statement.*
b. *CROSS I results give evidence against the statement.*
c. *CROSS II results give evidence supporting the statement.*
d. *CROSS II results give evidence against the statement.*
e. *Neither CROSS I nor CROSS II results support the statement.*

a
Application

14.35 Vestigial wings is a recessive trait.

c
Application

14.36 The genes for body color and wing shape are linked.

d
Application

14.37 An F_2 cross should produce flies that will fall into a Mendelian 9:3:3:1 ratio.

c
Application

14.38 There are 17 centimorgans between the genes for body color and wing shape.

b
Application

14.39 Male calico cats are the result of
   a. sex-linked inheritance.
   b. nondisjunction, where the male calico presumably has two X chromosomes.
   c. incomplete dominance of multiple alleles.
   d. recessive alleles retaining their fundamental natures even when expressed.
   e. a reciprocal translocation.

b
Application

14.40 Genes A and B are linked with 12 map units between them. A heterozygous individual Ab/aB would be expected to produce gametes in which of the following frequencies?
   a. 44% AB 6% Ab 6% aB 44% ab
   b. 6% AB 44% Ab 44% aB 6% ab
   c. 6% AB 6% Ab 44% aB 44% ab
   d. 12% AB 12% Ab 38% aB 38% ab
   e. 6% Ab 12% aB 50% AB 32% ab

a
Conceptual
Understanding

14.41 People who have red hair usually have freckles. This can best be explained by
   a. linkage.
   b. reciprocal translocation.
   c. independent assortment.
   d. sex-influenced inheritance.
   e. nondisjunction.

e
Application

14.42 Vermilion eyes is a sex-linked recessive characteristic in fruit flies. If a female having vermilion eyes is crossed with a wild-type male, what proportion of the F_1 males will have vermilion eyes?
   a. None
   b. 25%
   c. 50%
   d. 75%
   e. 100%

e
Application

14.43 In humans, male-pattern baldness is controlled by a gene that occurs in two allelic forms. Allele *Hn* determines nonbaldness and allele *Hb* determines pattern baldness. The interaction of these two alleles in the heterozygote condition is of special interest because in the presence of male hormone, allele *Hn* is dominant over *Hb*. If a man and woman both with genotype *Hn*/*Hb* have many children, approximately what percentage of their male children would be expected eventually to be bald?

a. 0%
b. 25%
c. 33%
d. 50%
e. 75%

For Questions 14.44–14.46, select the term from the list below that best fits each of the following descriptions. Each term may be used once, more than once, or not at all. There is only one correct answer for each question. Questions 14.44–14.46 refer to the sex chromosomes in different animals.

a. XX
b. XY
c. XO
d. ZZ
e. ZW

e
Factual Recall

14.44 The sex chromosome makeup of a hen.

c
Factual Recall

14.45 The sex chromosome makeup of a male grasshopper.

c
Factual Recall

14.46 The sex chromosome makeup of a human with Turner syndrome.

c
Factual Recall

14.47 The diploid chromosome number in honeybees is 32. What is the number of chromosomes in the somatic cells of a male honeybee?

a. 4
b. 8
c. 16
d. 32
e. 64

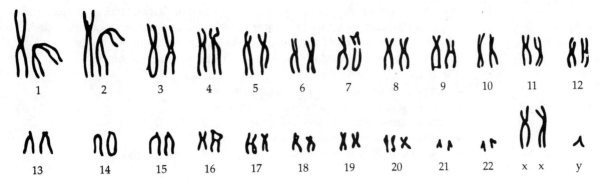

Figure 14.6

c
Factual Recall

14.48 The karyotype shown in Figure 14.6 is associated with which of the following genetic disorders?
  a. Turner syndrome
  b. Down syndrome
  c. Klinefelter syndrome
  d. hemophilia
  e. male-pattern baldness

Refer to the following information to answer Questions 14.49 and 14.50. The sequence of genes on the chromosome of a certain common fly is THMORGAN. You find a population of this fly at the top of a mountain. When you cross it with normal flies, the number of offspring is reduced. In studying its chromosome, you find that it has a sequence of genes of TGMORHAN.

e
Conceptual
Understanding

14.49 Which of the following best explains this?
  a. A deletion occurred.
  b. A duplication occurred.
  c. A single inversion occurred.
  d. A translocation occurred.
  e. Two inversions occurred.

e
Conceptual
Understanding

14.50 You find a second population that you believe is intermediate between the two. What should its sequence of genes be?
  a. THMORGAN
  b. TGROMHAN
  c. THROMGAN
  d. TGMORHAN
  e. either b or c

b
Conceptual
Understanding

14.51 A cassette tape or compact disk contains a series of distinct pieces of electrical or physical information that are interpreted by a tape player or compact disk player, resulting in music. If the information that produces a song is analogous to a gene, and the tape or compact disk is analogous to a chromosome, then the linkage of genes is analogous to
   a. the length of the songs on the tape or disk.
   b. the sequence of songs on the tape or disk.
   c. the volume of the music on the tape or disk.
   d. the tempo of the music on the tape or disk.
   e. the informational content of the songs on the tape or disk.

e
Conceptual
Understanding

14.52 The following is a map of four genes on a chromosome:

A   5   W   3   E   12   G
|___5___|___3___|_____12_____|

Between which two genes would you expect the highest frequency of recombination?
   a. *A* and *W*
   b. *W* and *E*
   c. *E* and *G*
   d. *A* and *E*
   e. *A* and *G*

a
Conceptual
Understanding

14.53 A certain type of grass has a diploid chromosome number of 8. A similar species of grass has a diploid chromosome number of 10. Interspecific hybridization between the two species results in sterile hybrids that can, nonetheless, reproduce vegetatively. The chromosome number of those hybrids would be
   a. 9.
   b. 16.
   c. 18.
   d. 20.
   e. 36.

c
Conceptual
Understanding

14.54 Blueprints contain the information to construct a building. Each floor of the building is represented by a set of blueprints (instructions). Thus, an individual blueprint page is analogous to a gene, and the blueprints for an entire floor would be analogous to a chromosome. That being the case, aneuploidy would result in
   a. an entire duplicate building.
   b. a building that has an extra floor.
   c. a building lacking a floor.
   d. a building in which one floor has an extra room.
   e. either b or c.

Chapter 15

e
Factual Recall

15.1 What does transformation involve in bacteria?
   a. the creation of a strand of DNA from an RNA molecule
   b. the creation of a strand of RNA from a DNA molecule
   c. the infection of cells by a phage DNA molecule
   d. the type of semiconservative replication shown by DNA
   e. the transfer of DNA from one strain to another

a
Conceptual
Understanding

15.2 What happens when T2 phages are grown with radioactive phosphorous?
   a. Their DNA becomes radioactive.
   b. Their proteins become radioactive.
   c. Their DNA is found to be of medium density in a centrifuge tube.
   d. They are no longer able to transform bacterial cells.
   e. They transfer their radioactivity to *E. coli* chromosomes during infection.

d
Conceptual
Understanding

15.3 In the following list of DNA properties, which one would be impossible for a single-stranded DNA molecule?
   a. replication
   b. information storage
   c. exchange with other organisms
   d. repair of thymine dimers
   e. mutation

e
Conceptual
Understanding

15.4 To function as the heritable genetic code, DNA molecules must have all of the following structural features EXCEPT
   a. the ability to form complementary base pairs with other DNA nucleotides.
   b. the ability to form complementary base pairs with RNA nucleotides.
   c. a very stable double-stranded form when not being transcribed or replicated.
   d. a sequence of nucleotides that can be decoded into a sequence of amino acids in a protein.
   e. histone proteins associated with the double helix.

Refer to the following list of enzymes to answer Questions 15.5–15.8. The answers may be used once, more than once, or not at all.
   a. *helicase*
   b. *exonuclease*
   c. *ligase*
   d. *polymerase*
   e. *primase*

d
Factual Recall

15.5 Catalyzes synthesis of a new strand of DNA.

a
Factual Recall

15.6 Enhances separation of DNA strands during replication.

c
Factual Recall

15.7 Covalently connects segments of DNA.

e
Factual Recall

15.8 Synthesizes short segments of RNA.

c
Application

15.9 If cytosine makes up 22% of the nucleotides in a sample of DNA from an organism, then adenine would make up what percent of the bases?
a. 22
b. 44
c. 28
d. 56
e. It cannot be determined from the information provided.

d
Factual Recall

15.10 The problem of replicating the lagging strand—that is, adding bases in the $3' \rightarrow 5'$ direction—is solved by DNA through the use of
a. base pairing.
b. replication forks.
c. the unwinding enzyme, helicase.
d. Okazaki fragments.
e. topoisomerases.

d
Factual Recall

15.11 All of the following elements are present in DNA EXCEPT
a. oxygen.
b. nitrogen.
c. carbon.
d. sulfur.
e. phosphorus.

c
Factual Recall

15.12 All of the following were determined directly from X-ray diffraction photographs of crystallized DNA EXCEPT
a. the diameter of the double helix.
b. the helical shape of DNA.
c. the specifity of base pairing.
d. the linear distance required for one full turn of the double helix.
e. the width of the helix.

a
Factual Recall

15.13 What kind of chemical bonds are found between paired bases of the DNA double helix?
a. hydrogen
b. ionic
c. covalent
d. sulfhydryl
e. phosphate

a
Factual Recall

15.14 What is the primer that is required to initiate the synthesis of a new DNA strand?
    a. RNA
    b. DNA
    c. protein
    d. ligase
    e. primase

c
Factual Recall

15.15 Which enzyme catalyzes the elongation of a DNA strand in the $5' \rightarrow 3'$ direction?
    a. primase
    b. DNA ligase
    c. DNA polymerase
    d. topoisomerase
    e. helicase

c
Factual Recall

15.16 Which of the following descriptions best fits the class of molecules known as nucleotides?
    a. a nitrogen base and a phosphate group only
    b. a nitrogen base and a five-carbon sugar only
    c. a nitrogen base, a phosphate group, and a five-carbon sugar
    d. a five-carbon sugar, a phosphate group, and a purine
    e. a pyrimidine, a purine, and a six-carbon sugar

Refer to the following information to answer Questions 15.17–15.19. For each of the important discoveries that led to our present knowledge of the nature of genes described below, select the investigator(s) associated with each.
    a. Griffith
    b. Hershey and Chase
    c. Avery, MacLeod, and McCarty
    d. Chargaff
    e. Meselson and Stahl

c
Factual Recall

15.17 Chemicals from heat-killed S cells were purified. The chemicals were tested for the ability to transform live R cells. The transforming agent was found to be DNA.

b
Factual Recall

15.18 The DNA of a phage was injected into the bacterial host, but the protein coat stayed outside. The viral DNA directed the host to replicate new phage viruses.

d
Factual Recall

15.19 In any DNA sample, the amount of adenine equals the amount of thymine and the amount of guanine equals the amount of cytosine.

b
Conceptual
Understanding

15.20 When T2 phage viruses that infect bacteria make more viruses in the presence of radioactive sulfur, which of the following results?
    a. The viral DNA is tagged by radioactivity.
    b. The viral proteins are tagged by radioactivity.
    c. The viral DNA is found to be of medium density in a centrifuge tube.
    d. They transfer their radioactivity to *E. coli* DNA.
    e. Both the viral DNA and the viral proteins are tagged by radioactivity.

e
Conceptual
Understanding

15.21 Suppose one were provided with an actively dividing culture of *E. coli* bacteria to which radioactive thymine had been added. What would happen if a cell replicated once in the presence of this radioactive base?
  a. One of the daughter cells, but not the other, would have radioactive DNA.
  b. Neither of the two daughter cells would be radioactive.
  c. All four bases of the DNA would be radioactive.
  d. Radioactive thymine would pair with nonradioactive guanine.
  e. DNA in both daughter cells would be radioactive.

a
Factual Recall

15.22 In DNA, the designations 3′ and 5′ refer to the
  a. bonds formed between phosphate groups and carbon atoms of deoxyribose.
  b. carbon or nitrogen atoms on the rings of purine or pyrimidine bases.
  c. cross-linking of the third and fifth carbon atoms of deoxyribose.
  d. bonding between purines and deoxyribose and between pyrimidines and deoxyribose.
  e. bonds that form between adenine and thymine and between guanine and cytosine.

Use Figure 15.1 to answer Questions 15.23 and 15.24.

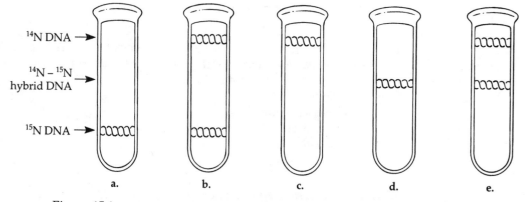

Figure 15.1

e
Factual Recall

15.23 In the Meselson-Stahl experiment, bacteria were grown in a medium containing ^{15}N and were then transferred to a medium containing ^{14}N. Which of the results in Figure 15.1 would be expected after two DNA replications in ^{14}N?

b
Conceptual
Understanding

15.24 A space probe returns with a culture of a microorganism found on a distant planet. Analysis shows that it is a carbon-based life form that has DNA. You grow the cells in ^{15}N medium for several generations and then transfer it to ^{14}N medium. Which pattern in Figure 15.1 would you expect if the DNA was replicated in a conservative manner?

e
Factual Recall

15.25 In trying to determine whether DNA or protein was the genetic material, Hershey and Chase made use of which of the following facts?
  a. DNA does not contain sulfur, whereas protein does.
  b. DNA contains phosphorus, but protein does not.
  c. DNA contains greater amounts of phosphorus than does protein.
  d. Protein contains greater amounts of sulfur than does DNA.
  e. Both a and b are correct.

d
Conceptual
Understanding

15.26 Which of the following statements does NOT apply to the Watson and Crick model of DNA?
  a. The two strands of the DNA helix are antiparallel.
  b. The distance between the strands of the helix is 20 angstroms.
  c. The framework of the helix consists of sugar-phosphate units of the nucleotides.
  d. The two strands of the helix are held together by covalent bonds.
  e. The purines are attracted to pyrimidines.

a
Conceptual
Understanding

15.27 It became apparent to Watson and Crick after completion of their model that the DNA molecule could carry a vast amount of hereditary information in its
  a. sequence of bases.
  b. phosphate-sugar backbones.
  c. complementary pairing of bases.
  d. side groups of nitrogenous bases.
  e. different five-carbon sugars.

c
Factual Recall

15.28 If radioactive sulfur (^{35}S) is used in the culture medium of bacteria that harbor phage viruses, it will later appear in the
  a. viral DNA.
  b. bacterial RNA.
  c. viral coats.
  d. viral RNA.
  e. bacterial cell wall.

c
Factual Recall

15.29 What is the function of DNA polymerase?
  a. to unwind the DNA helix during replication
  b. to seal together the broken ends of DNA strands
  c. to add nucleotides to the end of a growing DNA strand
  d. to repair damaged DNA molecules
  e. to rejoin the two DNA strands (one new and one old) after replication

c
Application

15.30 In an analysis of the nucleotide composition of DNA to see which bases are equivalent in concentration, which of the following would be true?
  a. A = C
  b. A = G and C = T
  c. A + C = G + T
  d. A + T = G + C
  e. Both b and c are true.

e
Application

15.31 DNA ligase functions in
  a. elongation of the $5' \rightarrow 3'$ strand.
  b. elongation of the $3' \rightarrow 5'$ strand.
  c. DNA repair.
  d. unwinding of the double helix.
  e. Both b and c are correct.

c
Conceptual
Understanding

15.32 When a double-stranded DNA molecule is heated, it denatures into two single-stranded molecules. The reason for this is that
  a. the proteins associated with the double helix are denatured and can no longer hold the DNA strands together.
  b. the heat causes the helix to straighten, breaking the connections between the bases.
  c. the heat breaks the hydrogen bonds holding the bases together in the center of the molecule but does not affect the covalent bonds of the backbone.
  d. the heat denatures the bases, preventing them from hydrogen-bonding with each other.
  e. the heat causes the phosphate groups to ionize, preventing them from hydrogen-bonding to the bases.

e
Conceptual
Understanding

15.33 Adenine and guanine have five nitrogen atoms; thymine has two, and cytosine has three. A DNA molecule from *E. coli* has about 5 million base pairs. If it were completely labeled with ^{15}N, how many additional neutrons would that DNA have, compared with a molecule composed of normal ^{14}N?
  a. about 1 million
  b. about 5 million
  c. about 7.5 million
  d. between 10 and 15 million
  e. about 37.5 million

a
Conceptual
Understanding

15.34 A DNA molecule consists of two strands of nucleotides. One strand is the information used by the cell, and the other strand is a complementary series of bases. This is analogous to
  a. a photograph and a photographic negative.
  b. two sides of a divided highway.
  c. a baseball and a bat.
  d. an up escalator and a down escalator.
  e. Both b and d are correct.

e
Conceptual
Understanding

15.35 The two strands of a DNA molecule run in opposite directions. The 3' and 5' ends of one strand are opposite the 5' and 3' ends of the complementary strand. This is analogous to
  a. a photograph and a photographic negative.
  b. two sides of a divided highway.
  c. a baseball and a bat.
  d. an up escalator and a down escalator.
  e. Both b and d are correct.

a
Conceptual
Understanding

15.36 For a couple of decades, we knew the nucleus contained DNA and proteins. The prevailing opinion was that the proteins were the genes and the DNA was a "string" that held them together. The reason for this belief was that
   a. All of the below are correct.
   b. proteins take a greater variety of three-dimensional forms.
   c. proteins have four different levels of structure; DNA has only two.
   d. proteins are made of 20 amino acids and DNA is made of four nucleotides.
   e. proteins can vary in their polarity and charge; DNA cannot.

e
Conceptual
Understanding

15.37 Beavis and a friend found out about Hershey and Chase's experiment. For a science fair project, they decide to repeat the experiment with modifications. They decide that labeling the phosphates of the DNA wasn't good enough. Each nucleotide has only one phosphate, whereas each has two to five nitrogens. Thus, labeling the nitrogens would provide a stronger label than labeling the phosphates. You must tell them that this will not work because
   a. there is no radioactive isotope of nitrogen.
   b. radioactive nitrogen has a half-life of 100,000 years and the material would be too dangerous for too long.
   c. Meselson and Stahl already did this experiment.
   d. although there are more nitrogens in a nucleotide, labeled phosphates actually have 16 extra neutrons, so they are more radioactive.
   e. amino acids (and thus proteins) also have nitrogen atoms, thus the radioactivity would not distinguish between DNA and proteins.

a
Conceptual
Understanding

15.38 Tobacco mosaic virus has RNA rather than DNA as its genetic material. If RNA from a tobacco mosaic virus is mixed with proteins from a DNA virus, the result is a mixed virus. If that virus infects a cell and reproduces, what would you expect the resulting viruses to be like?
   a. tobacco mosaic virus
   b. a DNA virus
   c. a hybrid—tobacco mosaic virus RNA and protein from the DNA virus
   d. a hybrid—tobacco mosaic virus protein and nucleic acid from the DNA virus
   e. I would not expect any viruses to result.

Chapter 16

b
Conceptual
Understanding

16.1 What is the relationship among DNA, a gene, and a chromosome?
  a. A chromosome contains hundreds of genes which are composed of protein.
  b. A chromosome contains hundreds of genes which are composed of DNA.
  c. A gene contains hundreds of chromosomes which are composed of protein.
  d. A gene is composed of DNA, but there is no relationship to a chromosome.
  e. A gene contains hundreds of chromosomes which are composed of DNA.

d
Factual Recall

16.2 What is one function of a signal sequence?
  a. to direct an mRNA molecule into the cisternal space of ER
  b. to bind RNA polymerase to DNA and initiate transcription
  c. to terminate translation of the messenger RNA
  d. to attach ribosomes synthesizing secretory proteins to the ER
  e. to signal the initiation of transcription

d
Application

16.3 Which of the following gene products, if absent or defective, would prevent the functioning of the others?
  a. transfer RNA
  b. ribosomal RNA
  c. messenger RNA
  d. RNA polymerase
  e. aminoacyl-tRNA synthetase

c
Application

16.4 The genetic code is essentially the same for all organisms. From this, one can logically assume all of the following EXCEPT:
  a. A gene from an organism could theoretically be expressed by any other organism.
  b. All organisms have a common ancestor.
  c. DNA was the first genetic material.
  d. All organisms must either manufacture nucleotides or obtain them from their environment.
  e. Related organisms have many similar genes.

c
Factual Recall

16.5 Where is the attachment site for RNA polymerase?
  a. structural gene region
  b. initiation region
  c. promoter region
  d. operator region
  e. regulator region

b
Factual Recall

16.6 What is an anticodon part of?
  a. DNA
  b. tRNA
  c. mRNA
  d. ribosome
  e. activating enzyme

b
Application

16.7 A part of an mRNA molecule with the following sequence is being read by a ribosome: 5' CCG-ACG 3' (mRNA). The following activated transfer RNA molecules are available. Two of them can correctly match the mRNA so that a dipeptide can form.

tRNA Anticodon	Amino Acid
GGC	Proline
CGU	Alanine
UGC	Threonine
CCG	Glycine
ACG	Cysteine
CGG	Alanine

The dipeptide that will form will be
a. cysteine–alanine.
b. proline–threonine.
c. glycine–cysteine.
d. alanine–alanine.
e. threonine–glycine.

b
Conceptual
Understanding

16.8 Which of the following is FALSE?
a. Transcriptionally produced gene products are molecules of RNA.
b. Proteins are translated in the cytoplasm.
c. Steroid hormones may bind directly to DNA and regulate expression.
d. Histones are found only in eukaryotic chromosomes.
e. RNA polymerase attaches to DNA at the promoter sequence.

b
Application

16.9 DNA has two functions: it can self-replicate and it can make non-DNA molecules. DNA is capable of these because
a. its two strands are held together by easily broken electrostatic interactions.
b. its nucleotides will form base pairs with both ribose and deoxyribose nucleotides.
c. both DNA and proteins can be synthesized directly at the DNA template.
d. its replication is semiconservative.
e. replication and expression are thermodynamically spontaneous and require no enzymes.

a
Factual Recall

16.10 Once transcribed, eukaryotic hnRNA typically undergoes substantial alteration that includes
a. excision of introns.
b. fusion into circular forms known as plasmids.
c. linkage to histone molecules.
d. union with ribosomes.
e. fusion with other newly transcribed mRNA.

c
Factual Recall

16.11 Which of the following is true for both prokaryotic and eukaryotic gene expression?
   a. After transcription, a 3' poly-A tail and a 5' cap are added to mRNA.
   b. Translation of mRNA can begin before transcription is complete.
   c. RNA polymerase may recognize a promoter region upstream from the gene.
   d. mRNA is synthesized in the 3' → 5' direction.
   e. The mRNA transcript is the exact complement of the gene from which it was copied.

e
Application

16.12 A particular eukaryotic protein is 300 amino acids long. Which of the following could be the number of nucleotides in the DNA that codes for this protein?
   a. 3
   b. 100
   c. 300
   d. 900
   e. 1800

c
Application

16.13 A particular triplet of bases in the coding sequence of DNA is AGT. What is the corresponding triplet in the complementary strand of DNA?
   a. AGT
   b. UCA
   c. TCA
   d. GAC
   e. TCA in eukaryotes, but UCA in prokaryotes

b
Application

16.14 The corresponding codon for the mRNA transcribed from the gene in Question 16.13 is
   a. AGT.
   b. UCA.
   c. TCA.
   d. AGU.
   e. Either UCA or TCA, depending on wobble in the first base.

d
Application

16.15 The anticodon on the tRNA that binds the mRNA codon in Question 16.14 is
   a. AGT.
   b. UCA.
   c. TCA.
   d. AGU.
   e. Either UCA or TCA, depending on wobble in the first base.

e
Conceptual
Understanding

16.16 Accuracy in the translation of mRNA into the primary structure of a protein depends on specificity in the
   a. binding of ribosomes to mRNA.
   b. shape of the A and P sites of ribosomes.
   c. bonding of the anticodon to the codon.
   d. attachment of amino acids to tRNAs.
   e. Both c and d are correct.

b
Factual Recall

16.17 What are the coding segments of a stretch of eukaryotic DNA called?
    a. introns
    b. exons
    c. codons
    d. replicons
    e. transposons

d
Factual Recall

16.18 All of the following are directly involved in translation EXCEPT
    a. mRNA.
    b. tRNA.
    c. ribosomes.
    d. DNA.
    e. amino acid-activating enzymes.

c
Factual Recall

16.19 The nitrogenous base adenine is found in all members of which of the following groups?
    a. proteins, triglycerides, and testosterone
    b. proteins, ATP, and DNA
    c. ATP, RNA, and genes
    d. alpha glucose, ATP, and DNA
    e. proteins, carbohydrates, and ATP

d
Factual Recall

16.20 RNA differs from DNA in that RNA
    a. contains ribose as its sugar.
    b. is found only in cytoplasm.
    c. contains uracil instead of thymine.
    d. a and c are correct.
    e. a, b, and c are correct.

b
Factual Recall

16.21 When a ribosome first attaches to an mRNA molecule, one tRNA binds to the ribosome. The tRNA that recognizes the initiation codon binds to the
    a. amino acid site (A site) of the ribosome only.
    b. peptide site (P site) of the ribosome only.
    c. large ribosomal subunit only.
    d. second tRNA before attaching to the ribosome.
    e. Both a and c are correct.

b
Application

16.22 If proteins were composed of only 12 different kinds of amino acids, what would be the smallest possible codon size in a genetic system with four different nucleotides?
    a. 1
    b. 2
    c. 3
    d. 4
    e. 12

b
Factual Recall

16.23 From the following list, which is the first event in translation in eukaryotes?
  a. elongation of the polypeptide
  b. base pairing of activated methionine-tRNA to AUG of the messenger
  c. binding of the larger ribosomal subunit to smaller ribosome subunits
  d. covalent bonding between the first two amino acids
  e. Both b and d occur simultaneously.

c
Factual Recall

16.24 What type of bonding is responsible for maintaining the shape of the tRNA molecule?
  a. covalent bonding between sulfur atoms
  b. ionic bonding between phosphates
  c. hydrogen bonding between base pairs
  d. van der Waals interactions between hydrogen atoms
  e. peptide bonding between amino acids

e
Factual Recall

16.25 As a ribosome translocates along an mRNA molecule by one codon, which of the following occurs?
  a. The transfer RNA that was in the A site moves into the P site.
  b. The tRNA that was in the P site moves into the A site.
  c. The tRNA that was in the P site departs from the ribosome.
  d. The tRNA that was in the A site departs from the ribosome.
  e. Both a and c are correct.

e
Factual Recall

16.26 A frameshift mutation could result from
  a. a base insertion only.
  b. a base deletion only.
  c. a base substitution only.
  d. deletion of three consecutive bases.
  e. either an insertion or a deletion of a base.

d
Factual Recall

16.27 If the triplet UUU codes for the amino acid phenylalanine in bacteria, then in plants UUU should code for
  a. leucine.
  b. valine.
  c. cystine.
  d. phenylalanine.
  e. proline.

b
Conceptual
Understanding

16.28 Which point mutation would be most likely to have a catastrophic effect on the functioning of a protein?
  a. a base substitution
  b. a base deletion near the start of the coding sequence
  c. a base deletion near the end of the coding sequence, but not in the terminator codon
  d. deletion of three bases near the start of the coding sequence, but not in the initiator codon
  e. a base insertion near the end of the coding sequence, but not in the terminator codon

b
Conceptual
Understanding

16.29 Choose the answer that has these events of protein synthesis in the proper sequence.
1. An aminoacyl-tRNA binds to the A site.
2. A peptide bond forms.
3. tRNA leaves the P site and the P site remains vacant.
4. A small ribosomal subunit associates with mRNA.
5. tRNA translocates to the P site.

a. 1, 3, 2, 4, 5
b. 4, 1, 2, 5, 3
c. 5, 4, 3, 2, 1
d. 4, 1, 3, 2, 5
e. 2, 4, 5, 1, 3

c
Factual Recall

16.30 Which of these statements represents a common misconception regarding point mutations?
a. They involve changes in one base pair.
b. They can cause drastic changes in polypeptide structure.
c. They always produce a change in the amino acid sequence of a protein.
d. They can lead to the shortening of the mutated polypeptide.
e. They could result in a frameshift mutation.

d
Application

16.31 A portion of the genetic code is UUU = phenylalanine, GCC = alanine, AAA = lysine, and CCC = proline. Assume the correct code places the amino acids phenylalanine, alanine, and lysine in a protein (in that order). Which of the following DNA sequences would substitute proline for alanine?
a. AAA-CGG-TTA
b. AAT-CGG-TTT
c. AAA-CCG-TTT
d. AAA-GGG-TTT
e. AAA-CCC-TTT

b
Application

16.32 The following DNA sequence shows a "gene" encoding a small peptide. The three "stop" codons are UAA, UAG, and UGA.

promoter

5' (ATGACGTATAA) TGACCGTACATGAGTAATACATAAATCAG 3'
3' (TACTGCATATT) ACTGGCATGTACTCATTATGTATTTAGTC 5'

How many animo acids long will the small protein encoded by this "gene" be?
a. 3
b. 4
c. 5
d. 6
e. 7

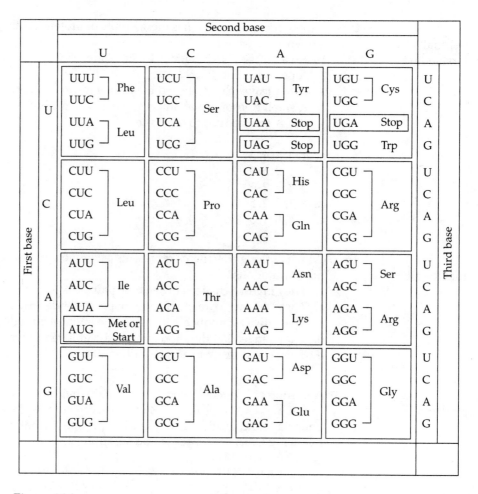

Figure 16.1

Questions 16.33–16.37 refer to Figure 16.1.

e
Application

16.33 A possible sequence of nucleotides in DNA that would code for the polypeptide sequence Phe-Leu-Ile-Val would be
a. 5′ TTG-CTA-CAG-TAG 3′.
b. 3′ AAC-GAC-GUC-AUA 5′.
c. 5′ AUG-CTG-CAG-TAT 3′.
d. 3′ AAA-AAT-ATA-ACA 5′.
e. 3′ AAA-GAA-TAA-CAA 5′.

d
Application

16.34 What amino acid sequence will be generated, based on the following mRNA codon sequence? 5′AUG-UCU-UCG-UUA-UCC-UUG
a. met-arg-glu-arg-glu-agr
b. met-glu-arg-arg-gln-leu
c. met-ser-leu-ser-leu-ser
d. met-ser-ser-leu-ser-leu
e. met-leu-phe-arg-glu-glu

c
Application

16.35 A peptide has the sequence NH$_2$-phe-pro-lys-gly-phe-pro-COOH. What is the sequence in DNA that codes for this peptide?
a. 3′ UUU-CCC-AAA-GGG-UUU-CCC
b. 3′ AUG-AAA-GGG-TTT-CCC-AAA-GGG
c. 3′ AAA-GGG-TTT-CCC-AAA-GGG
d. 5′ GGG-AAA-TTT-AAA-CCC-ACT-GGG
e. 5′ ACT-TAC-CAT-AAA-CAT-TAC-UGA

e
Application

16.36 What is the sequence of a peptide based on the mRNA sequence 5′UUUUCUUAUUGUCUU?
a. leu-cys-tyr-ser-phe
b. cyc-phe-tyr-cys-leu
c. phe-leu-ile-met-val
d. leu-pro-asp-lys-gly
e. phe-ser-tyr-cys-leu

d
Application

16.37 Suppose the following DNA sequence was mutated from AGAGAGAGAGAGAGAGAGA to AGAAGAGAGATCGAGAGA. What amino acid sequence will be generated based on this mutated DNA?
a. arg-glu-arg-glu-agr-glu
b. glu-arg-glu-leu-leu-leu
c. ser-leu-ser-leu-ser-leu
d. ser-ser-leu
e. leu-phe-arg-glu-glu-glu

a
Factual Recall

16.38 What are polysomes?
a. groups of ribosomes reading the same mRNA simultaneously
b. ribosomes containing more than two subunits
c. multiple copies of ribosomes found associated with giant chromosomes
d. aggregations of vesicles containing ribosomal RNA
e. ribosomes associated with more than one tRNA

c
Conceptual
Understanding

16.39 All of the following are found in prokaryotic messenger RNA EXCEPT
a. the AUG codon.
b. the UGA codon.
c. introns.
d. uracil.
e. cytosine.

b
Factual Recall

16.40 The first event in translation of eukaryotes (starting with methionine) is the
a. joining of the ribosomal subunits.
b. base pairing of met-tRNA to AUG of the messenger RNA.
c. binding of the large ribosomal subunit to AUG of mRNA.
d. covalent bonding between the first two amino acids.
e. forming of polysomes.

d
Factual Recall

16.41 During translation, chain elongation continues until what happens?
    a. No further amino acids are needed by the cell.
    b. All tRNAs are empty.
    c. The polypeptide is long enough.
    d. Chain terminator codons occur.
    e. The ribosomes run off the end of mRNA.

d
Conceptual
Understanding

16.42 Which of the following represents a similarity between RNA and DNA?
    a. the presence of a double-stranded helix
    b. the presence of uracil
    c. the presence of an OH group on the 2' carbon of the sugar
    d. nucleotides consisting of a phosphate, sugar, and nitrogen base
    e. repair systems that correct genetic code errors

c
Factual Recall

16.43 What are ribosomes composed of?
    a. two subunits, each consisting of rRNA only
    b. two subunits, each consisting of several proteins only
    c. both rRNA and protein
    d. mRNA, rRNA, and protein
    e. mRNA, tRNA, rRNA, and protein

a
Application

16.44 All of the following are transcribed from DNA EXCEPT
    a. protein.
    b. exons.
    c. rRNA.
    d. tRNA.
    e. mRNA.

c
Factual Recall

16.45 Where is eukaryotic ribosomal RNA transcribed?
    a. the Golgi apparatus
    b. ribosomes
    c. nucleoli
    d. X chromosomes
    e. prokaryotic cells only

a
Application

16.46 Sickle-cell anemia is probably the result of which kind of mutation?
    a. point only
    b. frameshift only
    c. nonsense only
    d. nondisjunction only
    e. both b and d

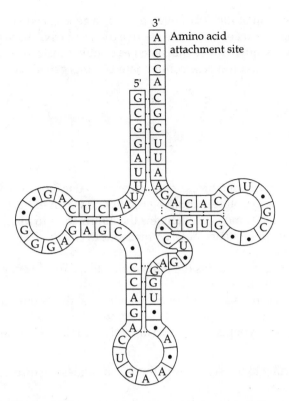

Figure 16.2

d
Application

16.47 Figure 16.2 represents tRNA that recognizes and binds a particular amino acid (in this instance, phenylalanine). Which of the following triplets of bases on the mRNA strand codes for this amino acid?
a. UGG
b. GUG
c. GUA
d. UUC
e. CAU

c
Factual Recall

16.48 Which of the following does NOT occur during the termination phase of translation?
a. A termination codon moves into the A site.
b. The newly formed polypeptide is released.
c. A tRNA with the next amino acid enters the P site.
d. The two ribosomal subunits separate.
e. Translation stops.

a
Application

16.49 Which of the following DNA mutations is the most potentially damaging to the protein it specifies?
a. a base-pair deletion
b. a codon substitution
c. a substitution in the last base of a codon
d. a codon deletion
e. a point mutation

c
Application

16.50 A new form of life is discovered. It has a genetic code much like that of other organisms except that there are five different DNA bases instead of four and the base sequences are translated as doublets instead of triplets. How many amino acids could be accommodated by this genetic code?
  a. 5
  b. 10
  c. 25
  d. 64
  e. 32

b
Application

16.51 Beadle and Tatum proposed the one gene–one enzyme concept. In its original form, this hypothesis could be restated in which of the following ways?
  a. One DNA molecule contains the information to make one enzyme.
  b. A given sequence of DNA nucleotides contains the information to make one enzyme.
  c. Each gene contains the information to make one enzyme, one lipid, and one carbohydrate.
  d. Each gene is actually an enzyme that catalyzes the production of one protein.
  e. Each polypeptide is the result of the activity of one enzyme.

b
Factual Recall

16.52 According to the signal hypothesis, ribosomes are directed to the ER membrane
  a. by a specific characteristic of the ribosome itself, which distinguishes free ribosomes from bound ribosomes.
  b. by a certain amino acid sequence at the beginning of the polypeptide chain being synthesized by the ribosome.
  c. by moving through a channel from the nucleus.
  d. by a chemical signal given off by the ER.
  e. by a signal sequence of RNA that precedes the start codon of the message.

Questions 16.53–16.55 refer to the following simple metabolic pathway:

$$A \xrightarrow{\text{enzyme a}} B \xrightarrow{\text{enzyme b}} C$$

c
Conceptual
Understanding

16.53 According to Beadle and Tatum's one gene–one polypeptide theory, at least _____ gene(s) is (are) necessary for this pathway.

  a. 0
  b. 1
  c. 2
  d. 3
  e. It cannot be determined from the pathway.

a
Conceptual
Understanding

16.54 A mutation results in a defective enzyme *a*. Which of the following would be a consequence?
  a. an accumulation of A and no production of B and C
  b. an accumulation of A and B and no production of C
  c. an accumulation of B and no production of A and C
  d. an accumulation of B and C and no production of A
  e. an accumulation of C and no production of A and B

a
Conceptual
Understanding

16.55 One strain of a diploid organism is homozygous for a recessive allele coding for a defective enzyme a. Another strain is homozygous for a recessive allele coding for a defective enzyme b. Crossing those two strains will result in a strain that would grow on which of the following?
  a. All of the below are correct.
  b. a minimal medium (supplying A)
  c. a minimal medium (supplying A), supplemented with B
  d. a minimal medium (supplying A), supplemented with C
  e. a minimal medium (supplying A), supplemented with B and C

a
Conceptual
Understanding

16.56 You are a member of the Official Junior Astronaut Club. The club has a code so that Martians don't intercept important messages being beamed back to Earth from your base on the moon. You use a simple number/letter substitution code in which the numbers 1 through 26 are substituted for the letters A through Z. However, to be clever, you have also used the substitution of numbers 31 through 56 for the letters A through Z and the substitution of numbers 61 through 86 for the letters A through Z. The Official Junior Astronaut Club code is
  a. degenerate.
  b. universal.
  c. triplet.
  d. ambiguous.
  e. overlapping.

For Questions 16.57 and 16.58, each of the following is a modification of the sentence THECATATETHERAT.
  a. THERATATETHECAT
  b. THETACATETHERAT
  c. THECATARETHERAT
  d. THECATATTHERAT
  e. CATATETHERAT

d
Application

16.57 Which of the above is analogous to a frameshift mutation?

c
Application

16.58 Which of the above is analogous to a single substitution mutation?

c
Conceptual
Understanding

16.59 The enzyme polynucleotide phosphorylase randomly assembles a polymer of nucleotides. You add polynucleotide phosphorylase to a solution of adenosine triphosphate and guanosine triphosphate. The resulting artificial mRNA molecule would have _____ possible different codons if the code involved two-base sequences and _____ possible different codons if the code involved three-base sequences.
  a. 2; 3
  b. 2; 4
  c. 4; 8
  d. 4; 16
  e. 16; 64

a
Conceptual
Understanding

16.60 We recently read about an outbreak of Ebola virus in Zaire. Originally, it was thought that the outbreak was caused by a new virus that resulted from a mutation in the original virus. Closer examination showed that the viruses were the same. How would you tell if a virus had mutated?
a. All of the below are correct.
b. Look for differences in its physical characteristics (with an electron microscope).
c. Look for differences in the amino acid sequence of the proteins the virus produces.
d. Look for differences in the nucleotide sequence of its DNA (or RNA).
e. Look for differences in its pattern of infection.

b
Conceptual
Understanding

16.61 Your friend Forrest wants to create a new "green" shrimp. He plans on growing normal shrimp in green light because, he says, the green light will cause mutations that make the shrimp green. You must tell him this is not a good idea because
a. if the green light is capable of causing such mutations, it will also turn him green.
b. mutations are random, and no agent (even green light) can cause a specific mutation like turning shrimp green.
c. it cannot work because everyone knows that water absorbs green light (that's why lakes and ponds are green).
d. the green shrimp would be mutagenic and anyone eating them would be in danger of turning green.
e. the color of a shrimp has nothing to do with its genes.

Chapter 17

b
Factual Recall

17.1 The function of reverse transcriptase in retroviruses is to
a. hydrolyze the host cell's DNA.
b. use viral RNA as a template for DNA synthesis.
c. convert host cell RNA into viral DNA.
d. translate viral RNA into proteins.
e. use viral RNA as a template for making complementary RNA strands.

e
Conceptual
Understanding

17.2 The role of a metabolite that controls a repressible operon is to
a. bind to the promoter region and decrease the affinity of RNA polymerase for the promoter.
b. bind to the operator region and block the attachment of RNA polymerase to the promoter.
c. increase the production of inactive repressor proteins.
d. bind to the repressor protein and inactivate it.
e. bind to the repressor protein and activate it.

e
Conceptual
Understanding

17.3 Viruses have some of the properties of living organisms. Which of the following is a characteristic of all organisms, but NOT of viruses?
a. genetic information stored as nucleic acid
b. ability to control metabolism
c. ability to reproduce
d. structure includes proteins
e. plasma membrane

d
Application

17.4 In a hospital, a bacterium is isolated that is resistant to an antibiotic previously used against other kinds of bacteria. This is most likely the result of
a. transposition.
b. reverse transcription.
c. transduction.
d. transformation.
e. insertion.

a
Factual Recall

17.5 Which of the following is a TRUE statement about viruses?
a. Viruses are classified below the cellular level of biological organization.
b. A virus particle contains both DNA and RNA.
c. Individual virus particles are visible with light microscopes.
d. Assembly of viral capsids from proteins requires host cell assistance.
e. After assembly of the capsid, growth of virus particles continues until they are released.

d
Factual Recall

17.6 In prokaryotes, the primary transcript of structural genes is
a. hnRNA.
b. tRNA.
c. rRNA.
d. mRNA.
e. DNA.

a
Application

17.7 A mutation that renders the regulator gene of a repressible operon inactive in an *E. coli* cell would result in
a. continuous transcription of the structural gene controlled by that regulator.
b. complete inhibition of transcription of the structural genes.
c. irreversible binding of the repressor to the operator.
d. inactivation of RNA polymerase.
e. Both b and c are correct.

d
Factual Recall

17.8 What is the function of the operator locus of an inducible operon?
a. producing repressor molecules
b. identifying the substrate lactose
c. producing messenger RNA
d. permitting transcription
e. binding steroid hormones

e
Conceptual
Understanding

17.9 The tryptophan synthetase operon uses glucose to synthesize tryptophan. Repressible operons such as this one are
a. permanently turned on.
b. turned on only when tryptophan is present in the growth medium.
c. turned off only when glucose is present in the growth medium.
d. turned on only when glucose is present in the growth medium.
e. turned off whenever tryptophan is added to the growth medium.

c
Conceptual
Understanding

17.10 For a repressible operon to be transcribed, which of the following must be TRUE?
a. Corepressor must be present.
b. RNA polymerase and the active repressor must be present.
c. RNA polymerase must bind to the promoter and the repressor must be inactive.
d. RNA polymerase cannot be present and the repressor must be inactive.
e. RNA polymerase must not occupy the promoter and the repressor must be inactive.

c
Conceptual
Understanding

17.11 The use of the isotope ^{32}P as a tracer element in the study of invasion and lysis of bacteria by bacteriophage viruses has shown that
a. ATP from bacteriophages is identical to ATP found in eukaryotic cells.
b. bacteriophage protein is infectious in bacteria.
c. bacteriophage nucleic acid enters bacteria prior to lysis of the bacteria.
d. ^{32}P accelerates the lytic effect of bacteriophage infection.
e. ^{32}P in an inactive form enters the bacterial genome as a plasmid.

c
Factual Recall

17.12 Bacteriophages that have become integrated into the host cell chromosome are called
a. intemperate bacteriophages.
b. transposons.
c. prophages.
d. T-even bacteriophages.
e. plasmids.

b
Conceptual
Understanding

17.13 Transcription of the structural genes in an inducible operon
   a. occurs all the time.
   b. starts when the pathway's substrate is present.
   c. starts when the pathway's product is present.
   d. stops when the pathway's product is present.
   e. does not produce enzymes.

e
Factual Recall

17.14 Which of the following is TRUE about tumor viruses?
   a. They integrate viral nucleic acid into the host cell genome.
   b. They transform cells growing in tissue culture into rounded cells that lose their contact inhibition.
   c. They may not contain oncogenes, but may turn on the host cell's oncogenes.
   d. Only a and c are correct.
   e. a, b, and c are correct.

For Questions 17.15–17.18, match the following terms with the appropriate phrase or description below. Each term can be used once, more than once, or not at all.
   a. operon
   b. operator
   c. promoter
   d. repressor
   e. corepressor

d
Factual Recall

17.15 A protein that is produced by a regulatory gene.

c
Conceptual
Understanding

17.16 A mutation in this gene could change the rate at which RNA polymerase binds to the DNA.

e
Conceptual
Understanding

17.17 A lack of this nonprotein molecule in the cellular environment would result in the inability of the cell to "turn off" genes.

b
Factual Recall

17.18 The binding of an active repressor molecule at this site prevents the binding of RNA polymerase.

e
Conceptual
Understanding

17.19 You would expect the lactose operon to be transcribed when
   a. there is more glucose in the cell than lactose.
   b. there is more lactose in the cell than glucose.
   c. there is lactose but no glucose in the cell.
   d. the cyclic AMP levels are high within the cell.
   e. Both c and d are correct.

c
Conceptual
Understanding

17.20 Which of the following is an example of positive control in prokaryotes?
   a. tryptophan binding to the repressor molecule
   b. lactose binding to the repressor molecule
   c. CAP binding to the promoter
   d. cAMP levels falling in the cell
   e. inducible enzymes are being synthesized

Use the following answers for Questions 17.21–17.33. The answers may be used once, more than once, or not at all.

 a. *transduction*
 b. *transposition*
 c. *translation*
 d. *transformation*
 e. *conjugation*

b
Factual Recall

17.21 A DNA segment is moved from one location to another.

a
Factual Recall

17.22 DNA is transferred from one bacterium to another by a virus.

d
Factual Recall

17.23 DNA from one strain of bacteria is assimilated by another strain.

e
Factual Recall

17.24 A plasmid is exchanged between bacteria through a pilus.

d
Factual Recall

17.25 DNA from pneumonia-causing bacteria is mixed with harmless bacteria. The bacteria are injected into mice. The mice develop pneumonia and die.

e
Conceptual
Understanding

17.26 A colony of antibiotic-resistant bacteria is mixed with a colony of antibiotic-sensitive bacteria. After several days, all the bacteria are found to be antibiotic resistant.

e
Factual Recall

17.27 A group of F$^+$ bacteria is mixed with a group of F$^-$ bacteria. After several days, all of the bacteria are F$^+$.

a
Conceptual
Understanding

17.28 Bacterial strains A and B are growing together in a colony that has been infected with viruses. After a short period of time, a new strain of bacteria is detected that is very similar to strain A but has a few characteristics of strain B.

d
Factual Recall

17.29 Bacteria have proteins on the surface that recognize and take in DNA from closely related species.

b
Conceptual
Understanding

17.30 A sequence of DNA that has inverted sequences on either end is found scattered throughout the chromosome of a bacterium.

b
Factual Recall

17.31 Antibiotic-resistant genes from different plasmids are found integrated into one large plasmid.

b
Conceptual
Understanding

17.32 "Selfish DNA" is found at several sites in the genome of the bacterium.

b
Factual Recall

17.33 DNA is present that does not provide any known benefit to the cell, yet is replicated each time the genome replicates.

b
Factual Recall

17.34 Which of the following represents a difference between viruses and viroids?
   a. Viruses infect many types of cells while viroids infect only prokaryotic cells.
   b. Viruses have capsids composed of protein while viroids have no capsids.
   c. Viruses contain introns while viroids have only exons.
   d. Viruses have genomes composed of DNA while viroids have genomes composed of RNA.
   e. Viruses cannot pass through plasmodesmata while viroids can.

b
Conceptual
Understanding

17.35 Beijerinck sprayed plants with filtered sap from infected plants. He then repeated the experiment for several generations. The fact that there was no decrease in the ability of the sap to infect plants RULED OUT which of the following hypotheses?
   a. The disease was caused by an agent small enough to pass through the filter.
   b. The disease was caused by a toxin that could pass through the filter.
   c. The disease was caused by an agent that could reproduce.
   d. The disease was caused by an agent that contained nucleic acid.
   e. The disease was caused by an agent that had a protein capsid.

c
Conceptual
Understanding

17.36 A researcher lyses a cell that contains nucleic acid molecules and capsid units of TMV. He leaves this sap in a covered test tube over night. The next day he sprays this fluid on tobacco plants. Which of the following would you expect to occur?
   a. The plants would develop some but not all of the symptoms of the TMV infection.
   b. The plants would develop symptoms typically produced by viroids.
   c. The plants would develop the typical symptoms of TMV infection.
   d. The plants would not show any disease symptoms.
   e. The plants would become infected, but the sap from these plants would be unable to infect other plants.

c
Factual Recall

17.37 Which of the following statements best describes oncogenes?
   a. They are found only in tumor cells.
   b. They are found only in tumor-causing viruses.
   c. They code for growth factors or proteins associated with growth factors.
   d. Activation of a single oncogene can transform a healthy cell.
   e. Both a and b accurately describe oncogenes.

e
Factual Recall

17.38 Most molecular biologists believe that viruses originated from fragments of cellular nucleic acid. Which of the following observations supports this theory?
   a. Viruses contain either DNA or RNA.
   b. Viruses are enclosed in protein capsids rather than plasma membranes.
   c. Viruses can reproduce only inside host cells.
   d. Viruses can infect both prokaryotic and eukaryotic cells.
   e. Viral genomes are usually more similar to the genome of the host cell than to the genome of other cells.

e
Factual Recall

17.39 What is the most common source of genetic diversity in a bacterial colony?
   a. transposons
   b. plasmids
   c. recombination
   d. crossing over
   e. mutation

a
Factual Recall

17.40 In which of the following cases would a mutation have the most significant impact on the genetic diversity of a species?
   a. The species reproduces only asexually.
   b. The species reproduces only sexually.
   c. The species usually reproduces asexually, but can reproduce sexually when conditions become unfavorable.
   d. The species has a relatively long reproductive cycle.
   e. The species' reproductive cycle is unpredictable.

d
Conceptual
Understanding

17.41 Reproduction in bacteria requires
   a. the production of a mitotic spindle.
   b. a plasmid.
   c. cyclic AMP.
   d. replication of DNA.
   e. both b and d.

Refer to the following information and Figure 17.1 to answer Questions 17.42–17.44. Beta-galactosidase, an enzyme that hydrolyzes the disaccharide lactose into its component monosaccharides, is only produced in bacteria in the presence of an "inducer" molecule. The following graph shows the production of this enzyme and its mRNA precursor.

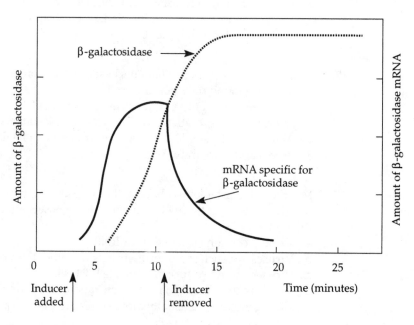

Figure 17.1

b
Application

17.42 The beta-galactosidase curve continues to rise after the inducer is removed because
   a. the mRNA and the protein it codes for are made by different procedures.
   b. the mRNA is broken down rapidly in the cytoplasm, but the enzymes are more resistant to decomposition.
   c. the inducer substance combines with the mRNA to activate it, and when it is removed, the mRNA decomposes.
   d. the beta-galactosidase feeds back negatively to inhibit the mRNA molecules that produce it.
   e. there is nothing to prevent its unrestrained growth.

d
Application

17.43 All of the following conclusions can be drawn from the data EXCEPT:
   a. The synthesis of mRNA specific for beta-galactosidase precedes the synthesis of the enzyme.
   b. The enzyme's increase in concentration levels off as the mRNA concentration approaches zero.
   c. The mRNA curve would probably have leveled off had the inducer remained present.
   d. The maximum concentration of beta-galactosidase would have been greater had the inducer been withdrawn sooner.
   e. The presence of beta-galactosidase is dependent upon the presence of the inducer.

b
Application

17.44 The inducer molecule is most likely
   a. a protein that binds to mRNA promotors.
   b. a molecule that is a substrate of the beta-galactosidase enzyme.
   c. a molecule that acts as an allosteric inhibitor of the enzyme beta-galactosidase.
   d. a molecule that is a competitive inhibitor of the enzymes that break down mRNA specific for beta-galactosidase.
   e. an enzyme that itself binds to the cell's DNA (operon) and therefore stimulates its transcription.

a
Factual Recall

17.45 Which of the following statements regarding transposons is FALSE?
   a. Transposons have specific target sites within the genome.
   b. Transposons are found in both prokaryotes and eukaryotes.
   c. Transposons can move from a plasmid to the chromosome of the bacterium.
   d. Transposons may replicate at the original site and insert the copy at another site.
   e. Transposons may carry only the genes necessary for insertion.

e
Factual Recall

17.46 An Hfr bacterium is one that has
   a. at least one plasmid present in the cytosol.
   b. a special recognition site that will take up closely related DNA from its environment.
   c. several insertion sequences scattered throughout its chromosome.
   d. several copies of a single transposon repeated randomly throughout its chromosome.
   e. a plasmid that has become integrated into its chromosome.

b
Application

17.47 Two strains of Hfr cells were allowed to conjugate. The conjugation was interrupted at 5-minute intervals and the following fragments were obtained: ZDKP, LYMG, AZDK, KPVQ, QLYM. What is the correct sequence for these genes in the chromosome?
a. DKPLYMQVPGA
b. AZDKPVQLYMG
c. YLQVMGAZDKP
d. GMLYQZVPKDA
e. GVMQYLAZDPK

c
Conceptual
Understanding

17.48 A virus injects its DNA into a cell. Some genes are transcribed quite rapidly. Those genes are probably involved in
a. producing DNA polymerase.
b. producing viral capsule proteins.
c. producing repressor proteins to control the bacterial cell.
d. producing proteins that lyse the bacterial cell.
e. producing various enzymes to alter cellular metabolism.

b
Application

17.49 When two viruses with different genotypes infect a cell at the same time, recombination between their DNA molecules can happen much like transformation in bacteria or crossing over in eukaryotes. The following recombination frequencies were found for four different genes:

	b	c	d
a	.05	.15	.10
b		.10	.15
c			.10

What do these data tell you about the nature of the viral DNA genome?
a. It is one linear DNA molecule with gene a at one end, and gene d at the other.
b. It is one circular molecule of DNA.
c. It is two DNA molecules with genes a and b on one, and c and d on the other.
d. It is two DNA molecules with genes a and c on one, and b and d on the other.
e. It is two DNA molecules with genes a and d on one, and b and c on the other.

d
Conceptual
Understanding

17.50 For many bacteriophages, infections are self-limiting. Each phage that infects a cell produces hundreds of new phages, and eventually cells are infected by several phages at different times. This results in a competition for regulatory control over the cell, with the eventual result of the death of the cell without the production of new viruses. This phenomenon is called *superinfection*. Superinfection can be prevented by which of the following changes in the bacteriophage?
a. restricting the host range of the phage
b. slowing down the speed of the lytic infection
c. making larger numbers of new phages
d. the virus becoming a prophage in the cell
e. evolving a more complex protein coat

a
Conceptual
Understanding

17.51 The "central dogma" of molecular genetics is a statement describing the flow of information in a cell. DNA makes RNA, which makes proteins. This path is not reversible. The exception to part of this statement seems to be
  a. retroviruses.
  b. temperate phages.
  c. herpesviruses.
  d. tumor viruses.
  e. all viruses.

Questions 17.52–17.55 refer to Figure 17.2, which represents a bacterial chromosome with six loci indicated.

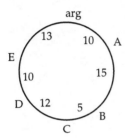

Figure 17.2

c
Conceptual
Understanding

17.52 Which two genes would show the greatest frequency of cotransformation?
  a. arg and A
  b. A and B
  c. B and C
  d. arg and C
  e. A and E

a
Conceptual
Understanding

17.53 When infected by a temperate phage, occasionally infected cells will show transformation for D. Which locus/loci could represent the site of insertion of the prophage?
  a. either C, D, or E
  b. B
  c. C
  d. D
  e. E

d
Conceptual
Understanding

17.54 An Hfr strain is formed when the F plasmid inserts itself. If the F plasmid inserts itself between B and C, which of the following would represent the order of transformation resulting from conjugation?
  a. arg A B C D E
  b. A B C D E arg
  c. B C D E arg A
  d. C D E arg A B
  e. D E arg A B C

b
Conceptual
Understanding

17.55 If the F plamid inserts itself between B and C, then which of the following
would be most likely to be transferred by specialized transduction involving
the F plasmid?
a. A
b. B
c. arg
d. D
e. E

a
Factual Recall

17.56 Which of the following does NOT consist of a sequence of bases?
a. repressor
b. structural gene
c. promoter
d. regulator gene
e. operator

Chapter 18

e
Conceptual
Understanding

18.1 Which of the following is an example of transcriptional control of gene expression?
   a. mRNA is stored in the cytoplasm and needs a control signal to initiate translation.
   b. mRNA exists for a specific time before it is degraded.
   c. There is an amplification of genes for rRNA.
   d. RNA processing occurs before mRNA exits the nucleus.
   e. Transcription factors bind to the enhancer and promoter region.

a
Conceptual
Understanding

18.2 The plasticity of the genome may result from chemical changes in chromosomes. All of the following are examples of chemical plasticity EXCEPT
   a. the compaction of chromosomes into heterochromatin.
   b. the cassette mechanism for altering mating types in yeast.
   c. selective gene loss, especially in certain insects.
   d. the rearrangements of immunoglobulin genes in differentiating B lymphocytes.
   e. gene amplification of rRNA genes.

c
Factual Recall

18.3 All of the following are potential control mechanisms for regulation of gene expression in eukaryotic organisms EXCEPT
   a. the degradation of mRNA.
   b. the transport of mRNA from the nucleus.
   c. the lactose operon.
   d. transcription.
   e. gene amplification.

c
Factual Recall

18.4 A eukaryotic gene typically has all of the following features EXCEPT
   a. introns.
   b. a promoter.
   c. an operator.
   d. a start base triplet.
   e. a transcriptional stop message.

d
Conceptual
Understanding

18.5 The gene that stimulates tumorogenesis in Burkitt's lymphoma is expressed when it is moved to chromosome 14 from chromosome 8. This is an example of gene expression regulated by
   a. diffusible factors.
   b. gene amplification.
   c. steroid hormones.
   d. translocation.
   e. point mutations.

e
Conceptual
Understanding

18.6 Gene expression in eukaryotes may depend upon
a. the position of the gene on the chromosome.
b. the state of the external environment.
c. the stage of development of the organism.
d. a and c.
e. a, b, and c.

a
Conceptual
Understanding

18.7 Which of the following is true of gene regulation in BOTH prokaryotes and eukaryotes?
a. DNA binding proteins interact with other proteins and environmental factors to control gene expression.
b. Nucleosomes regulate gene expression by preventing transcriptional factors from associating with DNA.
c. Heterochromatin remains coiled, thus preventing its transcription.
d. 10–25% of the cell's DNA forms "satellite" DNA, thus limiting the amount of DNA available for transcription.
e. Noncoding sequences are transcribed into RNA from the DNA, thus slowing down the rate at which the genes are transcribed.

b
Conceptual
Understanding

18.8 If a cell were unable to produce histone proteins, which of the following would be expected?
a. An increase in the amount of "satellite" DNA produced during centrifugation.
b. Chromosomes would not form during prophase.
c. Spindle fibers would not form during prophase.
d. The amplification of other protein genes would compensate for the lack of histones.
e. Pseudogenes would be transcribed to compensate for the decreased protein in the cell.

a
Conceptual
Understanding

18.9 The globin pseudogenes lack introns. This supports which of the following hypotheses?
a. Some pseudogenes are transposons involving reverse transcription.
b. Some pseudogenes arose from mutations in duplicated genes.
c. Some pseudogenes arose from RNA introduced into the cell by a virus.
d. Some pseudogenes arose from mistakes in DNA replication and recombination.
e. Some pseudogenes arose from mistakes in transcription of functional genes.

e
Conceptual
Understanding

18.10 If a pseudogene were transposed between a functioning gene and its "upstream" regulatory components, which of the following would most likely occur?
a. The functioning gene would not be transcribed.
b. The pseudogene would not be transcribed.
c. The pseudogene would be transcribed.
d. Both genes would be transcribed.
e. Both a and c would probably occur.

e
Factual Recall

18.11 The processing of the RNA transcript involves
   a. the removal of introns and the splicing together of exons.
   b. the removal of exons and the splicing together of introns.
   c. the addition of a guanine cap and a poly-A tail.
   d. the attachment of introns to ribosomal RNA.
   e. Both a and c are correct.

c
Factual Recall

18.12 Which of the following is a plausible mechanism proposed for coordinating the expression of the genes in a metabolic pathway found in eukaryotic cells?
   a. The genes coding for the enzymes in the pathway are usually grouped closely together on the chromosomes.
   b. The genes coding for the enzymes are all under the control of a single promoter.
   c. A regulator protein recognizes a specific nucleotide sequence in every gene that codes for an enzyme in the pathway.
   d. The genes coding for the enzymes in the pathway are amplified and rearranged during development of the cell.
   e. Both a and b are proposed mechanisms of coordination.

e
Factual Recall

18.13 Which of the following supports the statement that an organism's genome is plastic?
   a. The DNA in a cell carries the complete instructions for making the proteins for the entire organism.
   b. Only a fraction of the DNA is expressed in any one cell at a particular time.
   c. Certain genes may increase in number at certain times in some cells.
   d. Certain genes are selectively lost in some tissues of an organism.
   e. Both c and d support the statement.

e
Conceptual
Understanding

18.14 In which of the following cell types would you observe gene amplification?
   a. white blood cells that produce antibodies
   b. red blood cells that produce hemoglobin
   c. cells of a developing fetus
   d. cells that produce silk in silk worms
   e. a developing amphibian ovum

c
Factual Recall

18.15 In which of the following would you expect to find the most methylation of the DNA?
   a. tandem arrays for ribosomal genes
   b. pseudogenes
   c. Barr bodies
   d. globin genes
   e. transposons

d
Factual Recall

18.16 Which of the following statements concerning transposons is FALSE?
   a. Transposons may increase the production of a particular protein.
   b. Transposons may prevent the normal functioning of a gene.
   c. Transposons may decrease the production of a particular protein.
   d. Transposons may reduce the amount of DNA within certain cells.
   e. Both a and c are false.

d
Factual Recall

18.17 Steroid hormones produce their effects in cells by
    a. activating key enzymes in metabolic pathways.
    b. activating translation of certain mRNAs.
    c. promoting the degradation of specific mRNAs.
    d. promoting transcription of certain regions of DNA.
    e. promoting the formation of looped domains in certain regions of DNA.

e
Factual Recall

18.18 Eukaryotes use all of the following as a means of controlling gene expression EXCEPT the
    a. binding of regulatory proteins to DNA.
    b. degradation of mRNA molecules.
    c. processing of the mRNA transcript before it can be transcribed.
    d. modification of the amino acid sequence of a protein after it has been translated.
    e. modification of the RNA nucleotides to enhance transcription.

e
Factual Recall

18.19 Which of the following events is necessary for the production of a full-blown malignant tumor?
    a. activation of an oncogene in the cell
    b. the inactivation of tumor suppressor genes within the cell
    c. the presence of mutagenic substances within the cell's environment
    d. the presence of a retrovirus within the cell
    e. Both a and b are necessary.

c
Factual Recall

18.20 Which of the following statements concerning proto-oncogenes is FALSE?
    a. They code for proteins associated with cell growth.
    b. They are similar to oncogenes found in retroviruses.
    c. They are produced by somatic mutations induced by carcinogenic substances.
    d. They are involved in producing proteins for cell adhesion.
    e. They are genes that code for proteins involved in cell division.

a
Factual Recall

18.21 What percentage of the DNA in a typical eukaryotic cell is expressed at any given time?
    a. 3–5%
    b. 5–20%
    c. 20–40%
    d. 40–60%
    e. 60–90%

d
Factual Recall

18.22 In a nucleosome, what is the DNA wrapped around?
    a. polymerase molecules
    b. ribosomes
    c. mRNA
    d. histones
    e. nucleolus protein

a
Conceptual
Understanding

18.23 In eukaryotes, what is the active transcription generally associated with?
a. euchromatin only
b. heterochromatin only
c. very tightly packed DNA only
d. highly methylated DNA only
e. both euchromatin and highly methylated DNA

b
Factual Recall

18.24 Chromosome puffs are thought to represent chromosomal regions where
a. genes are inactivated by repressor proteins.
b. genes are especially active in transcription.
c. hormones are produced.
d. regulatory genes are located.
e. genes have been damaged.

d
Conceptual
Understanding

18.25 Muscle cells and nerve cells in one kind of animal owe their differences in structure to
a. having different genes.
b. having different chromosomes.
c. using different genetic codes.
d. expressing different genes.
e. having unique ribosomes.

a
Factual Recall

18.26 Which of the following represents an order of increasingly higher levels of organization?
a. nucleosome, 30-nanometer chromatin fiber, looped domain
b. looped domain, 30-nanometer chromatin fiber, nucleosome
c. looped domain, nucleosome, 30-nanometer chromatin fiber
d. nucleosome, looped domain, 30-nanometer chromatin fiber
e. 30-nanometer chromatin fiber, nucleosome, looped domain

c
Factual Recall

18.27 A cell that remains flexible in its developmental possibilities is said to be
a. differentiated.
b. determined.
c. totipotent.
d. genomically equivalent.
e. epigenic.

a
Conceptual
Understanding

18.28 The cloning of a plant from somatic cells is consistent with the view that
a. differentiated cells retain all the genes of the zygote.
b. genes are lost during differentiation.
c. the differentiated state is normally very unstable.
d. differentiated cells contain masked mRNA.
e. cells can be easily reprogrammed to differentiate and develop into another kind of cell.

c
Factual Recall

18.29 What is meant by the word *metastasis*?
    a. the transformation of a normal cell to a cancer cell
    b. a mutation that causes cancer
    c. the spread of cancer cells from their site of origin
    d. the activation of an oncogene
    e. the development of contact inhibition

c
Conceptual
Understanding

18.30 If one were to observe the activity of methylated DNA, it would be expected that it would
    a. be replicating.
    b. be unwinding in preparation for protein synthesis.
    c. have turned off or slowed down the process of transcription.
    d. be very active in translation.
    e. induce protein synthesis by not allowing repressors to bind with it.

b
Conceptual
Understanding

18.31 A difference between prokaryote and eukaryote RNA is that
    a. prokaryote RNA has uracil, eukaroyte RNA has thymine.
    b. eukaryote RNA lasts much longer before being degraded.
    c. prokaryote RNA never leaves the cell nucleus.
    d. prokaryote RNA contains deoxyribose.
    e. eukaryote RNA is in the form of a double helix.

e
Conceptual
Understanding

18.32 What do pseudogenes and introns have in common?
    a. They code for RNA end products, rather than proteins.
    b. They both contain uracil.
    c. They have multiple promoter sites.
    d. The both code for histones.
    e. They are not expressed nor do they code for functional proteins.

d
Conceptual
Understanding

18.33 All of the following statements concerning gene expression in eukaryotes are true EXCEPT:
    a. Chromosome puffs are sites of active mRNA synthesis.
    b. The mRNA synthesized in the nucleus is modified before it goes to the ribosome and becomes translated.
    c. The euchromatic regions of a chromosome contain active genes while the heterochromatic regions contain inactive genes.
    d. Gene expression in the nucleus is controlled by histone proteins.
    e. Promoter regions function as regulatory sites that influence the binding of RNA polymerase.

a
Conceptual
Understanding

18.34 Which of the following statements is true about control mechanisms in eukaryotic cells?
    a. Methylation of DNA may cause inactivity in part or all of a chromosome.
    b. Cytoplasmic inductive influences act by altering what a particular gene makes.
    c. Eukaryotic genes are organized in large operon systems.
    d. Active gene transcription occurs in the heterochromatic regions of the nucleus.
    e. Lampbrush chromosomes are areas of active tRNA synthesis.

c
Conceptual
Understanding

18.35 All of the following statements concerning the eukaryotic chromosome are
true EXCEPT that
a. it is composed of DNA and protein.
b. the nucleosome is the structural subunit.
c. gene expression is controlled by the histones.
d. it consists of a single molecule of DNA wound around nucleosomes.
e. active transcription occurs on euchromatin.

b
Conceptual
Understanding

18.36 Lampbrush chromosomes are involved in
a. mRNA synthesis in spermatocytes.
b. mRNA synthesis in oocytes.
c. the transcription of rRNA.
d. the transcription of tRNA.
e. DNA replication.

d
Factual Recall

18.37 There is good evidence that which of the following occurs in the vicinity of
chromosomal puffs?
a. tRNA is being synthesized.
b. Amino acids are being synthesized.
c. Proteins are being synthesized.
d. mRNA is being synthesized.
e. DNA is being synthesized.

d
Factual Recall

18.38 Most of the DNA in eukaryotic chromosomes is
a. organized into operons.
b. highly repetitive DNA.
c. moderately repetitive DNA.
d. single-copy DNA sequences that are never transcribed.
e. single-copy sequences that are repeatedly transcribed.

*Questions 18.39–18.42 refer to the following terms. Each term may be used once, more
than once, or not at all.*
  a. enhancer sequence
  b. promoter region
  c. RNA polymerase III
  d. pseudogene
  e. intron

c
Factual Recall

18.39 Transcribes the genes for small RNA molecules, including tRNA.

a
Factual Recall

18.40 Recognition sites for proteins that make the DNA more accessible to RNA
polymerase, thereby boosting the activity of nearby genes several
hundredfold.

b
Factual Recall

18.41 Site important for controlling the initiation of transcription in eukaryotic DNA.

d
Factual Recall

18.42 Has sequences very similar to functional genes but lacks the signals for gene expression.

a
Factual Recall

18.43 The numerous copies of rRNA genes in a salamander are an example of
   a. eukaryotic multigene families.
   b. prokaryotic multigene families.
   c. a highly repetitive sequence.
   d. enhanced promoter regions.
   e. satellite DNA.

e
Application

18.44 All of the following are usually associated with transposons EXCEPT
   a. inverted repeats.
   b. probable presence in all organisms.
   c. an enzyme that catalyzes insertion into new sites.
   d. acquisition of resistance to antibiotics by bacteria.
   e. insertion into DNA by recombination of homologs.

Questions 18.45–18.49 refer to the following terms. Each term may be used once, more than once, or not at all.
   a. highly repetitive DNA
   b. moderately repetitive DNA
   c. unique sequence DNA
   d. methylated DNA
   e. pseudogenes

a
Conceptual
Understanding

18.45 When DNA is heated, the two strands separate. As the DNA cools, the strands come back together ("reanneal") as a function of the complementation of the sequences of bases. Which type of DNA would reanneal first?

c
Conceptual
Understanding

18.46 When DNA is heated, the two strands separate. As the DNA cools, the strands come back together (reanneal) as a function of the complementation of the sequences of bases. Which type of DNA would reanneal last?

a
Factual Recall

18.47 Many transposons fall into this class of DNA.

b
Application

18.48 Control regions such as promoters would fall into which class of DNA?

c
Factual Recall

18.49 Most genes coding for proteins would fall into which class of DNA?

e
Conceptual
Understanding

18.50 Which of the following would be considered evidence that chromosome puffs are a physical result of gene activity?
   a. Different tissues show puffs in different locations.
   b. As tissues develop, the puffs occur in different locations at different times.
   c. When RNA nucleotides have radioactive labels, the puffs show a great deal of radioactivity.
   d. both a and b
   e. all of the above

a
Factual Recall

18.51 Which of the following would you expect NOT to be part of a multigene family?
   a. genes coding for the enzymes used in glycolysis
   b. rRNA genes
   c. tRNA genes
   d. genes for histone proteins
   e. genes for globin subunits

b
Conceptual
Understanding

18.52 When an advertiser wants to send out an ad, a computer prints an address label for everyone on a mailing list. This can be considered analogous to eukaryotic control of genes. The list might be considered to be the array of genes that need to be turned on. Using that analogy, the mailing labels would be analogous to
   a. methylated bases.
   b. transcription factors.
   c. enhancer regions.
   d. introns.
   e. RNA polymerase.

c
Conceptual
Understanding

18.53 If the structure of a TV show is analogous to the structure of a gene, then the introns of a gene would be analogous to
   a. the opening theme music.
   b. the segments of the show.
   c. the commercials between segments of the show.
   d. the commercials between shows.
   e. the closing credits.

a
Factual Recall

18.54 Which of the following is NOT a mechanism whereby a proto-oncogene is converted to an oncogene?
   a. methylation of bases
   b. point mutation
   c. gene transposition
   d. gene amplification
   e. chromosome translocation

Chapter 19

d
Factual Recall

19.1 Biotechnology is presently being used to do which of the following?
  a. produce vaccines
  b. correct defects in human germ cells
  c. produce human gene products
  d. Only a and c are correct.
  e. a, b, and c are correct.

e
Conceptual
Understanding

19.2 PCR could be used to amplify DNA from which of the following?
  a. a fossil
  b. a fetal cell
  c. a virus
  d. Only b and c are correct.
  e. a, b, and c are correct.

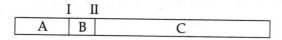

Figure 19.1

b
Application

19.3 The segment of DNA shown in Figure 19.1 has restriction sites I and II, which create restriction fragments A, B, and C. Which of the gels produced by electrophoresis and shown in Figure 19.2 would represent the separation and identity of these fragments?

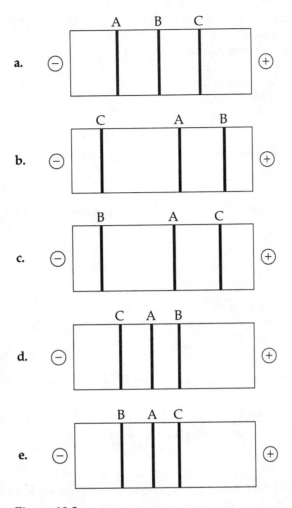

Figure 19.2

a
Conceptual
Understanding

19.4 It is theoretically possible for a gene from any organism to function in any other organism. Why is this possible?
    a. All organisms have the same genetic code.
    b. All organisms are made up of cells.
    c. All organisms have similar nuclei.
    d. All organisms have ribosomes.
    e. All organisms have transfer RNA.

d
Factual Recall

19.5 The polymerase chain reaction is important because it allows us to
    a. insert eukaryotic genes into prokaryotic plasmids.
    b. incorporate genes into viruses.
    c. make DNA from RNA transcripts.
    d. make many copies of DNA.
    e. insert regulatory sequences into eukaryotic genes.

Use the following choices to answer Questions 19.6–19.11. Each choice may be used once, more than once, or not at all.
    a. restriction endonuclease
    b. DNA ligase
    c. reverse transcriptase
    d. RNA polymerase
    e. DNA polymerase

b
Factual Recall

19.6 Which enzyme permanently seals together DNA fragments that have complementary sticky ends?

c
Factual Recall

19.7 Which enzyme is used to make complementary DNA (cDNA)?

d
Factual Recall

19.8 Which enzyme joins a phosphate group to ribose?

e
Factual Recall

19.9 Which enzyme is used to make multiple copies of genes in the polymerase chain reaction (PCR)?

a
Factual Recall

19.10 Which enzyme is used to produce RFLPs?

a
Factual Recall

19.11 *Eco* RI is an example of which type of enzyme?

c
Factual Recall

19.12 Which of the following procedures would produce RFLPs?
    a. incubating a mixture of single-strand DNA from two closely related species
    b. incubating DNA nucleotides with DNA polymerase
    c. incubating DNA with restriction endonucleases
    d. incubating RNA with DNA nucleotides and reverse transcriptase
    e. incubating DNA fragments with "sticky ends" with ligase

c
Conceptual
Understanding

19.13 If you discovered a bacterial cell that contained no restriction endonuclease, which of the following would you expect to happen?
    a. The cell would be unable to replicate its DNA.
    b. The cell would create incomplete plasmids.
    c. The cell would be easily infected and lysed by bacteriophages.
    d. The cell would become an obligate parasite.
    e. Both a and d would occur.

Use the following information to answer Questions 19.14–19.17.
A eukaryotic gene has sticky ends produced by the restriction endonuclease Eco R1. It is added to a mixture containing Eco RI and a bacterial plasmid that carries two genes which make it resistant to ampicillin and tetracycline. The plasmid has one recognition site for Eco RI located in the tetracycline-resistance gene. This mixture is incubated for several hours and then added to bacteria growing in nutrient broth. The bacteria are allowed to grow overnight and are streaked on a plate using a technique that produces isolated colonies that are clones of the original. Samples of these colonies are then grown in four different media: nutrient broth plus ampicillin, nutrient broth plus tetracycline, nutrient broth plus ampicillin and tetracycline, and nutrient broth containing no antibiotics.

e
Conceptual
Understanding

19.14 The bacteria containing the engineered plasmid would grow in
    a. the nutrient broth only.
    b. the nutrient broth and the tetracycline broth only.
    c. the nutrient broth, the ampicillin broth, and the tetracycline broth.
    d. the ampicillin and tetracycline broth only.
    e. the ampicillin and the nutrient broth.

d
Conceptual
Understanding

19.15 The bacteria that contained the plasmid, but not the eukaryotic gene, would grow
    a. in the nutrient broth plus ampicillin, but not in the broth containing tetracycline.
    b. only in the broth containing both antibiotics.
    c. in the broth containing tetracycline, but not in the broth containing ampicillin.
    d. in all four types of broth.
    e. only in the broth that contained no antibiotics.

d
Conceptual
Understanding

19.16 Why was the gene inserted in the plasmid before it was mixed with the bacteria?
    a. The plasmid acted as a vector to introduce the gene into the bacteria.
    b. The plasmid contains control regions necessary for the replication of the gene.
    c. The eukaryotic gene contains introns which must be removed by the plasmid.
    d. Only a and b are correct.
    e. a, b, and c are correct.

a
Conceptual
Understanding

19.17 Bacteria that did not take up any plasmids would grow on which media?
    a. the nutrient broth only
    b. the nutrient broth and the tetracycline broth only
    c. the nutrient broth and the ampicillin broth only
    d. the tetracycline and ampicillin broth only
    e. all four broths

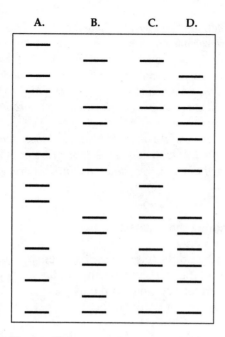

Figure 19.3

Use the following information and Figure 19.3 to answer Questions 19.18–19.20. The DNA fingerprints above represent four different individuals.

b
Application

19.18 Which of the following statements is consistent with the results?
    a. B is the child of A and C.
    b. C is the child of A and B.
    c. D is the child of B and C.
    d. A is the child of B and C.
    e. A is the child of C and D.

b
Application

19.19 Which of the following statements is most likely TRUE?
    a. D is the child of A and C.
    b. D is the child of A and B.
    c. D is the child of B and C.
    d. A is the child of C and D.
    e. B is the child of A and C.

d
Application

19.20 Which of the following are probably siblings?
    a. A and B
    b. A and C
    c. A and D
    d. C and D
    e. B and D

e
Factual Recall

19.21 A DNA fingerprint is produced by
    a. treating selected segments of DNA with restriction enzymes.
    b. electrophoresis of restriction fragments.
    c. oligonucleotides from PCRs.
    d. electroporation of cDNAs.
    e. Both a and b are correct.

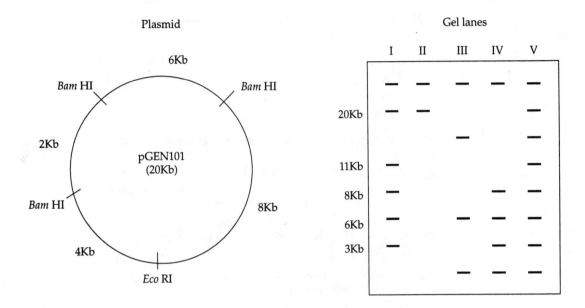

Figure 19.4

Use the following information and Figure 19.4 to answer Questions 19.22–19.24. The plasmid pGEN101 shown in Figure 19.4 was treated with various mixtures of restriction enzymes. The electrophoresis gel shows the results of each of those digestions.

c
Application

19.22. Which lane represents the fragments produced using *Bam* HI only?
    a. I
    b. II
    c. III
    d. IV
    e. V

b
Application

19.23 Which lane represents the fragments produced using *Eco* RI only?
    a. I
    b. II
    c. III
    d. IV
    e. V

d
Application

19.24 Which lane represents the fragments produced when the plasmid was cut with both *Eco* RI and *Bam* HI?
    a. I
    b. II
    c. III
    d. IV
    e. V

e
Factual Recall

19.25 Which of the following can be used as a cloning vector?
    a. *Eco* RI
    b. lambda phage
    c. *E. coli*
    d. bacterial plasmid
    e. Both b and d are correct.

c
Factual Recall

19.26 Why is it difficult to get bacteria to express genes directly from eukaryotic DNA?
    a. Eukaryotic genes are not transcribed in a single transcript.
    b. Eukaryotic genes do not contain enhancer sequences.
    c. Eukaryotic genes contain introns.
    d. Eukaryotic genes lack controlling regions.
    e. Eukaryotic genes may contain transposons.

b
Factual Recall

19.27 Reverse transcriptase is important in genetic engineering because
    a. it is found in retroviruses.
    b. it allows us to make DNA from RNA.
    c. it allows bacteria to translate eukaryotic RNA.
    d. it removes exons from eukaryotic genes.
    e. both c and d are correct.

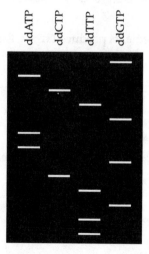

Figure 19.5

c
Application

19.28 A DNA fragment of unknown sequence is divided into four portions and mixed with all the elements necessary to synthesize the complementary strand. To each of these four mixtures a different dideoxynucleotide is added in addition to the normal nucleotides. Dideoxynucleotides compete with the normal nucleotides for insertion into the synthesizing strand of DNA. When a dideoxynucleotide is added, the synthesis of the strand stops. Each mixture is then separated by electrophoresis and the gel in Figure 19.5 is obtained. What is the sequence of the new strand being synthesized?
   a. ACTGAACTGTTGG
   b. CTGACTTCGACAA
   c. TTGTCGAAGTCAG
   d. GACTGAAGCTGTT
   e. AAGGCTTAGCTTA

Questions 19.29–19.32 refer to the techniques, tools, or substances below. Answers may be used once, more than once, or not at all.
   a. *restriction enzymes*
   b. *gene cloning*
   c. *DNA ligase*
   d. *gel electrophoresis*
   e. *reverse transcriptase*

b
Factual Recall

19.29 Produces many copies of a gene for basic research or for large-scale production of a gene product.

e
Factual Recall

19.30 Enables one to create complementary DNA (cDNA) from mRNA; results in a smaller gene product (RNA processed—no introns) that is more easily translated by bacteria.

d
Factual Recall

19.31 Separates molecules by movement due to size and electrical charge.

c
Factual Recall

19.32 Seals the sticky ends of restriction fragments to make recombinant DNA.

c
Factual Recall

19.33 Restriction fragments of DNA are separated from one another by which process?
   a. filtering
   b. centrifugation
   c. gel electrophoresis
   d. chromatography
   e. electron microscopy

c
Conceptual
Understanding

19.34 What is a cloning vector?
   a. the enzyme that cuts DNA into restriction fragments
   b. a DNA probe used to locate a particular gene in the genome
   c. an agent, such as a plasmid, used to transfer DNA from an *in vitro* solution into a living cell
   d. the laboratory apparatus used to clone genes
   e. the sticky end of a DNA fragment

d
Conceptual
Understanding

19.35 Specific DNA fragments of a gene library are contained in
  a. recombinant plasmids of bacteria.
  b. recombinant viral DNA.
  c. eukaryotic chromosomes.
  d. Only a and b are correct.
  e. a, b, and c are correct.

c
Factual Recall

19.36 Bacteria containing recombinant plasmids are often identified by which process?
  a. examining the cells with an electron microscope
  b. using radioactive tracers to locate the plasmids
  c. exposing the bacteria to an antibiotic that kills the cells lacking the plasmid
  d. removing the DNA of all cells in a culture to see which cells have plasmids
  e. producing antibodies specific for each bacterium containing a recombinant plasmid

c
Factual Recall

19.37 The yeast cell has been referred to as the *E. coli* of eukaryotes because it
  a. is actually more similar to a prokaryotic cell than a true eukaryotic cell.
  b. resides in the human colon along with *E. coli.*
  c. is a relatively simple eukaryotic cell that is easily cultured in the laboratory.
  d. lacks a nucleus.
  e. is the simplest multicellular organism.

a
Factual Recall

19.38 Plasmids are important in biotechnology because they are
  a. a vehicle for the insertion of recombinant DNA into bacteria.
  b. recognition sites on recombinant DNA strands.
  c. surfaces for protein synthesis in eukaryotic recombinants.
  d. surfaces for respiratory processes in bacteria.
  e. proviruses incorporated into the host DNA.

d
Factual Recall

19.39 What is the genetic function of restriction endonuclease?
  a. adds new nucleotides to the growing strand of DNA
  b. joins nucleotides during replication
  c. joins nucleotides during transcription
  d. cleaves nucleic acids at specific sites
  e. repairs breaks in sugar-phosphate backbones

a
Factual Recall

19.40 The *Eco* RI enzyme used in constructing hybrid molecules of certain gene sequences and plasmid DNA acts by
  a. opening DNA molecules at specific sites, leaving sticky ends exposed.
  b. sealing plasmid DNA and foreign DNA into a closed circle.
  c. transcribing plasmid DNA into a transformed molecule.
  d. allowing a hybrid plasmid DNA into a transformed molecule.
  e. binding human genes to bacterial plasmids.

I. transform *E. coli* cells
II. cleave by endonuclease
III. extract plasmid DNA from bacterial cells
IV. join plasmid DNA with foreign DNA by hydrogen bonds
V. seal with DNA ligase

c
Application

19.41 Which of the following is the most logical sequence of the steps shown above for splicing foreign DNA into a plasmid and inserting the plasmid into a bacterium?
a. I, II, IV, III, V
b. II, III, V, IV, I
c. III, II, IV, V, I
d. III, IV, V, I, II
e. IV, V, I, II, III

d
Application

19.42 Restriction endonucleases are used to produce fragments of double-stranded DNA that are joined to carriers or vectors. Which of the following pieces of DNA would be a part of a double strand that could be cleaved by a restriction endonuclease?
a. AAAGGG
b. CAGCAG
c. CCCTTT
d. GAATTC
e. GCTTAC

b
Conceptual
Understanding

19.43 What two enzymes are needed to produce recombinant DNA?
a. endonuclease, transcriptase
b. endonuclease, ligase
c. polymerase, ligase
d. transcriptase, ligase
e. DNA polymerase, topoisomerase

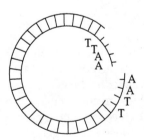

Figure 19.6

c
Factual Recall

19.44 Which enzyme was used to produce the molecule in Figure 19.6?
a. ligase
b. transcriptase
c. endonuclease
d. RNA polymerase
e. DNA polymerase

d
Application

19.45 *Eco* RI and *Hin* dIII are two different restriction endonucleases. If the DNA of different animals were cut with either *Eco* RI or *Hin* dIII, which of the cut DNAs would not join together easily so that they could be sealed with ligase?
  a. human DNA cut with *Eco* RI and chimp DNA cut with *Eco* RI
  b. prokaryotic DNA cut with *Hin* dIII and eukaryotic DNA cut with *Hin* dIII
  c. mouse liver DNA cut with *Eco* RI and mouse kidney DNA cut with *Eco* RI
  d. *E. coli* DNA cut with *Eco* RI and mouse DNA cut with *Hin* dIII
  e. mouse DNA cut with *Hin* dIII and chimp DNA cut with *Hin* dIII

c
Application

19.46 The principal problem with inserting an unmodified mammalian gene into the bacterial chromosome, and then getting that gene expressed, is that
  a. prokaryotes use a different genetic code from that of eukaryotes.
  b. bacteria translate polycistronic messages only.
  c. bacteria cannot remove eukaryotic introns.
  d. bacterial RNA polymerase cannot make RNA complementary to mammalian DNA.
  e. bacterial DNA is not found in a membrane-bound nucleus and is therefore incompatible with mammalian DNA.

c
Application

19.47 Assume that you are trying to insert a gene into a plasmid and someone gives you a preparation of DNA cut with restriction endonuclease X. The gene you wish to insert has sites on both ends for cutting by restriction endonuclease Y. You have a plasmid with a single site for Y, but not for X. Your strategy should be to
  a. insert the fragments cut with X directly into the plasmid without cutting the plasmid.
  b. cut the plasmid with restriction endonuclease X and insert the fragments cut with Y into the plasmid.
  c. cut the DNA again with restriction endonuclease Y and insert these fragments into the plasmid cut with the same enzyme.
  d. cut the plasmid twice with restriction endonuclease Y and ligate the two fragments onto the ends of the human DNA fragments cut with restriction endonuclease X.
  e. cut the plasmid with endonuclease X and then insert the gene into the plasmid.

c
Conceptual
Understanding

19.48 A DNA gene that contains introns can be made shorter (but remain functional) for genetic engineering purposes by
  a. using RNA polymerase to transcribe the gene.
  b. using a restriction endonuclease to cut the gene into shorter pieces.
  c. using reverse transcriptase to reconstruct the gene from its mRNA.
  d. using DNA polymerase to reconstruct the gene from its polypeptide product.
  e. using DNA ligase to put together fragments of the DNA that codes for a particular polypeptide.

d
Factual Recall

19.49 After being digested with a restriction enzyme, DNA fragments are separated by gel electrophoresis. Specific fragments are then identified through the use of a
a. plasmid.
b. restriction enzyme.
c. sticky end.
d. probe.
e. RFLP.

e
Factual Recall

19.50 Probes are short, single-stranded DNA segments that are used to identify DNA fragments with a particular sequence. Before the probe can identify a specific restriction fragment, what must be done?
a. The fragments must be separated by electrophoresis.
b. The double-stranded DNA must be heated to make it single stranded.
c. The DNA must be permanently attached to a suitable substrate so it doesn't migrate.
d. both a and b
e. a, b, and c

d
Conceptual
Understanding

19.51 Which of the following statements about probes is FALSE?
a. They are single-stranded segments of DNA.
b. Shorter probes adhere to more fragments than do longer probes.
c. In order to be useful, the probe must be labeled.
d. They must be produced with the same restriction endonuclease as the fragments.
e. In many cases, a probe from one organism can be used to locate a homologous DNA segment in another organism.

a
Factual Recall

19.52 DNA fragments from a gel are transferred to a membrane via a procedure called Southern blotting. The purpose of Southern blotting is to
a. permanently attach the DNA fragments to a substrate.
b. separate the two complementary DNA strands.
c. transfer only the DNA that is of interest.
d. analyze the RFLPs in the DNA.
e. separate out the PCRs.

Questions 19.53 and 19.54 use the following information: RFLP analysis has identified the frequencies of six alleles in the Caucasian population as follows:

Allele	Frequency
A_1	.25
A_2	.15
A_3	.10
A_4	.05
A_5	.20
A_6	.25

d
Application

19.53 A man is found to have genotpe A_3/A_3. With what proportion of the Causasian population would he share that genotype?
a. .20
b. .1
c. .05
d. .01
e. .001

d
Application

19.54 A man is believed to have committed a rape. His genotype is A_2/A_4 and the victim's is A_3/A_3. The DNA from a vaginal swab is found to have alleles A_2, A_3, and A_4. The RFLP analysis proves the man is
a. not guilty.
b. probably not guilty.
c. certainly guilty.
d. possibly guilty.
e. triploid.

e
Application

19.55 RFLP analysis of some blood from a crime scene shows that the blood belongs to someone with genotype D_1/D_3, E_2/E_5, F_5/F_5, G_1/G_3. The frequencies of these genotypes in the Caucasian population are $D_1/D_3 = 0.01$, $E_1/E_5 = 0.02$, $F_5/F_5 = 0.001$, $G_1/G_3 = 0.00001$. With what proportion of the Causasian population does this man share a genotype?
a. .0311
b. .0001
c. .00005
d. .000001
e. 2×10^{-11}

b
Conceptual
Understanding

19.56 Dideoxyribose is a modified ribose sugar that does not have a hydroxyl group attached to either the 2' or the 3' carbon. This makes it useful in sequencing DNA molecules because DNA polymerase cannot attach a nucleotide to it. This attachment cannot be made because
a. the hydroxyl group is missing from the 2' carbon.
b. the hydroxyl group is missing from the 3' carbon.
c. it is unable to form hydrogen bonds with complementary bases because of the hydroxyl group missing from the 2' carbon.
d. the nucleotide would lack a phosphate group because of the missing hydroxyl group at the 3' carbon.
e. the fact that two hydroxyl groups are missing causes the base to become unstable.

e
Conceptual
Understanding

19.57 The DNA of a cell is like a library. The books of a library are analogous to genes, and the sections of a library are analogous to chromosomes. Which of the following would NOT be a library activity analogous to a function of biotechnology?
a. finding a particular book in the library
b. moving a book from one library to another
c. reading and understanding the contents of a book
d. identifying a library by the books that it has
e. returning books that had been checked out

Chapter 20

b
Factual Recall

20.1 All of the following statements are part of the Darwin-Wallace theory of natural selection EXCEPT:
   a. Heritable variations occur in natural populations.
   b. Characteristics which are acquired during the life of an individual are passed on to offspring.
   c. Organisms tend to increase in numbers at a rate more rapid than the environment can support.
   d. On average, the best adapted individuals leave more offspring.
   e. There exists in nature a constant struggle for survival.

a
Conceptual
Understanding

20.2 During a study session about evolution, one of your fellow students remarks, "The giraffe stretched its neck while reaching for higher leaves—its offspring inherited longer necks as a result." To correct your fellow student's misconception, what would you say?
   a. Characteristics acquired during an organism's life are not passed on through genes.
   b. Spontaneous mutations can result in the appearance of new traits.
   c. Only favorable adaptations have survival value.
   d. Disuse of an organ may lead to its eventual disappearance.
   e. Overproduction of offspring leads to a struggle for survival.

d
Application

20.3 Insects with wing mutations that prevent flight (e.g., the "vestigial wing" mutation in fruit flies) usually can't survive long in nature. Flightlessness is selected against. But in four of the following environments the trait could actually be selected for. In which environment would useless wings NOT be selected for?
   a. an island where stiff winds blow some flying insects out to sea, never to return
   b. a swamp full of frogs that can see and catch flying insects better than crawling insects
   c. a forest full of bats that catch and eat insects while in flight
   d. a cage with no predators, in which food is provided in low dishes
   e. a cage with slippery walls that insects can't climb and an electrified screen on top that electrocutes insects that touch it

e
Factual Recall

20.4 Darwin differed from Lamarck in his proposal that
   a. species are not fixed.
   b. evolution leads to adaptation.
   c. life on Earth has had a long evolutionary history.
   d. life on Earth did not evolve abruptly but rather through a gradual process of minute changes.
   e. inherent variations in the population are more important in evolution than variations acquired during individual lifetimes.

e
Factual Recall

20.5 What would be the best technique for determining the phylogenetic relationship among several closely related species?
a. examining the fossil record
b. comparison of homologous structures
c. comparative embryology
d. comparative anatomy
e. DNA analysis and protein comparison

e
Factual Recall

20.6 Darwin was able to formulate his theory of evolution based on several facts. Which of the following facts was unavailable to Darwin in the mid-nineteenth century?
a. Most populations are stable in size.
b. Individual organisms in a population are not alike.
c. All populations have the potential to increase.
d. Natural resources are limited.
e. Characteristics are inherited as genes on chromosomes.

a
Factual Recall

20.7 What was the prevailing notion before Lyell and Darwin? The Earth is
a. 6000 years old and populations are unchanging.
b. 6000 years old and populations gradually change.
c. 6000 years old and populations changed radically after periodic catastrophes.
d. very old and populations are unchanging.
e. very old and populations gradually change.

b
Conceptual
Understanding

20.8 Which of the following is an acceptable definition of evolution?
a. a change in the phenotypic makeup of a population
b. a change in the genetic makeup of a population
c. a change in the environmental conditions
d. a change in the genotypic makeup of an individual
e. a change in the species composition of a community

e
Conceptual
Understanding

20.9 Of the following anatomical structures, which is homologous to the wing of a bat?
a. the dorsal fin of a shark
b. the tail of a kangaroo
c. the wing of a butterfly
d. the tail fin of a fish
e. the arm of a human

c
Conceptual
Understanding

20.10 Anatomical structures that show similar function but dissimilar embryonic and evolutionary background are said to be
a. homologous.
b. primitive.
c. analogous.
d. monophyletic.
e. polyphyletic.

a
Factual Recall

20.11 Single-stranded DNA extracted from human cells is added to single-stranded DNA from chimpanzee cells. The two different DNA strands bind tightly to each other, indicating much similarity in base sequences. This technique is called
a. DNA hybridization.
b. restriction mapping.
c. DNA sequencing.
d. DNA electrophoresis.
e. RFLP analysis.

a
Factual Recall

20.12 Which of the following ideas that Darwin incorporated into his theory was proposed by Hutton?
a. gradual geological processes
b. extinctions evident in the fossil record
c. adaptation of species to the environment
d. a hierarchical classification of organisms
e. the inheritance of acquired characteristics

e
Factual Recall

20.13 Natural selection is based on all of the following EXCEPT:
a. Variation exists within populations.
b. The fittest individuals leave the most offspring.
c. There is differential reproductive success within populations.
d. Populations tend to produce more individuals than the environment can support.
e. Individuals must adapt to their environment.

b
Factual Recall

20.14 Which of the following represents an idea Darwin took from the writings of Thomas Malthus?
a. All species are fixed in the form in which they are created.
b. Populations tend to increase at a rate greater than their food supply.
c. The Earth changed over the years through a series of catastrophic upheavals.
d. The environment is responsible for natural selection.
e. The Earth is more than 10,000 years old.

a
Factual Recall

20.15 On which of the following did Linnaeus base his classification system?
a. morphology and anatomy
b. evolutionary history
c. the fossil record
d. Only a and b are correct.
e. a, b, and c are correct.

b
Factual Recall

20.16 Which of the following disciplines has contributed LEAST to the body of evidence for evolution?
a. biogeography
b. mycology
c. molecular biology
d. taxonomy
e. paleontology

a
Factual Recall

20.17 Which of the following did Darwin NOT understand about natural selection?
    a. the source of genetic variation
    b. that organisms became extinct
    c. that variation is common in populations
    d. that competition exists in populations
    e. that populations overproduce offspring

d
Factual Recall

20.18 Which of the following was NOT part of Darwin's explanation of natural selection?
    a. Organisms commonly produce more offspring than can possibly survive.
    b. Variations exist within each species.
    c. Members of a species compete with each other for food and space.
    d. New variations continually arise by mutation.
    e. Usually the better-adapted of each generation survive to reproduce.

b
Factual Recall

20.19 The statement "Improving the intelligence of an adult through education will result in that adult's descendants being born with a greater native intelligence" is an example of
    a. Darwinism.
    b. Lamarckism.
    c. neo-Darwinism.
    d. *scala naturae.*
    e. natural theology.

e
Factual Recall

20.20 Darwin's and Lamarck's theories of evolution both suggest that
    a. species are fixed.
    b. the Earth is 6000 years old.
    c. the environment creates favorable characteristics on demand.
    d. the main mechanism of evolution is the inheritance of acquired characteristics.
    e. the interaction of organisms with their environment is important in the evolutionary process.

e
Factual Recall

20.21 What did Charles Darwin publish in 1859?
    a. *Vestiges of Creation*
    b. *Philosophie Zoologique*
    c. *On the Nature of Things*
    d. *The Growth of Biological Thought*
    e. *On the Origin of Species by Means of Natural Selection*

c
Factual Recall

20.22 The taxonomic system developed by Linnaeus is best described as a
    a. binary scheme of groupings.
    b. branching diagram of interrelationships.
    c. hierarchy of increasingly general categories.
    d. map that distinguishes kinship among animals.
    e. decimal plan for sorting all living organisms.

c
Factual Recall

20.23 In his publications, Darwin primarily wrote most about
    a. phylogeny.
    b. speciation.
    c. adaptation.
    d. macroevolution.
    e. the origin of life.

e
Factual Recall

20.24 Idealism, or essentialism, is an idea most associated with
    a. Cuvier.
    b. Darwin.
    c. Lamarck.
    d. Lyell.
    e. Plato.

b
Factual Recall

20.25 Charles Lyell was an advocate of
    a. use and disuse.
    b. uniformitarianism.
    c. industrial melanism.
    d. the modern synthesis.
    e. the inheritance of acquired characteristics.

b
Factual Recall

20.26 Which of the following has provided an abundance of evidence that the Earth has had a succession of flora and fauna?
    a. population genetics
    b. the fossil record
    c. natural selection
    d. creationism
    e. catastrophism

a
Conceptual
Understanding

20.27 How would one describe Darwin?
    a. gradualist
    b. creationist
    c. essentialist
    d. catastrophist
    e. geneticist

a
Conceptual
Understanding

20.28 All of the following influenced Darwin as he synthesized the concept of natural selection EXCEPT
    a. Mendel's laws of inheritance.
    b. the finches of the Galapagos.
    c. Lyell's *Principles of Geology*.
    d. Malthus' *Essays of Populations*.
    e. the results of artificial selection.

c
Factual Recall

20.29 Who was the naturalist who synthesized a concept of natural selection independently of Darwin?
    a. Lyell
    b. Mendel
    c. Wallace
    d. Henslow
    e. Malthus

c
Factual Recall

20.30 Which of the following elements of Darwinism is associated with Malthus?
  a. Artificial selection improves plant and animal breeds.
  b. Differential reproductive success is the cornerstone of natural selection.
  c. The potential for population growth exceeds what the environment can support.
  d. Species become better adapted to their local environments through natural selection.
  e. Favorable variations accumulate in a population after many generations of natural selection.

a
Factual Recall

20.31 Current arguments among evolutionists about evolution are mainly concerned with the
  a. mechanism of evolutionary change.
  b. existence of vestigial organs.
  c. importance of homologous structures.
  d. effects of ontogeny versus phylogeny.
  e. significance of natural versus artificial selection.

e
Factual Recall

20.32 Natural selection is based on all of the following aspects EXCEPT:
  a. Variation exists within populations.
  b. The fittest individuals leave the most offspring.
  c. There is differential reproductive success with populations.
  d. Populations tend to produce more individuals than the environment can support.
  e. The environment tends to create favorable characteristics within populations.

a
Conceptual
Understanding

20.33 The effect of natural selection in the case of the English peppered moth, *Biston betularia*, illustrates that the advantage of inherited traits depends on the
  a. environment.
  b. presence of cytochrome *c*.
  c. intensity of melanin.
  d. principle of common descent.
  e. presence of homologies among moths.

b
Factual Recall

20.34 All of the following statements are inferences of natural selection EXCEPT:
  a. There is heritable variation among individuals.
  b. Production of offspring is unrelated to the abundance of essential resources.
  c. Since only a fraction of offspring survive, there is a struggle for limited resources.
  d. Individuals whose inherited characteristics best fit them to the environment will leave more offspring.
  e. Unequal reproductive success leads to adaptations.

d
Conceptual
Understanding

20.35 When single-stranded DNA from a human is mixed with single-stranded DNA from a chimpanzee, we find that about 99% of the DNA is homologous. This can be taken as evidence that
   a. humans and chimpanzees originated in similar environments.
   b. humans evolved from chimpanzees.
   c. chimpanzees evolved from humans.
   d. humans and chimpanzees are closely related.
   e. all organisms have similar DNA.

a
Conceptual
Understanding

20.36 One finds that organisms on islands are different from, but closely related to, similar forms found on the nearest continent. This is taken as evidence that
   a. island forms and mainland forms descended from common ancestors.
   b. common environments are inhabited by the same organisms.
   c. the islands were originally part of the continent.
   d. the island forms and mainland forms are converging.
   e. island forms and mainland forms share the same gene pool.

b
Conceptual
Understanding

20.37 Which of the following best describes the fossil record?
   a. Similar fossils are found in varying environments at different times.
   b. There is a progression, with older fossils being primitive and younger fossils being advanced.
   c. Life has remained essentially unchanged since it began 6000 years ago.
   d. All vertebrate classes make their first appearance in the fossil record in rocks of the same age.
   e. The fossil record proves Darwin's hypothesized genetic variation in populations.

d
Conceptual
Understanding

20.38 Linnaeus' concept of taxonomy is that the more closely two organisms resemble each other, the more closely related they are phylogenetically. In evolutionary terms, the more closely related two organisms are,
   a. All of the below are correct.
   b. the more similar their DNA sequences are.
   c. the more recently they shared a common ancestor.
   d. the less likely they are to be related to fossil forms.
   e. Both a and b are correct.

c
Conceptual
Understanding

20.39 "Ontogeny recapitulates phylogeny" means that development passes through stages that represent the evolutionary ancestors of the organism. Our modern view of this statement is best described by which of the following?
   a. It is generally valid.
   b. It is interesting only from an historical perspective.
   c. It is more accurate to say that development passes through stages that are shared with the embryonic stages of evolutionary ancestors.
   d. It is more accurate to say that development passes through stages that are determined by adaptations to the environment.
   e. It is more accurate to say that development passes through stages that are shared with the adult stages of evolutionary ancestors.

e
Conceptual
Understanding

20.40 Ichthyosaur was an aquatic dinosaur. Fossils show us that it had a dorsal fin and a tail just as fish do, even though its closest relatives were terrestrial reptiles that had neither dorsal fins nor aquatic tails. The dorsal fins and tails of ichthyosaurs and fish are
   a. homologous.
   b. analogous.
   c. adaptations to a common environment.
   d. Both a and c are correct.
   e. Both b and c are correct.

a
Application

20.41 A biologist studied a population of squirrels for 15 years. Over that time, the population was never fewer than 30 squirrels and never more than 45. Her data showed that over half of the squirrels born did not survive to reproduce, because of competition for food and predation. Suddenly, the population increased to 80. In a single generation, 90% of the squirrels that were born lived to reproduce. What inferences might you make about that population?
   a. All of the below are reasonable inferences.
   b. The amount of available food probably increased.
   c. The number of predators probably decreased.
   d. The young squirrels in the next generation will show greater levels of variation than in the previous generations because squirrels that would not have survived in the past are now surviving.
   e. Both b and c are reasonable inferences.

d
Conceptual
Understanding

20.42 Evolution is a general concept. TV shows evolve in ways similar to biological systems. Given this similarity, which would NOT be a good analogy between TV shows and evolution?
   a. Shows (organisms) that are successful last for a long time; those that are not successful are canceled (become extinct).
   b. Over a period of years, as public opinion (environment) changes, shows (organisms) change.
   c. Successful shows (organisms) often generate (reproduce) other shows (organisms) via a process called *spin-offs*.
   d. The characteristics of successful shows (organisms) are copied by other new shows (organisms).
   e. Over time, different shows (organisms) may develop similar characteristics.

Chapter 21

a
Factual Recall

21.1 Which of the following is NOT a requirement for maintenance of Hardy-Weinberg equilibrium?
  a. an increasing mutation rate
  b. random mating
  c. large population size
  d. no migration
  e. no natural selection

b
Application

21.2 In a population that is in Hardy-Weinberg equilibrium, the frequency of the allele a is 0.3. What is the percentage of the population that is homozygous for this allele?
  a. 3
  b. 9
  c. 21
  d. 30
  e. 42

e
Application

21.3 In a population that is in Hardy-Weinberg equilibrium, the frequency of the allele a is 0.3. What is the percentage of the population that is heterozygous for this allele?
  a. 3
  b. 9
  c. 21
  d. 30
  e. 42

c
Factual Recall

21.4 The gene pool can best be described as the
  a. group of genes not described by the Hardy-Weinberg theorem.
  b. total number of gene loci that occur in each species.
  c. total aggregate of genes in a population at any time.
  d. group of genes responsible for polygenic traits.
  e. genes only found in isolated populations.

d
Factual Recall

21.5 Gene frequencies in a gene pool may shift randomly and by chance. This is called
  a. artificial selection.
  b. adaptive radiation.
  c. climatic shift.
  d. genetic drift.
  e. natural selection.

e
Factual Recall

21.6 Through time, the movement of people on Earth has steadily increased. This has altered the course of human evolution by increasing
    a. nonrandom reproduction.
    b. geographical isolation.
    c. genetic drift.
    d. mutations.
    e. gene flow.

e
Factual Recall

21.7 Natural selection is most closely related to
    a. diploidy.
    b. gene flow.
    c. genetic drift.
    d. assortative mating.
    e. differential reproductive success.

a
Factual Recall

21.8 Which of the following is likely to have been produced by sexual selection?
    a. a male lion's mane
    b. bright colors of female flowers
    c. the ability of desert animals to concentrate their urine
    d. different sizes of male and female pine cones
    e. camouflage coloration in animals

c
Application

21.9 In a Hardy-Weinberg population, the frequency of the a allele is 0.4. What is the frequency of individuals with Aa genotype?
    a. 0.16
    b. 0.20
    c. 0.48
    d. 0.60
    e. Cannot tell from the information provided.

d
Conceptual
Understanding

21.10 Which of the following is one important evolutionary feature of the diploid condition?
    a. Only diploid organisms can reproduce sexually.
    b. Recombination can only occur in diploid organisms.
    c. Genes are more resistant to mutation in diploid cells.
    d. Diploid organisms express less of their genetic variability than haploid organisms.
    e. Diploid organisms are more likely to successfully clone than are haploid organisms.

d
Application

21.11 In a population with two alleles, A and a, the frequency of a is 0.6. What would be the frequency of heterozygotes if the population is in Hardy-Weinberg equilibrium?
    a. 0.16
    b. 0.36
    c. 0.4
    d. 0.48
    e. 0.64

e
Conceptual
Understanding

21.12 You sample a population of butterflies and find that 42% are heterozygous for a particular gene. What would be the frequency of the recessive allele in this population?
a. 0.09
b. 0.3
c. 0.49
d. 0.7
e. Allele frequency cannot be estimated from this information.

a
Factual Recall

21.13 The decrease in the size of plants on the slopes of mountains as altitudes increase is an example of
a. a cline.
b. a bottleneck.
c. relative fitness.
d. genetic drift.
e. speciation.

d
Factual Recall

21.14 Which of the following is the unit of evolution? In other words, which of the following can evolve in the Darwinian sense?
a. gene
b. chromosome
c. individual
d. population
e. species

c
Factual Recall

21.15 Natural selection tends to reduce variation in gene pools. Which process serves to balance natural selection by creating new alleles?
a. meiosis
b. sex
c. mutation
d. migration
e. reproduction

e
Conceptual
Understanding

21.16 In a large, sexually reproducing population, the frequency of an allele changes from 60% to 20%. From this change, one can most logically assume
a. that the allele is linked to a detrimental allele.
b. that the allele mutates readily.
c. that random processes have changed allelic frequencies.
d. that there is no sexual selection.
e. that the allele reduces fitness.

b
Conceptual
Understanding

21.17 Which factor is the most important in producing the variability that occurs in each generation of humans?
a. diploidy
b. genetic recombination
c. genetic drift
d. nonrandom mating
e. natural selection

d
Factual Recall

21.18 Most copies of harmful recessive alleles in a population are carried by
     individuals that are
     a. haploid.
     b. polymorphic.
     c. homozygous for the allele.
     d. heterozygous for the allele.
     e. afflicted with the disorder caused by the allele.

a
Factual Recall

21.19 The Darwinian fitness of an individual is measured best by
     a. the number of its offspring that survive to reproduce.
     b. the number of supergenes in the genotype.
     c. the number of mates it attracts.
     d. its physical strength.
     e. how long it lives.

e
Conceptual
Understanding

21.20 All of the following statements about balanced polymorphism are correct
     EXCEPT:
     a. It is maintained by natural selection.
     b. It is associated with heterogeneous environments.
     c. It can be caused by frequency-dependent selection.
     d. It results from the perpetuation of genetic variation.
     e. It occurs in populations at a Hardy-Weinberg equilibruim.

*Use the following information to answer Questions 21.21–21.24. In a hypothetical popula-
tion of 1000 people, tests of blood type genes show that 100 have the genotype AA, 600
have the genotype AB, and 300 have the genotype BB.*

d
Application

21.21 What is the frequency of the *A* allele?
     a. .001
     b. .002
     c. .100
     d. .400
     e. .600

e
Application

21.22 What is the frequency of the *B* allele?
     a. .001
     b. .002
     c. .100
     d. .400
     e. .600

a
Conceptual
Understanding

21.23 What percentage of the population will have type O blood?
     a. 0
     b. 10
     c. 24
     d. 48
     e. 60

c
Application

21.24 If there are 4000 children produced by this generation, how many would be expected to have AB blood?
  a. 960
  b. 100
  c. 1920
  d. 2000
  e. 2400

Choose among the following options to answer Questions 21.25–21.31. Each option may be used once, more than once, or not at all.
  a. *random selection*
  b. *directional selection*
  c. *stabilizing selection*
  d. *diversifying selection*
  e. *sexual selection*

d
Conceptual
Understanding

21.25 An African butterfly species exists in two strikingly different color patterns, each of which closely resembles other species that are distasteful to birds.

e
Conceptual
Understanding

21.26 Brightly colored peacocks mate more frequently than do drab colored peacocks.

c
Conceptual
Understanding

21.27 Most Swiss starlings produce 4 to 5 eggs in each clutch.

b
Conceptual
Understanding

21.28 Fossil evidence indicates that horses have gradually increased in size over geological time.

c
Conceptual
Understanding

21.29 The average birth weight for human babies is about 7 pounds.

d
Conceptual
Understanding

21.30 A certain species of land snail exists as either a cream color or a solid brown color. Intermediate individuals are relatively rare.

b
Conceptual
Understanding

21.31 Pathogenic bacteria found in many hospitals are antibiotic-resistant.

e
Factual Recall

21.32 When we say that one organism has a greater fitness than another organism, we specifically mean that it
   a. lives longer than others of its species.
   b. competes for resources more successfully than others of its species.
   c. mates more frequently than others of its species.
   d. utilizes resources more efficiently than other species occupying similar niches.
   e. leaves more viable offspring than others of its species.

e
Application

21.33 In peas, a gene controls flower color such that R = red and r = white. In an isolated pea patch, there were 36 red flowers and 64 white flowers. Assuming a Hardy-Weinberg equilibrium, what is the value of q for this population?
   a. .36
   b. .60
   c. .64
   d. .75
   e. .80

b
Factual Recall

21.34 Recessive alleles in a genetic equilibrium
   a. are not significant.
   b. remain stable indefinitely.
   c. are constantly selected against.
   d. are on a steady increase.
   e. are on a steady decrease.

a
Factual Recall

21.35 Which term best describes a change in allelic frequencies due to an influx of new members into a population?
   a. gene flow
   b. genetic drift
   c. founder effect
   d. selection
   e. convergent evolution

a
Application

21.36 In modern terminology, diversity is understood to be a result of genetic variation. Sources of variation for evolution include all of the following EXCEPT
   a. mistakes in translation of structural genes.
   b. mistakes in DNA replication.
   c. translocations and mistakes in meiosis.
   d. recombination at fertilization.
   e. recombination by crossing over in meiosis.

c
Factual Recall

21.37 What is the measure of Darwinian fitness in a population?
   a. longevity in a species
   b. survival under adverse conditions
   c. the number of fertile offspring
   d. strength, in a predator
   e. fleetness, in a prey animal

Use the following information to answer questions 21.38–21.40. A large population of laboratory animals has been allowed to breed randomly for a number of generations. After several generations, 49 percent of the animals display a recessive trait (aa), the same percentage as at the beginning of the breeding program. The rest of the animals show the dominant phenotype, with heterozygotes indistinguishable from the homozygous dominants.

b
Conceptual
Understanding

21.38 What is the most reasonable conclusion that can be drawn from the fact that the frequency of the recessive trait (*aa*) has not changed over time?
    a. The population is undergoing genetic drift.
    b. The two phenotypes are about equally adaptive under laboratory conditions.
    c. The genotype *AA* is lethal.
    d. There has been a high rate of mutation of allele *A* to allele *a*.
    e. There has been sexual selection favoring allele *a*.

a
Application

21.39 What is the estimated frequency of allele *a* in the gene pool?
    a. 0.70
    b. 0.51
    c. 0.49
    d. 0.30
    e. 0.07

b
Application

21.40 What proportion of the population is probably heterozygous (*Aa*) for this trait?
    a. 0.51
    b. 0.42
    c. 0.21
    d. 0.09
    e. 0.07

c
Factual Recall

21.41 A change in the frequencies of alleles in the gene pool of a small population arising from chance events is called
    a. gene flow.
    b. selection.
    c. genetic drift.
    d. mutation pressure.
    e. differential reproduction.

a
Factual Recall

21.42 The most important missing evidence in Darwin's theory in 1859 was
    a. the source of genetic variation.
    b. the evidence of overproduction of offspring.
    c. the evidence that some organisms became extinct.
    d. the observation that variation is common in populations.
    e. the observation that competition exists in populations.

d
Conceptual
Understanding

21.43 Cattle breeders have improved the quality of meat over the years by which process?
   a. artificial selection
   b. directional selection
   c. stabilizing selection
   d. Only a and b are correct.
   e. a, b, and c are correct.

d
Factual Recall

21.44 Which of the following is correct about a population that has a lethal recessive allele?
   a. The allele inevitably will mutate to a nonlethal state.
   b. This allele will ultimately bring about the extinction of this population.
   c. The allele eventually will be removed from the population by natural selection.
   d. The homozygous recessive genotype of this allele has a relative fitness of 0 and a selection coefficient of 1.
   e. The heterozygous genotype of this allele has a relative fitness of 0 and a selection coefficient of 0.5.

e
Conceptual
Understanding

21.45 Which of the following statements best summarizes organic evolution as it is viewed today?
   a. It is goal directed.
   b. It represents the result of selection for acquired characteristics.
   c. It is synonymous with the process of gene flow.
   d. It is the descent of humans from the present-day great apes.
   e. It is the differential survival and reproduction of the most fit phenotypes.

Questions 21.46–21.48 refer to the following events. In 2468, two male space colonists and three female space colonists settle on an uninhabited Earth-like planet in the Andromeda galaxy. The colonists and their offspring randomly mate for many generations. All five of the original colonists had free ear lobes, and two are heterozygous for that trait. The allele for free ear lobes is dominant to the allele for attached ear lobes.

b
Application

21.46 If one assumes that the Hardy-Weinberg equilibrium applies to the population of colonists on this planet, about how many people will have attached ear lobes when the planet's population reaches 10,000?
   a. 0
   b. 400
   c. 800
   d. 1,000
   e. 10,000

c
Factual Recall

21.47 If two of the original colonists died before they produced offspring, the ratios of genotypes could be quite different in the subsequent generations. This is an example of
   a. diploidy.
   b. gene flow.
   c. genetic drift.
   d. diversifying selection.
   e. stabilizing selection.

a
Factual Recall

21.48 After many generations, the population on this planet has an unusually high frequency for the incidence of color blindness. This is most likely due to
   a. the founder effect.
   b. sexual selection.
   c. coadapted genes.
   d. mutations.
   e. pleiotropy.

e
Conceptual
Understanding

21.49 Over a period of time, the frequency of an allele in a population decreases from 0.01 to 0.003. Which of the following might explain this change?
   a. The allele is disadvantageous when homozygous.
   b. Migrants entering the population had a lower frequency of that allele than did current members of the population.
   c. One member of the population who was homozygous recessive for the allele accidentally died.
   d. a and b are correct.
   e. a, b, and c are correct.

e
Conceptual
Understanding

21.50 Genetic recombination is a critical process in evolution. This statement is supported by the continuous existence of which of the following in evolving populations?
   a. sex
   b. bacterial conjugation
   c. exchange of chromosome regions in meiosis (crossing over)
   d. a and c
   e. a, b, and c

e
Conceptual
Understanding

21.51 What effect do sexual processes (meiosis and fertilization) have on the allelic frequencies in a population?
   a. They tend to reduce the frequencies of deleterious alleles and increase the frequencies of advantageous ones.
   b. They tend to increase the frequencies of deleterious alleles and decrease the frequencies of advantageous ones.
   c. They tend to selectively combine favorable alleles into the same zygote but do not change allelic frequencies.
   d. They tend to increase the frequency of new alleles and decrease the frequency of old ones.
   e. They have no effect on allelic frequencies.

d
Application

21.52 The frequency of the *a* allele in Population 1 is 0.3. The frequency of the *a* allele in Population 2 is 0.8, and the frequency of the *a* allele in Population 3 is 0.4. For which of the following would gene flow cause the greatest change in alleleic frequencies?
   a. if some members of Population 1 joined Population 3
   b. if some members of Population 3 joined Population 1
   c. if some members of Population 2 joined Population 3
   d. if some members of Population 2 joined Population 1
   e. if some members of Population 3 joined Population 2

Questions 21.53–21.55 utilize the following information: You are studying three popula-
tions of cardinals. Population 1 has ten birds, of which one is brown (a recessive trait)
rather than red. Population 2 has 100 birds. In that population, ten of the birds are brown.
Population 3 has 30 birds, and three of them are brown. Use the following options to an-
swer the questions:
  a. Population 1
  b. Population 2
  c. Population 3
  d. They are all the same.
  e. It is impossible to tell from the information given.

d
Application

21.53 In which population is the frequency of the allele for brown feathers highest?

b
Application

21.54 In which population would it be least likely that an accident would significantly alter the frequency of the brown allele?

a
Conceptual
Understanding

21.55 Which population is most likely to be subject to the bottleneck effect?

e
Conceptual
Understanding

21.56 Variations in populations can be demonstrated by studying which of the following?
  a. morphological characteristics
  b. proteins
  c. DNA sequences
  d. b and c are correct
  e. a, b, and c are correct

b
Conceptual
Understanding

21.57 Your friend Forrest says he is having trouble with this Hardy-Weinberg idea, and in particular the idea of assortative mating. He says, "How can you ever have an equilibrium in humans? The vast majority of mating is highly assortative. We prefer mates of our own race, social standing, educational level, and ethnicity." Which of the following would be your response to Forrest?
  a. "You are correct. That is why the Hardy-Weinberg concept is only a theorem—it doesn't apply to all populations."
  b. "Although you are correct that much mating is assortative, when it comes to the Hardy-Weinberg theorem, we are concerned only with the specific genotype. So, mating can be assortative for one trait like race, while still being random for other traits like blood type."
  c. "You are correct. However, the Hardy-Weinberg theorem is relevant only to a population, and mating within populations is still random."
  d. "You may be correct. However, those things you mentioned are all nongenetic. When it comes to genetic traits, mating is not assortative."
  e. "Although it may appear that you are correct, the human population is so large that mating can be considered random."

Chapter 22

b
Factual Recall

22.1 Which of the following is the most likely pattern for the origin of species?
a. anagenesis
b. cladogenesis
c. phyletic evolution
d. spontaneous generation
e. inheritance of acquired characteristics

e
Factual Recall

22.2 To a punctuationalist, the "sudden" appearance of a new species in the fossil record means that
a. the species is now extinct.
b. the Earth is only 6000 years old.
c. speciation occurred instantaneously.
d. speciation occurred in one generation.
e. speciation occurred over many thousand years.

a
Factual Recall

22.3 The biologist who proposed the biological species concept is
a. Mayr.
b. Gould.
c. Wright.
d. Sheldon.
e. Eldridge.

a
Factual Recall

22.4 The only taxonomic category that actually exists as a discrete unit in nature is the
a. species.
b. genus.
c. family.
d. class.
e. phylum.

b
Conceptual
Understanding

22.5 Which of the following statements is consistent with the punctuated equilibrium interpretation of speciation?
a. Evolution proceeds at a slow, steady pace.
b. Long periods of minor change are interrupted by short bursts of significant change.
c. Rapid speciation is caused by population explosions.
d. There is an equilibrium between living and extinct species.
e. Large populations evolve more quickly than small ones.

e
Factual Recall

22.6 A rapid method of speciation that has been important in the history of flowering plants is
a. genetic drift.
b. parapatric speciation.
c. a mutation in the gene controlling the timing of flowering.
d. behavioral isolation.
e. polyploidy.

d
Factual Recall

22.7 If two species are able to interbreed but produce sterile hybrids, their species integrity is maintained by
a. gametic isolation.
b. a prezygotic barrier.
c. hybrid inviability.
d. a postzygotic barrier.
e. introgression.

b
Application

22.8 A new plant species formed from hybridization of a plant with a diploid number of 16 with a plant with a diploid number of 12 would probably have a gamete chromosome number of
a. 12
b. 14
c. 16
d. 22
e. 28

Questions 22.9 and 22.10 are based on this paragraph. A botanist discovers a large population of annual plants. The plants all look basically the same, but seem to be of two different size classes. The larger and smaller plants inhabit the same areas and are visited by the same pollinating insects.

a
Application

22.9 What is the most likely reason for the size differences?
a. The larger plants are polyploids derived from the smaller plants.
b. The larger plants happen to be in areas with more nutrients.
c. The smaller plants are haploids which developed from unfertilized eggs.
d. The larger plants germinated in the winter and the smaller plants germinated in the summer.
e. The larger plants are dominant and the smaller plants are recessive.

c
Application

22.10 What would be the LEAST productive research to discover the relationship between the plants?
a. electrophoretic studies to see if they have the same enzymes
b. chromosome counts
c. growing seeds of one size of plant with various nutrient concentrations
d. growing seeds from the two sizes of plants under identical conditions
e. careful measurement of anatomical features

c
Factual Recall

22.11 Although different species of warblers often migrate together and use the same habitats for mating and feeding, they rarely hybridize. The isolating mechanism most likely to be operating is
a. ecological isolation.
b. temporal isolation.
c. behavioral isolation.
d. mechanical isolation.
e. gametic isolation.

c
Factual Recall

22.12 Which of the following is NOT considered to be a reproductive isolating mechanism?
   a. sterile offspring
   b. ecological isolation
   c. feeding behavior
   d. gametic incompatibility
   e. timing of courtship display

c
Factual Recall

22.13 Which of the following reproductive isolating mechanisms is postzygotic?
   a. habitat isolation
   b. temporal isolation
   c. hybrid sterility
   d. behavioral isolation
   e. gamete incompatibility

a
Conceptual
Understanding

22.14 Some species of *Anopheles* mosquito live in brackish water, some in running fresh water, and others in stagnant water. What type of reproductive barrier is most obviously separating these different species?
   a. ecological isolation
   b. temporal isolation
   c. behavioral isolation
   d. gametic isolation
   e. postzygotic isolation

d
Factual Recall

22.15 The reproductive barrier that maintains the species boundary between horses and donkeys is
   a. mechanical isolation.
   b. gametic isolation.
   c. hybrid inviability.
   d. hybrid sterility.
   e. hybrid breakdown.

c
Conceptual
Understanding

22.16 According to advocates of the punctuated equilibrium theory,
   a. natural selection is unimportant as a mechanism of evolution.
   b. given enough time, most existing species will branch gradually into new species.
   c. a new species accumulates most of its unique features as it comes into existence.
   d. most evolution results from disruption of a Hardy-Weinberg equilibrium.
   e. transitional fossils are intermediate between newer species and their parent species.

Use the following options to answer Questions 22.17–22.23. For each description of reproductive isolation, select the option which best describes it. Options may be used once, more than once, or not at all.
   a. *gametic*
   b. *temporal*
   c. *behavioral*
   d. *habitat*
   e. *mechanical*

e
Conceptual
Understanding

22.17 Two species of orchids with different floral anatomy.

b
Conceptual
Understanding

22.18 Two species of trout that breed in different seasons.

c
Conceptual
Understanding

22.19 Two species of meadowlarks with different mating songs.

d
Conceptual
Understanding

22.20 Two species of garter snakes living in the same region, but one lives in water and the other lives on land.

b
Conceptual
Understanding

22.21 Two species of pine shed their pollen at different times.

c
Conceptual
Understanding

22.22 Mating fruit flies recognize the appearance, odor, tapping motions, and sounds of members of their own species, but not of other species.

d
Conceptual
Understanding

22.23 The scarlet oak is adapted to moist bottomland, whereas the black oak is adapted to dry upland soils.

c
Conceptual
Understanding

22.24 The biological species concept is inadequate for grouping
    a. plants.
    b. parasites.
    c. asexual organisms.
    d. endemic populations.
    e. sympatric populations.

d
Conceptual
Understanding

22.25 The most important factor in preserving horses and donkeys as distinct biological species is
    a. allopolyploidy.
    b. a geographic barrier.
    c. a prezygotic barrier.
    d. a postzygotic barrier.
    e. an allopatric barrier.

b
Conceptual
Understanding

22.26 Races of humans are unlikely to evolve extensive differences in the future for which of the following reasons?

    I. The environment is unlikely to change.
    II. Humans are essentially perfect.
    III. The human races are incompletely isolated.

    a. I only
    b. III only
    c. I and II only
    d. II and III only
    e. I, II, and III

e
Factual Recall

22.27 The only way that two populations can assure their integrity as distinct biological species is by
a. sympatry.
b. allopatry.
c. introgression.
d. geographic isolation from one another.
e. reproductive isolation from one another.

a
Factual Recall

22.28 Two species of frogs belonging to the same genus occasionally mate, but the offspring do not complete development. This is an example of
a. the postzygotic barrier called hybrid inviability.
b. the postzygotic barrier called hybrid breakdown.
c. the prezygotic barrier called hybrid sterility.
d. gametic isolation.
e. adaptation.

c
Conceptual
Understanding

22.29 If two subspecies, A and B, are not considered separate species even though they cannot interbreed, then
a. they are groups that are endemic to isolated geographic regions.
b. they have eliminated postzygotic barriers but not prezygotic barriers.
c. gene flow between A and B may exist through other related subspecies.
d. gene flow has ceased and genetic isolation is complete.
e. their diploid gametes are produced by nondisjunction.

c
Factual Recall

22.30 A characteristic of allopatric speciation is
a. the appearance of new species in the midst of old ones.
b. asexually reproducing populations.
c. geographic isolation.
d. artificial selection.
e. large populations.

c
Factual Recall

22.31 The process of a new species arising within the range of the parent populations is termed
a. semispeciation.
b. adaptive radiation.
c. sympatric speciation.
d. parapatric speciation.
e. allopatric speciation.

e
Application

22.32 The formation of a land bridge between North and South America about three million years ago resulted in which of the following?

I. allopatry of marine populations that were previously sympatric
II. sympatry of marine populations that were previously allopatric
III. sympatry of terrestrial populations that were previously allopatric

a. I only
b. II only
c. III only
d. I and II
e. I and III

d
Factual Recall

22.33 The biologists who proposed the theory of punctuated equilibrium are
  a. Mayr and Wright.
  b. Watson and Crick.
  c. Darwin and Wallace.
  d. Eldridge and Gould.
  e. Sheldon and Templeton.

d
Application

22.34 Plant species A has a diploid number of 28. Plant species B has a diploid number of 14. A new, sexually reproducing species C arises as an allopolyploid from hybridization of A and B. The diploid number of C would probably be
  a. 14
  b. 21
  c. 28
  d. 42
  e. 63

b
Conceptual
Understanding

22.35 The origin of a new plant species by hybridization coupled with nondisjunction is an example of
  a. allopatric speciation.
  b. sympatric speciation.
  c. autopolyploidy.
  d. introgression.
  e. a peak shift.

d
Conceptual
Understanding

22.36 All of the following statements about splinter populations, or peripheral isolates, are correct EXCEPT:
  a. The gene pool may represent the extremes of genotypic and phenotypic clines.
  b. Many peripheral isolates have an increased likelihood of experiencing a founder effect.
  c. Life on the frontier is usually harsh for the peripheral isolates, and most become extinct.
  d. They undergo speciation readily because they are large populations with immense gene pools.
  e. The selective factors operating on peripheral isolates may be quite different from those operating on the parent population.

c
Conceptual
Understanding

22.37 According to one hypothesis, the production of sterile mules in nature by the mating of horses and donkeys tends to
  a. result in the extinction of one of the two species.
  b. decrease character displacement between horses and donkeys.
  c. reinforce prezygotic isolating mechanisms between horses and donkeys.
  d. weaken the intrinsic reproductive isolating mechanisms between horses and donkeys.
  e. eventually result in the formation of a single species from the two parental species (horses and donkeys).

a
Factual Recall

22.38 Which of the following would be a position held by a punctuationalist?
  a. A new species forms most of its unique features as it comes into existence and then changes little for the duration of its existence.
  b. One should expect to find many transitional fossils left by organisms in the process of forming new species.
  c. Given enough time, most existing species will gradually evolve into new species.
  d. Natural selection is unimportant as a mechanism of evolution.
  e. Most speciation is anagenic.

b
Conceptual
Understanding

22.39 The Hawaiian Islands are a great showcase of evolution because of intense
  a. ecological isolation and parapatric speciation.
  b. adaptive radiation and allopatric speciation.
  c. allopolyploidy and sympatric speciation.
  d. cross-specific mating and reinforcement.
  e. hybrid vigor and allopatric speciation.

c
Conceptual
Understanding

22.40 Which of the following statements about biological species is (are) correct?
  I. Biological species are defined by reproductive isolation.
  II. Biological species are the model used for grouping extinct forms of life.
  III. The biological species is the largest unit of population in which gene flow is possible.

  a. I only
  b. II only
  c. I and III
  d. II and III
  e. I, II, and III

c
Application

22.41 Plant species A has a diploid number of 8. A new species, B, arises as an autopolyploid from A. The diploid number of B would probably be
  a. 4
  b. 8
  c. 16
  d. 32
  e. 64

a
Conceptual
Understanding

22.42 Which of the following best describes what occurs when two species hybridize and a fraction of the hybrids manage to backcross with one of the parent species?
  a. introgression
  b. genetic drift
  c. random mating
  d. heterozygote advantage
  e. differential reproduction

b
Conceptual
Understanding

22.43 Which of the following statements about speciation is correct?
   a. The goal of natural selection is speciation.
   b. When reunited, two allopatric populations will not interbreed.
   c. Natural selection chooses the reproductive barriers for populations.
   d. Prezygotic reproductive barriers usually evolve before postzygotic barriers.
   e. Natural selection amplifies sexual adaptations that lead to reproductive success.

d
Conceptual
Understanding

22.44 Differences in all of the following would be useful in distinguishing one biological species from another EXCEPT
   a. physiology.
   b. biochemistry.
   c. behavior.
   d. fossil morphology.
   e. genetic makeup.

d
Conceptual
Understanding

22.45 Which of the following would apply to *both* anagenesis and cladogenesis?
   a. branching
   b. phyletic
   c. increased diversity
   d. speciation
   e. more species

a
Conceptual
Understanding

22.46 For which of the following organisms is the concept of a "biological species" not applicable?
   a. triploid, asexual plants
   b. humans of different ethnic origins
   c. dogs of different breeds
   d. haploid male honeybees
   e. *Peromyscus maniculatus*

a
Factual Recall

22.47 Which of the following is a postzygotic barrier to hybridization?
   a. hybrid sterility
   b. habitat isolation
   c. mechanical isolation
   d. genetic isolation
   e. behavioral isolation

d
Conceptual
Understanding

22.48 Dog breeders perpetuate breeds of dogs by controlling mating. This is analogous to which of the following natural isolating mechanisms?
   a. reduced hybrid fertility
   b. hybrid breakdown
   c. mechanical isolation
   d. habitat isolation
   e. gametic isolation

b
Conceptual
Understanding

22.49 All of the following would increase the rate of allopatric speciation EXCEPT
   a. genetic drift
   b. introgression
   c. founder effect
   d. natural selection
   e. geotypic clines

e
Conceptual
Understanding

22.50 All of the following have contributed to the diversity of organisms on the Hawaiian archipelago EXCEPT that
   a. the islands have a volcanic origin.
   b. the islands are distant from the mainland.
   c. multiple invasions have occurred.
   d. adaptive radiation has occurred.
   e. the islands are very young.

b
Conceptual
Understanding

22.51 Which of the following is a way that polyploidy can most directly influence speciation?
   a. It can improve success in island habitats.
   b. It can overcome hybrid sterility.
   c. It can change the mating behavior of animals.
   d. It can generate new adaptive peaks.
   e. It can enhance the rate of stabilizing selection.

a
Conceptual
Understanding

22.52 Which of the following is a correct statement about the concept of punctuated equilibrium?
   a. It explains variation in the tempo of speciation.
   b. It contradicts much of the evidence for evolution.
   c. It explains gradual changes in the fossil record.
   d. It applies only to trilobites.
   e. It argues against the possibility of morphological stasis.

e
Conceptual
Understanding

22.53 An adaptive peak represents
   a. a symbolic landscape with valleys and rivers.
   b. a place where alpine organisms can flourish.
   c. a stable environment for a given popoulation.
   d. a place where microevolution can cause a founder effect.
   e. a successful combination of allele frequencies.

Chapter 23

b
Conceptual
Understanding

23.1 A randomly selected group of organisms from a family would show more
genetic variation than a randomly selected group from a
a. class.
b. genus.
c. kingdom.
d. order.
e. phylum.

a
Factual Recall

23.2 Which of the following can be a mechanism of macroevolution?
a. a change in a regulatory gene, which has a major impact on morphology
b. a change of the classification protocol from phenetic to cladistic
c. DNA–DNA hybridization
d. introgression
e. genetic drift

d
Application

23.3 The half-life of carbon-14 is 5,600 years. A fossil that has one-eighth the normal
proportion of carbon-14 to carbon-12 is probably
a. 1,400 years old.
b. 2,800 years old.
c. 11,200 years old.
d. 16,800 years old.
e. 22,400 years old.

d
Application

23.4 A biologist discovers two new species of organisms, one in Africa and one in
South America. The organisms resemble one another closely. Which type of
evidence would be LEAST useful in determining whether these organisms
are closely related or are the products of convergent evolution?
a. the history and timing of continental drift
b. a comparison of DNA from the two species
c. the fossil record of the two species
d. analysis of the behavior of the two species
e. comparative embryology

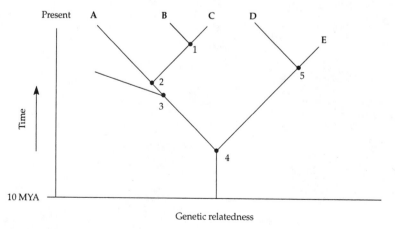

Figure 23.1

Use Figure 23.1 to answer Questions 23.5–23.7.

d
Application

23.5 A common ancestor for species C and E could be at position number
   a. 1.
   b. 2.
   c. 3.
   d. 4.
   e. 5.

c
Application

23.6 The most closely related species are
   a. A and B.
   b. B and D.
   c. C and B.
   d. D and E.
   e. E and A.

e
Application

23.7 Which species is extinct?
   a. A
   b. B
   c. C
   d. D
   e. E

b
Conceptual
Understanding

23.8 The ostrich and the emu look very similar and live in similar habitats,
      although they are not very closely related. This is an example of
   a. divergent evolution.
   b. convergent evolution.
   c. coevolution.
   d. adaptive radiation.
   e. sympatric speciation.

c
Conceptual
Understanding

23.9 Macroevolution includes the study of all of the following EXCEPT
   a. mass extinctions.
   b. evolutionary novelties.
   c. speciation.
   d. the study of evolutionary trends.
   e. global episodes of major adaptive radiations.

b
Factual Recall

23.10 Phylogeny is the
   a. life history of an organism.
   b. evolutionary history of a species.
   c. theory of evolution by natural selection.
   d. branch of biology concerned with the diversity of life.
   e. branch of biology concerned with the relations between plants and the
       Earth's surface.

e
Factual Recall

23.11 The only taxon that actually exists as a natural unit is the
   a. class.
   b. family.
   c. genus.
   d. phylum.
   e. species.

e
Conceptual
Understanding

23.12 According to the hypothesis known as species selection, an evolutionary trend results from
    a. a drive toward perfection.
    b. paraphyletic groupings of endemic species.
    c. stepwise progression of a single unbranched lineage.
    d. phyletic evolution that gradually transforms a single population.
    e. varying longevity of species and varying speciation rates.

d
Factual Recall

23.13 The taxonomic school that places greatest emphasis on the chronology of phylogenetic branching is
    a. systematic taxonomy.
    b. classical taxonomy.
    c. numerical taxonomy.
    d. cladistics.
    e. phenetics.

c
Conceptual
Understanding

23.14 A major evolutionary episode that corresponded in time most closely with the formation of Pangaea was the
    a. origin of humans.
    b. Cambrian explosion.
    c. Permian extinctions.
    d. Pleistocene ice ages.
    e. Cretaceous extinctions.

a
Factual Recall

23.15 All of the following are usual methods for dating fossils EXCEPT
    a. molecular clocks.
    b. carbon-14.
    c. potassium-40.
    d. L- and D-amino acids.
    e. superimposition of sedimentary rock.

c
Factual Recall

23.16 The changing facial features of a maturing child are an example of
    a. phylogeny.
    b. preadaptation.
    c. allometric growth.
    d. paedogenesis.
    e. homologies.

b
Factual Recall

23.17 The anomalous layer of iridium clay that has figured so prominently in the asteroid theory is located at which geological boundary?
    a. Laurasia–Gondwana
    b. Mesozoic–Cenozoic
    c. Paleozoic–Mesozoic
    d. Precambrian–Paleozoic
    e. North American–Pacific plates

e
Factual Recall

23.18 The asteroid hypothesis is associated most prominently with which of the
following events in the history of life?
   a. origin of life
   b. origin of humans
   c. origin of eukaryotes
   d. Permian extinctions
   e. Cretaceous extinctions

b
Factual Recall

23.19 The correct sequence from the most to the least comprehensive taxonomic
level is
   a. family, phylum, class, kingdom, order, species, and genus
   b. kingdom, phylum, class, order, family, genus, and species
   c. kingdom, phylum, order, class, family, genus, and species
   d. phylum, kingdom, order, class, species, family, and genus
   e. phylum, family, class, order, kingdom, genus, and species

b
Factual Recall

23.20 The vertebrate eye and the eye of squids are similar in structure and function.
Which of the following processes accounts for most of this similarity?
   a. cladogenesis
   b. convergent evolution
   c. the iridium anomaly
   d. parapatric speciation
   e. balanced polymorphism

d
Factual Recall

23.21 A technique used in molecular systematics relies on the comparison of
cytochrome c in different animals. This technique is referred to as
   a. DNA–DNA hybridization.
   b. restriction mapping.
   c. electron transport.
   d. protein comparison.
   e. translation.

c
Factual Recall

23.22 The two names of a species' binomial are its specific name and its
   a. class.
   b. family.
   c. genus.
   d. order.
   e. phylum.

b
Application

23.23 All of the following statements about macroevolution are correct EXCEPT:
   a. Long stable periods have been interrupted by brief intervals of extensive
      species extinction.
   b. Most evolutionary trends appear to be the result of gradual phyletic change
      in an unbranched lineage.
   c. Major adaptive radiations have often followed the evolution of novel
      features.
   d. Continental drift has had a significant impact on macroevolution.
   e. Differential speciation is probably a driving force behind macroevolution.

Questions 23.24–23.28 refer to the following geological time periods. Each answer may be used once, more than once, or not at all.

  a. Cretaceous
  b. Precambrian
  c. Pleistocene
  d. Paleocene
  e. Permian

c
Factual Recall

23.24 Epoch when humans appeared.

d
Factual Recall

23.25 Epoch when mammals first flourished.

b
Factual Recall

23.26 Era of the origin of the first animals.

a
Factual Recall

23.27 Period when the dinosaurs became extinct.

e
Factual Recall

23.28 Period of mass extinctions about 250 million years ago.

Questions 23.29–23.33 refer to the following terms. Each term may be used once, more than once, or not at all.

  a. nonadaptive
  b. analogous
  c. homologous
  d. paraphyletic
  e. preadaptational

c
Factual Recall

23.29 Shared derived characters.

c
Factual Recall

23.30 Shared primitive characters.

e
Factual Recall

23.31 Feathers and hollow bones.

b
Factual Recall

23.32 Human hand and lobster claw.

c
Factual Recall

23.33 Whale flipper and horse foreleg.

Questions 23.34–23.36 refer to the hypothetical patterns of taxonomic hierarchy shown in Figure 23.2.

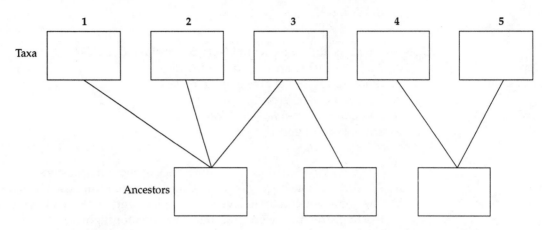

Figure 23.2

e
Application

23.34 Which of the following numbers could represent monophyletic taxa?
  a. 2 only
  b. 4 only
  c. 5 only
  d. 1, 2, and 3
  e. 1, 2, 4, and 5

b
Application

23.35 Which of the following numbers could represent polyphyletic taxa?
  a. 2 only
  b. 3 only
  c. 5 only
  d. 1, 2, and 3
  e. 1, 2, 3, 4, and 5

c
Application

23.36 If taxon 4 represents modern reptiles and taxon 5 represents birds, then these taxa should be considered
  a. monophyletic.
  b. synapomorphic.
  c. paraphyletic.
  d. polyphyletic.
  e. phylogenetic.

a
Conceptual
Understanding

23.37 Which combination of the following species characteristics would cause the greatest likelihood of fossilization?

  I.   aquatic
  II. Arctic
  III. hard body parts
  IV. presence of organic material
  V.   flight
  VI. long lasting

a. I, III, and VI
b. I, II, and VI
c. III only
d. III and VI
e. II, IV, and V

a
Conceptual
Understanding

23.38 Which of the following would NOT be considered fossils?
a. All of the below are considered fossils.
b. dinosaur footprints preserved in rocks
c. insects enclosed in amber
d. mammoths frozen in Arctic ice
e. ancient pollen grains

c
Conceptual
Understanding

23.39 How could flying vertebrates evolve from earthbound ancestors?
a. A structure somehow evolved in anticipation of future use for flight.
b. The flying vertebrates evolved from aquatic vertebrates with winglike fins.
c. Existing structures were gradually modified for flight.
d. Natural selection improved a structure in the context of its later utility for flight.
e. The vertebrates were destined to fly because there have been flying reptiles, birds, and mammals.

c
Conceptual
Understanding

23.40 In all of the following pairs, BOTH processes have played important roles in causing macroevolution EXCEPT
a. chance and natural selection
b. homeosis and heterochrony
c. mutation and molecular systematics
d. extinction and adaptive radiation
e. allometry and paedomorphosis

d
Factual Recall

23.41 Which of the following terms best describes the process in which organisms reach sexual maturity while retaining some juvenile characteristics?
a. homeosis
b. allometry
c. cladogenesis
d. paedomorphosis
e. convergent evolution

d
Factual Recall

23.42 Which of the following is believed to be the most likely cause of the Cretaceous extinction?
a. adaptive radiation
b. continental drift
c. the eruption of Mount Pinatubo
d. the impact of an asteroid or a comet
e. the collapse of food chains

b
Conceptual
Understanding

23.43 Which of the following can best be explained by continental drift?
   a. the relative age of fossils
   b. the scarcity of placental mammals in Australia
   c.   the Chicxulub crater
   d. the survival of *Purgatorius*
   e. polyphyletic taxa

e
Factual Recall

23.44 Which of the following can be used to determine the absolute age of fossils?
   a. index fossils
   b. the "DNA clock"
   c. cladistics
   d. sedimentary strata
   e. the half-life of isotopes

c
Conceptual
Understanding

23.45 When using a cladistic approach to systematics, which of the following is considered most important for classification?
   a. shared primitive characters
   b. analogous primitive characters
   c. shared derived characters
   d. a common five-toed ancestor
   e. overall phenotypic similarity

a
Application

23.46 Which of the following technologies would you use when trying to determine the evolutionary relationship between horses and zebras?
   a. All of the techniques below might prove useful.
   b. DNA–DNA hybridization
   c. analysis of cytochrome *c* differences
   d. restriction mapping of DNA
   e. analysis of fossil DNA

e
Conceptual
Understanding

23.47 Which of the following best describes our current understanding of the role of natural selection in evolutionary theory?
   a. Natural selection has been discarded as an important concept in evolution.
   b. Changes in gene pools due to natural selection are now seen to be unimportant.
   c. Microevolution has replaced natural selection as an organizing concept.
   d. Natural selection is able to explain virtually all changes in gene pools.
   e. Natural selection produces adaptations that are essential to the survival of organisms.

Chapter 24

b
Conceptual
Understanding

24.1 What is the evidence that protobionts may have formed spontaneously?
a. the discovery of ribozymes, showing that prebiotic RNA molecules may have been autocatalytic
b. the laboratory synthesis of microspheres, liposomes, and coacervates
c. the fossil record found in the Fig Tree Chert
d. the abiotic synthesis of polymers
e. the production of organic compounds within a laboratory apparatus simulating conditions on early Earth

b
Conceptual
Understanding

24.2 In what way is the complex process of photosynthesis similar to the simple reactions that are thought to have led to the origin of life on Earth?
a. Both involve an electron transport chain.
b. In both, simple molecules are reduced to form complex organic molecules.
c. Both require light energy in order to proceed.
d. Oxygen is a by-product of both types of reactions.
e. Both must occur within membrane-bound structures.

c
Conceptual
Understanding

24.3 In which of the following ways are ALL living things alike?
a. They are all composed of cells with nuclei.
b. They all breathe.
c. They all contain complex, reduced molecules.
d. They all have circulatory systems.
e. They all undergo mitosis.

b
Conceptual
Understanding

24.4 In classifying organisms into five kingdoms, scientists try to select groups of organisms with common characteristics due to common ancestry. Of the five kingdoms, which one includes organisms with the greatest variety of life styles, structures, and functions? In other words, organisms in which kingdom are least likely to have common ancestry?
a. Monera
b. Protista
c. Fungi
d. Plantae
e. Animalia

d
Factual Recall

24.5 Current theories of prebiotic evolution are based on evidence for all of the following EXCEPT
a. abiotic production of small organic molecules.
b. abiotic polymerization of amino acids.
c. abiotic replication of oligopeptides.
d. abiotic origin of DNA–protein interactions.
e. abiotic production of proteinoid microspheres.

a
Conceptual
Understanding

24.6 While studying a cell with an electron microscope, a scientist notes the following: numerous ribosomes, a well-developed endoplasmic reticulum, two nuclei, and a chitinous cell wall. Which of the following could be the source of this cell?
   a. a fungus
   b. an animal
   c. a bacterium
   d. a plant
   e. a virus

c
Factual Recall

24.7 The first genetic material was most likely
   a. a DNA polymer.
   b. a DNA oligonucleotide.
   c. an RNA polymer.
   d. a protein.
   e. a protein enzyme.

Use the following information to answer Questions 24.8–24.12. According to the Miller-Urey experimental results, chemical evolution leading up to and including the formation of living matter is believed to have occurred during the early history of the Earth. Below are five pairs of events that might have occurred during this period. Judge the relative time of each of these pairs of events according to the key below.
   a. Event I occurred before Event II
   b. Event II occurred before Event I
   c. Events I and II occurred simultaneously

a
Conceptual
Understanding

24.8 Event I
      nitrogen oxides and carbon dioxide
      in the atmosphere

Event II
free oxygen in the atmosphere

b
Conceptual
Understanding

24.9 Event I
      formation of photosynthetic organisms

Event II
formation of heterotrophic organisms

a
Conceptual
Understanding

24.10 Event I
       formation of amino acids

Event II
formation of enzymes

c
Conceptual
Understanding

24.11 Event I
       atmosphere of water, methane, and
       ammonia

Event II
reducing atmosphere

b
Conceptual
Understanding

24.12 Event I
       appearance of life on land

Event II
formation of ozone layer in the
atmosphere

d
Factual Recall

24.13 Approximately how far does the fossil record extend back in time?
a. 6,000 years
b. 3,500,000 years
c. 6,000,000 years
d. 3,500,000,000 years
e. 5,000,000,000,000 years

e
Factual Recall

24.14 What are the banded rocks called that are believed to be the fossils of bacterial mat communities?
a. coacervates
b. stalactites
c. stalagmites
d. strata
e. stromatolites

b
Factual Recall

24.15 Which gas was probably LEAST abundant in the Earth's early atmosphere?
a. H_2
b. O_2
c. CH_4
d. H_2O
e. NH_3

a
Factual Recall

24.16 In their laboratory simulation of the early Earth, Miller and Urey observed the abiotic synthesis of
a. amino acids
b. coacervates
c. DNA
d. liposomes
e. microspheres

a
Conceptual
Understanding

24.17 Which of the following factors was most important in the very early origin of life?
a. natural selection
b. competition for oxygen
c. low levels of solar energy
d. biotic synthesis of organic molecules
e. high levels of freon in the atmosphere

d
Conceptual
Understanding

24.18 Which of the following has not yet been synthesized in laboratory experiments studying the origin of life?
a. lipids
b. microspheres with selectively permeable membranes
c. proteins and other polymers
d. protobionts that use DNA to program protein synthesis
e. purines and pyrimidines

c
Conceptual
Understanding

24.19 Why was the primitive atmosphere of Earth more conducive to the origin of life than the modern atmosphere of Earth?
    a. The primitive atmosphere had a layer of ozone that shielded the first fragile cells.
    b. The primitive atmosphere removed electrons that impeded the formation of protobionts.
    c. The primitive atmosphere was a reducing one that facilitated the formation of complex substances from simple molecules.
    d. The primitive atmosphere had more oxygen than the modern atmosphere, and thus it successfully sustained the first living organisms.
    e. The primitive atmosphere had less free energy than the modern atmosphere, and thus newly formed organisms were less likely to be destroyed.

d
Conceptual
Understanding

24.20 A key role that clay may have played in the origin of life is the tendency for clay to
    a. form microspheres.
    b. assemble into liposome membranes.
    c. generate life through spontaneous processes.
    d. provide a catalytic surface for the polymerization of organic monomers.
    e. accumulate components necessary for life through the mechanism of panspermia.

b
Factual Recall

24.21 How many kingdoms were recognized by Linnaeus?
    a. one
    b. two
    c. three
    d. four
    e. five

e
Factual Recall

24.22 In what way were conditions on the early Earth different from those on modern Earth?
    a. The early Earth had no water.
    b. The early Earth was much cooler.
    c. The early Earth had an oxidizing atmosphere.
    d. Less ultraviolet radiation penetrated the early atmosphere.
    e. The early atmosphere had significant quantities of CH_4 and NH_3.

c
Conceptual
Understanding

24.23 Which of the following is the correct sequence of events in the origin of life?
    I. Formation of protobionts
    II. Synthesis of organic monomers
    III. Synthesis of organic polymers

    a. I, II, III
    b. I, III, II
    c. II, III, I
    d. III, I, II
    e. III, II, I

c
Conceptual
Understanding

24.24 Which of the following statements about the origin of genetic material is most probably correct?
  a. The first genes were DNA produced by reverse transcriptase from abiotically produced RNA.
  b. The first genes were DNA whose information was transcribed to RNA and later translated in polypeptides.
  c. The first genes were autocatalytic RNA molecules bound to clay surfaces.
  d. The first genes were RNA produced by autocatalytic, proteinaceous enzymes called ribozymes.
  e. The first genes were protobionts produced by dehydration syntheses of nucleic acids.

Questions 24.25–24.29 refer to the following scientists.
  a. *Fox*
  b. *Cech*
  c. *Oparin*
  d. *Miller and Urey*
  e. *Whittaker and Margulis*

b
Factual Recall

24.25 Discovered ribozymes.

a
Factual Recall

24.26 Synthesized proteinoids.

c
Factual Recall

24.27 Synthesized coacervates.

e
Factual Recall

24.28 Supported the five-kingdom scheme for classification.

c
Factual Recall

24.29 Postulated that the spontaneous synthesis of complex molecules could not occur today because oxygen attacks chemical bonds, extracting electrons.

Questions 24.30–24.36 refer to the following terms. Each term may be used once, more than once, or not at all.
  a. *coacervates*
  b. *liposomes*
  c. *mesosomes*
  d. *microspheres*
  e. *proteinoids*

a
Factual Recall

24.30 Have been shown to absorb substrates from their surroundings.

d
Factual Recall

24.31 Droplets that self-assemble from only polypeptides in cool water.

e
Factual Recall

24.32 Formation catalyzed by hot lava or metal atoms such as iron and zinc.

b
Factual Recall

24.33 Droplets that self-assemble from organic ingredients that include lipids.

a
Factual Recall

24.34 Colloidal droplets that form when polypeptides, nucleic acids, and polysaccharides are shaken together.

d
Factual Recall

24.35 Form an electrical potential across their boundary membrane.

b
Factual Recall

24.36 Droplets that self-assemble and form a bilayer membrane.

Questions 24.37–24.40 refer to the following list:
1. *Plantae*
2. *Fungi*
3. *Animalia*
4. *Protista*
5. *Monera*

e
Factual Recall

24.37 Which kingdom includes prokaryotic organisms?
   a. 1
   b. 2
   c. 3
   d. 4
   e. 5

a
Factual Recall

24.38 Which kingdom includes mosses and ferns?
   a. 1
   b. 2
   c. 3
   d. 4
   e. 5

b
Factual Recall

24.39 Which kingdom includes organisms that are dikaryotic and mycelial?
   a. 1
   b. 2
   c. 3
   d. 4
   e. 5

d
Factual Recall

24.40 Which kingdoms include photosynthetic organisms?
    a. 1 and 5 only
    b. 2 and 4 only
    c. 1, 2, and 5 only
    d. 1, 4, and 5 only
    e. 1, 2, 3, 4, and 5

a
Conceptual
Understanding

24.41 If the first genes were made of RNA, why do most organisms today have genes made of DNA?
    a. DNA is chemically more stable and replicates with fewer errors (mutations) than RNA.
    b. Only DNA can replicate during cell division.
    c. RNA is too involved with translation of proteins and cannot provide multiple functions.
    d. DNA forms the rod-shaped chromosomes necessary for cell division.
    e. Replication of RNA occurs too quickly.

a
Factual Recall

24.42 In an experiment, zinc and RNA nucleotides are placed in a test tube. If a short RNA molecule is added to the test tube, what happens?
    a. RNA becomes autocatalytic and replicates portions of itself.
    b. RNA degrades into nucleotides.
    c. An autocatalytic protein, amino adenosine triacid ester (AATE), is produced.
    d. A protein enzyme is formed that assists RNA in replication.
    e. DNA is formed.

d
Factual Recall

24.43 Why do some scientists think that the kingdom Monera should be subdivided into two kingdoms, Eubacteria and Archaebacteria?
    a. Eubacteria contain organisms that are mostly pathogenic to humans.
    b. Archaebacteria contain colonial forms and should not be classified with Eubacteria.
    c. The five-kingdom system was proposed in 1969 and is old.
    d. Molecular evidence suggests that the two groups diverged very early in evolutionary history.
    e. Only Eubacteria have rigid cell walls.

b
Conceptual
Understanding

24.44 What characteristic do all protobionts have in common?
    a. the ability to synthesize enzymes
    b. a boundary membrane
    c. RNA genes
    d. a nucleus
    e. the ability to replicate RNA

d
Conceptual
Understanding

24.45 Although absolute distinctions between the "most evolved" protobiont and the first living cell are fuzzy, all scientists agree that one major difference is that all protobionts
a. do not possess a selectively permeable membrane boundary.
b. are not capable of osmosis.
c. do not grow in size.
d. are not capable of controlled, precise reproduction.
e. do not absorb compounds from the external environment.

b
Conceptual
Understanding

24.46 What is the most important function of the ozone layer for organisms living on Earth today?
a. reducing the amount of pollution in the atmosphere
b. reducing the amount of ultraviolet radiation that reaches the Earth's surface
c. increasing the amount of sunlight that reaches the Earth's surface
d. increasing the circulation of free oxygen in the atmosphere
e. moderating the temperature of the Earth's surface

b
Conceptual
Understanding

24.47 High temperatures, the presence of metal ions such as zinc, and the presence of a substrate such as clay all directly facilitate the formation of
a. coacervates.
b. organic polymers.
c. genes.
d. cells.
e. liposomes.

e
Conceptual
Understanding

24.48 Which of the following is the strongest evidence that prokaryotes evolved before eukaryotes?
a. the primitive structure of plants
b. meterorites that have struck the Earth
c. abiotic experiments that constructed microspheres in the laboratory
d. Liposomes and coacervates look like prokaryotic cells.
e. The oldest fossilized cells resemble prokaryotes.

e
Conceptual
Understanding

24.49 How could RNA have become involved in the mechanism for protein translation?
a. Only ribozymes were available as catalysts.
b. RNA replication is enhanced if proteins are produced.
c. Natural selection acted against autocatalyic protein formation.
d. DNA was not available for protein translation.
e. Natural selection favored RNA molecules that sequenced proteins that enhanced the replication of more of the same RNA.

a
Factual Recall

24.50 Some scientists think that panspermia played a role in the origin of life on Earth. What is the idea of panspermia?
a. The first organic compounds came from meteorites striking the Earth.
b. Meteorites or comets struck the Earth and caused mass extinctions.
c. Extraterrestrial visitors brought living organisms to the Earth.
d. Spermlike cells were the first living organisms on early Earth.
e. Living organisms from other planets were carried to the Earth via comets.

e
Conceptual
Understanding

24.51 Why is the abiotic origin of the organic compounds essential for life considered a testable hypothesis?
a. Scientists can prove how life began.
b. Organic compounds can be tested for life in the laboratory.
c. Simple cells can be constructed.
d. Scientists can prove the importance of these compounds to living organisms.
e. Scientists can simulate the early Earth environment.

Chapter 25

a
Conceptual
Understanding

25.1 Prokaryotes have ribosomes different from those of eukaryotes. Because of this, which of the following is true?
   a. Some selective antibiotics can block protein synthesis of bacteria without harming the eukaryotic host.
   b. It is believed that eukaryotes did not evolve from prokaryotes.
   c. Protein synthesis can occur at the same time as transcription in prokaryotes.
   d. Some antibiotics can block the formation of cross-links in the peptidoglycan walls of bacteria.
   e. Prokaryotes are able to use a much greater variety of molecules as food sources.

b
Factual Recall

25.2 The first form of nutrition to evolve was probably that of
   a. photoautotrophs that used light energy to reduce CO_2 with electrons from H_2O.
   b. chemoheterotrophs that used abiotically made organic compounds.
   c. anaerobic chemoautotrophs.
   d. photoheterotrophs that used light for energy and abiotically made organic compounds for a carbon source.
   e. photoautotrophs (such as the cyanobacteria) that used water as a source of electrons and protons.

a
Factual Recall

25.3 Of all the organisms, the prokaryotes have the greatest range of metabolic diversity. Which category of prokaryotes is currently the most important ecologically?
   a. nitrogen fixers
   b. obligate anaerobes
   c. thermoacidophiles
   d. chemoautotrophs
   e. extreme halophiles

b
Factual Recall

25.4 In the following list of major metabolic pathways, which one must have been the most recent to evolve?
   a. glycolysis
   b. oxidative phosphorylation
   c. fermentation
   d. O_2-producing photosynthesis
   e. sulfur-producing photosynthesis

c
Application

25.5 Which of the following statements is TRUE for chemoautotrophs?
   a. They use hydrogen sulfide as their hydrogen source for the photosynthesis of their organic compounds.
   b. They "feed themselves" by obtaining energy from the chemical bonds of organic molecules.
   c. They oxidize inorganic compounds to obtain energy to drive the synthesis of their organic compounds.
   d. They live as decomposers of inorganic chemicals in organic litter.
   e. They obtain their energy from oxidizing chemical compounds and get their carbon skeletons from organic compounds.

d
Factual Recall

25.6 Which of the following is a FALSE statement about the ways that prokaryotic cells differ from eukaryotic cells?
    a. The prokaryotic genome has a unique organization.
    b. Prokaryotes have a relatively simple organization of their cytoplasm.
    c. Prokaryotes have a cell wall with unique components.
    d. Prokaryotes lack specialized membranes.
    e. Prokaryotes are typically much smaller than eukaryotes.

e
Factual Recall

25.7 All of the following are TRUE statements about prokaryotes EXCEPT:
    a. Prokaryotes dominate the biosphere.
    b. Prokaryotes are the most numerous organisms on Earth.
    c. Some prokaryotes can live in extreme habitats.
    d. Some prokaryotes are important as decomposers.
    e. Prokaryotes are the most important photosynthesizers.

d
Application

25.8 If all the bacteria on Earth suddenly disappeared, which of the following would be the most likely direct result?
    a. The number of organisms on earth would decrease by 10 to 20 percent.
    b. Human populations would thrive in the absence of disease.
    c. There would be little change in the Earth's ecosystems.
    d. Recycling of nutrients would be greatly reduced.
    e. The Earth's total photosynthesis would decline markedly.

Using the following list of types of bacterial metabolism, pick the one that best matches each of the following statements 25.9–25.14.
    a. *photoautotrophs*
    b. *photoheterotrophs*
    c. *chemoautotrophs*
    d. *saprobic chemoheterotrophs*
    e. *parasitic chemoheterotrophs*

e
Factual Recall

25.9 Responsible for many human diseases.

d
Factual Recall

25.10 Nutritional mode of the earliest prokaryotes.

a
Factual Recall

25.11 Use light energy to synthesize organic compounds from CO_2.

c
Factual Recall

25.12 Obtain energy by oxidizing inorganic substances.

b
Factual Recall

25.13 Use light energy to generate ATP only.

a
Factual Recall

25.14 Responsible for high levels of O_2 in Earth's present atmosphere.

b
Application

25.15 If ancient prokaryotes had not evolved a way to use water as a source of electrons and protons, which of the following processes is least likely to have evolved later?
a. enzyme catabolism
b. the Krebs cycle
c. protein synthesis
d. fermentation
e. glycolysis

a
Factual Recall

25.16 Which of the following statements best characterizes cell structure and function prior to the evolution of eukaryotic cells?
a. All forms of nutrition and metabolism had evolved in prokaryotes.
b. The basic types of locomotion by whiplash flagella evolved in prokaryotes.
c. The evolutionary advantages of diploidy were exploited by prokaryotes.
d. Precursors of rough endoplasmic reticula had evolved within prokaryotes.
e. Mitosis developed as a process in more recent prokaryotes.

b
Factual Recall

25.17 What were the earliest bacteria like?
a. aerobic heterotrophs
b. anaerobic heterotrophs
c. photoautotrophs
d. chemoautotrophs
e. parasites

e
Factual Recall

25.18 The antibiotics known as penicillins inhibit the ability of bacteria to
a. form spores.
b. perform respiration.
c. replicate DNA.
d. synthesize proteins.
e. synthesize cell walls.

e
Factual Recall

25.19 What are bacteria that are poisoned by oxygen called?
a. aerobes
b. aestivating bacteria
c. cyanobacteria
d. facultative anaerobes
e. obligate anaerobes

b
Application

25.20 Which of the following would most likely occur if all prokaryotes were suddenly to perish?
a. All life would eventually perish due to disease.
b. All life would eventually perish because chemical cycling would stop.
c. All life would eventually perish because of increased global warming due to the greenhouse effect.
d. Only the organisms that feed directly on prokaryotes would suffer any deleterious effects.
e. Very little change would occur because prokaryotes are not of significant ecological importance.

a
Factual Recall

25.21 Which of the following types of bacteria have chlorophyll *a*?
a. cyanobacteria
b. archaebacteria
c. pathogenic bacteria
d. bacteria that decompose organic matter
e. bacteria that oxidize H_2S during photosynthesis

e
Conceptual
Understanding

25.22 Which term applies to bacteria that oxidize organic matter as a source of energy?
a. photoautotrophs
b. photoheterotrophs
c. chemoautotrophs
d. obligate anerobes
e. saprobes

b
Factual Recall

25.23 Heterocysts are structures characteristic of some
a. fungi.
b. cyanobacteria.
c. spore-forming bacteria.
d. heterotrophic cyst-forming archaebacteria.
e. nitrogen-fixing bacteria in root nodules of legumes.

d
Factual Recall

25.24 In an aerobic prokaryotic cell, the molecules of the respiratory chain are located in the
a. cytosol.
b. cristae.
c. cell wall.
d. plasma membrane.
e. mitochondrial matrix.

d
Factual Recall

25.25 Which of the following is the antibiotic mechanism of tetracycline?
a. It oxidizes the chemical bonds of organic macromolecules.
b. It uncouples the rotary motor of prokaryotic flagella.
c. It plasmolyzes the plasma membrane.
d. It binds to prokaryotic ribosomes.
e. It disrupts the cell wall.

d
Conceptual
Understanding

25.26 All of the following statements about prokaryotes are correct EXCEPT:
a. The gradual accumulation of oxygen caused the extinction of many prokaryotes.
b. Glycolysis probably evolved in prokaryotes to regenerate ATP in anaerobic environments.
c. Early photosynthetic prokaryotes probably used pigments and light-powered photosystems to fix carbon dioxide.
d. The first prokaryotes were likely photoautotrophs that could utilize the abundant light energy and inorganic minerals of early Earth.
e. The gradual accumulation of oxygen led to the evolution of respiratory mechanisms to either tolerate or capitalize on rising oxygen levels.

a
Factual Recall

25.27 Proton pumps of bacteria probably functioned first for
    a. pH regulation.
    b. ATP synthesis.
    c. photosynthesis.
    d. reduction of O_2.
    e. oxidation of food.

e
Factual Recall

25.28 Coordination of two photosystems occurs during photosynthesis in
    a. chemoautotrophic bacteria.
    b. purple sulfur bacteria.
    c. green sulfur bacteria.
    d. anaerobic bacteria.
    e. cyanobacteria.

b
Factual Recall

25.29 Which of the following statements is correct about gram-negative bacteria?
    a. Penicillins are effective antibiotics to use against them.
    b. They often possess an outer cell membrane containing toxic lipopoly-saccharides.
    c. On a cell-to-cell basis, they possess more DNA than do the cells of any taxonomically higher organism.
    d. Their chromosomes are composed of DNA tightly wrapped around histone proteins.
    e. Their cell walls are primarily composed of peptidoglycan.

d
Conceptual
Understanding

25.30 The oxygen revolution probably began with the origin of
    a. plants.
    b. eukaryotes.
    c. prokaryotes.
    d. cyanobacteria.
    e. cellular respiration.

b
Factual Recall

25.31 The botulism toxin is an example of
    a. an antibiotic.
    b. an exotoxin.
    c. an endotoxin.
    d. a nitrogenase.
    e. a bacteriorhodopsin.

a
Factual Recall

25.32 Flagellated bacteria will move away from toxic substances. This phenomenon is termed
    a. chemotaxis.
    b. chemotropism.
    c. gliding.
    d. magnetotaxis.
    e. toxoplasmosis.

c
Factual Recall

25.33 Which of the following statements about prokaryotes is CORRECT?
    a. Bacterial cells conjugate to mutually exchange genetic material.
    b. Their genetic material is confined within a nuclear envelope.
    c. They divide by binary fission without mitosis or meiosis.
    d. The persistence of bacteria through time is due to metabolic similarity.
    e. Genetic variation in bacteria arises from their geometric growth rates.

b
Conceptual
Understanding

25.34 If present in a solution, members of which group could not be filtered out of the solution by a filter with pores 250 nm in diameter?
    a. archaebacteria
    b. mycoplasmas
    c. myxobacteria
    d. pseudomonads
    e. spirochetes

Questions 25.35–25.39 refer to the following categories of nutrition. Each answer may be used once, more than once, or not at all.
    a. saprobes
    b. chemoheterotrophs
    c. chemoautotrophs
    d. photoautotrophs
    e. photoheterotrophs

c
Conceptual
Understanding

25.35 Bacteria that oxidize NH_3 to NO_2.

c
Conceptual
Understanding

25.36 Bacteria that oxidize sulfur to sulfate.

e
Conceptual
Understanding

25.37 Bacteria that use light for energy and organic matter for a carbon source.

d
Conceptual
Understanding

25.38 Green sulfur bacteria.

c
Conceptual
Understanding

25.39 Methanogens.

Questions 25.40–25.45 refer to the following prokaryotic groups. Each answer may be used once, more than once, or not at all.
    a. Actinomycetes
    b. Chlamydias
    c. Enteric bacteria
    d. Pseudomonads
    e. Spirochetes

c
Factual Recall

25.40 *E. coli* and *Salmonella*.

e
Factual Recall

25.41 *Treponema pallidum* and *Borrelia burgdorferi*.

b
Factual Recall

25.42 Nongonococcal urethritis (NGU).

a
Factual Recall

25.43 Tuberculosis and leprosy.

a
Factual Recall

25.44 The antibiotic streptomycin.

d
Factual Recall

25.45 Most versatile chemoheterotrophs.

Questions 25.46–25.47 are based on the following information. A team of biologists is trying to determine the identity of a new organism. All characteristics have not been fully investigated, but, based on the data available, the species has the following four characteristics:
1. *presence of some unbranched hydrocarbons*
2. *presence of introns in the few genes that have been analyzed*
3. *methionine present and used as initiator for protein synthesis*
4. *absence of peptidoglycan in outer portions of the cell wall*

d
Application

25.46 Based on these characteristics, what domain might the organism belong to?
   a. Eucarya
   b. Eubacteria or Eucarya
   c. Archaebacteria
   d. Archaebacteria or Eucarya
   e. Eubacteria

e
Application

25.47 What additional characteristic would be most useful in order to confirm or refine your identification?
   a. whether the organism is pathogenic
   b. presence of formylmethionine
   c. number of RNA polymerases present
   d. whether growth is inhibited by streptomycin
   e. presence of a nuclear membrane

b
Factual Recall

25.48 Why do systematic biologists now reject the use of a single kingdom Monera for all prokaryotic organisms?
   a. Only eukaryotic organisms have membrane-bound organelles.
   b. Genetic data show that archaebacteria and eukaryotes share a more recent common ancestor.
   c. Structural data show that eubacteria are more closely related to eukaryotes and that archaebacteria differ in a greater number of characteristics.
   d. Only prokaryotic organisms show growth inhibition in the presence of antibiotics.
   e. Only species of eubacteria lack the noncoding parts of genes.

d
Application

25.49 A new pathogenic bacterium has been obtained from a number of individuals exhibiting the same symptoms, and it has been isolated and grown in pure culture. What additional steps, if any, are necessary to establish that this bacterium is the cause of the disease?
   a. Sufficient data have been accumulated and no further research need be done.
   b. Gram stains must be applied, and the appropriate exotoxins or endotoxins must be isolated and their chemical structures analyzed.
   c. The method of transfer of the infection must be identified and the bacteria must be shown to infect a variety of different species.
   d. Cultured bacteria must be introduced to uninfected organisms, cause the same symptoms, and then be isolated from these test organisms.
   e. An antibody must be located, isolated, and shown to provide immunization against further infection by the bacteria.

a
Conceptual
Understanding

25.50 A particular virus acts to transfer genes during bacterial transduction but does not harm the bacteria in any way. If the virus also benefits from the bacteria, the term that best describes the relationship between virus and bacteria is
   a. mutualism.
   b. parasitism.
   c. commensalism.
   d. symbiosis.
   e. predation.

c
Factual Recall

25.51 All of the following statements about bacterial cell walls are true EXCEPT:
   a. They differ in molecular composition from plant cell walls.
   b. They prevent cells from bursting in hypotonic environments.
   c. They prevent cells from dying in hypertonic conditions.
   d. They are analogous to cell walls of protists, fungi, and plants.
   e. They provide the cell with physical protection from the environment.

b
Application

25.52 A biologist discovers a species of bacteria that stains purple with Gram stain and causes symptoms similar to food poisoning in infected animals. What is the type of bacterium and the source of the symptoms produced?
   a. gram-positive bacterium with endotoxins imbedded in peptidoglycans
   b. gram-positive bacterium that produce exotoxins
   c. gram-negative bacterium with endotoxins imbedded in lipopolysaccharides
   d. gram-negative bacterium that produce exotoxins
   e. gram-positive bacterium that produce endospores

c
Factual Recall

25.53 Many physicians administer antibiotics to patients at the first sign of any disease symptoms. Why can this practice cause more problems for these patients and for others not yet infected?
   a. The antibiotic administered may not be effective for the particular type of bacterium.
   b. Antibiotics may cause other side effects in patients.
   c. Overuse of antibiotics can select for antibiotic-resistant strains of bacteria.
   d. Particular patients may be allergic to the antibiotic.
   e. Antibiotics may interfere with the ability to identify the bacteria present.

Chapter 26

b
Factual Recall

26.1 Which of the following cause red tides?
   a. red algae (Rhodophyta)
   b. dinoflagellates
   c. diatoms
   d. Only a and c are correct.
   e. a, b, and c are correct

b
Factual Recall

26.2 A certain unicellular eukaryote has a siliceous (glasslike) shell and autotrophic nutrition. It is most likely a
   a. dinoflagellate.
   b. diatom.
   c. radiozoan.
   d. foraminifera.
   e. slime mold.

e
Factual Recall

26.3 In what ways are all protists alike?
   a. They are all multicellular.
   b. They are all photosynthetic.
   c. They are all marine.
   d. They are all nonparasitic.
   e. They are all eukaryotic.

b
Conceptual
Understanding

26.4 The strongest evidence for the endosymbiotic origin of eukaryotic organelles is the homology between extant prokaryotes and
   a. nuclei and chloroplasts.
   b. mitochondria and chloroplasts.
   c. cilia and mitochondria.
   d. ribosomes and nuclei.
   e. ribosomes and cilia.

b
Factual Recall

26.5 According to the endosymbiont theory of the origin of eukaryotic cells, how did mitochondria originate?
   a. infoldings of the plasma membrane
   b. engulfed, originally free-living prokaryotes
   c. mutations of genes for oxygen-using metabolism
   d. the nuclear envelope folding outward
   e. a protoeukaryote becoming symbiotic with a protobiont

e
Factual Recall

26.6 Which feature of protists is probably endosymbiotic in origin?
   a. cysts
   b. microtubules
   c. acritarchs
   d. a nucleus
   e. mitochondria

e
Application

26.7 If eukaryotic cells had first evolved in an environment much lower in O$_2$ than was actually the case, how might eukaryotes be different today?
    a. They would all be unicellular.
    b. They would be unable to photosynthesize.
    c. They would be more motile.
    d. They would lack ribosomes.
    e. They would lack mitochondria.

d
Factual Recall

26.8 Evidence for an endosymbiotic origin of chloroplasts and mitochondria includes which of the following?
    a. Both have circular DNA.
    b. Both have prokaryoticlike ribosomes.
    c. Both have histone proteins associated with DNA.
    d. Only a and b are correct.
    e. a, b, and c are correct.

c
Application

26.9 A biologist collects a previously unknown organism in a marine habitat. It is relatively large (about 0.5 meter long), photosynthetic, and has no vascular tissue. Should it be classified as a protist in a five-kingdom classification scheme?
    a. No, because it is too large.
    b. No, because it is photosynthetic.
    c. Yes, because it has a simple morphology.
    d. Yes, because it is multicellular.
    e. More information is necessary.

b
Application

26.10 Why are red algae red?
    a. They live in warm coastal waters.
    b. They absorb blue and green light.
    c. They use red light for photosynthesis.
    d. They are related to cyanobacteria.
    e. They lack chlorophyll.

a
Factual Recall

26.11 Each of the following groups includes many planktonic species EXCEPT
    a. sporozoans.
    b. chrysophytes.
    c. bacillariophytes.
    d. dinoflagellates.
    e. actinopods.

a
Factual Recall

26.12 In which group would you find organisms with the most complex cell structure?
    a. Ciliophora
    b. Zoomastigophora
    c. Euglenophyta
    d. Phaeophyta
    e. Myxomycota

c
Factual Recall

26.13 Which of the following correctly pairs a protist with one of its characteristics?
a. Zoomastigophora—slender pseudopodia
b. Rhizopoda—flagellated stages
c. Apicomplexa—all parasitic
d. Actinopoda—calcium carbonate shell
e. Foraminifera—abundant in soils

a
Factual Recall

26.14 Which of the following includes unicellular, colonial, and multicellular members?
a. Chlorophyta only
b. Rhizopoda only
c. Euglenophyta only
d. Phaeophyta only
e. both Chlorophyta and Phaeophyta

d
Factual Recall

26.15 The largest seaweeds belong to which group?
a. Cyanobacteria
b. Rhodophyta
c. Chlorophyta
d. Phaeophyta
e. Euglenophyta

a
Application

26.16 A biologist discovers an alga that is marine, multicellular, lives in fairly deep water, and has phycoerythrin. It probably belongs to which group?
a. Rhodophyta
b. Phaeophyta
c. Chlorophyta
d. Dinoflagellata
e. Chrysophyta

b
Factual Recall

26.17 The Irish potato famine was caused by what kind of organism?
a. bacterium
b. oomycete
c. sporozoan
d. plasmodial slime mold
e. virus

b
Factual Recall

26.18 According to the endosymbiotic hypothesis, chloroplasts are most likely the descendants of
a. aerobic, heterotrophic prokaryotes.
b. photosynthetic prokaryotes.
c. photoautotrophic eukaryotes.
d. Only a and b are correct.
e. a, b, and c are correct.

b
Factual Recall

26.19 According to the endosymbiotic theory, the ancestors of mitochondria were probably
a. aerobic eukaryotes.
b. aerobic bacteria.
c. anaerobic bacteria.
d. cyanobacteria.
e. chloroplasts.

d
Conceptual
Understanding

26.20 Which process results in genetic recombination but is separate from reproduction in *Paramecium*?
  a. budding
  b. meiotic division
  c. mitotic division
  d. conjugation
  e. fission

d
Factual Recall

26.21 Protozoan protists are generally classified according to
  a. nutrition.
  b. cell shape.
  c. size.
  d. locomotion.
  e. type of reproduction.

b
Factual Recall

26.22 A snail-like, coiled shell of calcium carbonate is characteristic of
  a. zooflagellates.
  b. forams.
  c. heliozoans.
  d. amoebas.
  e. ciliates.

c
Factual Recall

26.23 Which of the following produces the dense glassy ooze of the ocean floor?
  a. heliozoans
  b. dinoflagellates
  c. radiozoans
  d. ciliates
  e. sporozoans

b
Conceptual
Understanding

26.24 Members of the Chlorophyta often differ from members of Plantae in that some chlorophytans
  a. are heterotrophs.
  b. are unicellular.
  c. have chlorophyll *a*.
  d. store carbohydrates as starch.
  e. have cellulose cell walls.

c
Conceptual
Understanding

26.25 *Chlamydomonas* reproduces asexually unless
  a. male and female zoospores are produced.
  b. growth conditions are very favorable.
  c. + and – strains are present.
  d. four haploid cells are produced by mitosis.
  e. the antheridium hooks around the oogonium to deposit sperm.

d
Conceptual
Understanding

26.26 Ways in which *Volvox* has become advanced over *Chlamydomonas* include which of the following?
   I.   Mature organism is a single haploid cell.
   II. Movement is coordinated among cells.
   III. Colonial organization borders on multicellularity.

   a. I only
   b. II only
   c. III only
   d. II and III
   e. I, II, and III

e
Conceptual
Understanding

26.27 All of the following are characteristic of the water molds EXCEPT
   a. coenocytic hyphae.
   b. flagellated zoospores.
   c. haploid antheridia and oogonia.
   d. large egg cells.
   e. feeding plasmodium.

b
Factual Recall

26.28 The chloroplast structure and biochemistry of the red algae are most like which of the following organisms?
   a. Chrysophyta
   b. Cyanobacteria
   c. Bryophyta
   d. Chlorophyta
   e. Phaeophyta

b
Factual Recall

26.29 Considering photosynthetic organisms, which characteristic, usually found only in vascular plants, is sometimes seen in the brown algae?
   a. cellulose cell walls
   b. conducting tissue
   c. chlorophyll *a*
   d. alternation of generations
   e. carbohydrates as a food reserve

e
Factual Recall

26.30 Which of the following is a characteristic pigment of the brown algae?
   a. laminarin
   b. algin
   c. phycocyanin
   d. leucosin
   e. fucoxanthin

b
Factual Recall

26.31 All of the following statements concerning protists are true EXCEPT:
   a. All Protista are eukaryotic organisms; many are unicellular or colonial.
   b. The organism that causes malaria is transmitted to humans by the bite of the tsetse fly.
   c. All apicomplexans (sporozoans) are parasitic.
   d. All slime molds have an amoeboid stage that is followed by a sedentary stage during which spores are produced. ⸱
   e. The Euglenophyta have a pigment system similar to that of green algae and higher land plants.

d
Conceptual
Understanding

26.32 All Protista are alike in that they are
   a. autotrophic.
   b. heterotrophic.
   c. unicellular.
   d. eukaryotic.
   e. flagellated.

e
Factual Recall

26.33 Which phylum containing eukaryotic organisms is believed to be ancestral to the plant kingdom?
   a. Chrysophyta
   b. Actinopoda
   c. Foraminifera
   d. Apicomplexa
   e. Chlorophyta

d
Factual Recall

26.34 Which of the following is mismatched?
   a. Apicomplexa—internal parasites
   b. Chrysophta—planktonic producers
   c. Euglenophyta—unicellular flagellates
   d. Ciliophora—freshwater producers
   e. Rhizopoda—ingestive heterotrophs

c
Application

26.35 You are given an unknown organism to identify. It is unicellular and heterotrophic. It is motile, with well-developed organelles and three nuclei—one large and two small. You conclude that this organism is most likely to be a member of which phylum?
   a. Rhizopoda
   b. Actinopoda
   c. Ciliophora
   d. Zoomastigophora
   e. Oomycota

e
Factual Recall

26.36 Diatomaceous earth consists of the shells of members of which phylum?
   a. Chrysophyta
   b. Ciliophora
   c. Myxomycota
   d. Chlorophyta
   e. Bacillariophyta

e
Factual Recall

26.37 Which of the following statements is NOT true about the dinoflagellates?
   a. They possess two unequal flagella.
   b. Some cause red tides.
   c. They are unicellular.
   d. They have chlorophyll.
   e. Their fossil remains form limestone deposits.

b
Conceptual
Understanding

26.38 All of the following statements provide evidence that chloroplasts and mitochondria originated as prokaryotic endosymbionts EXCEPT that they
   a. are the same size as bacteria.
   b. can be cultured on agar since they make all their own proteins.
   c. contain circular DNA molecules not associated with histones.
   d. have membranes that are similar to those found in the plasma membranes of prokaryotes.
   e. have ribosomes that are similar to those of bacteria.

e
Factual Recall

26.39 Which of the following characteristics of chloroplasts and mitochondria are more similar to prokaryote cells than to eukaryote cells?
  a. enzymes and transport systems of inner membranes
  b. DNA not associated with histone proteins
  c. single, circular chromosome
  d. Only a and c are correct.
  e. a, b, and c are correct.

b
Factual Recall

26.40 The oldest possible fossil representatives of the kingdom Protista currently known are
  a. stromatolites that are colonial and 3.5 billion years old.
  b. acritarchs that are parts of cysts and 2.1 billion years old.
  c. chromistans that are unicellular and 1.8 billion years old.
  d. plasmodia that are coenocytic and 1.8 billion years old.
  e. diplomonads with two nuclei that are 1.5 billion years old.

d
Factual Recall

26.41 Which of the major groups of alga-like protists are contained in the new group, Chromista?
  a. plasmodial and cellular slime molds and the water molds
  b. radiozoans, sporozoans, and amoebas
  c. green algae, euglenas, and ciliates
  d. brown and golden algae and the diatoms
  e. zooflagellates, dinoflagellates, and forams

b
Application

26.42 A biologist finds a new unicellular organism that possesses an endoplasmic reticulum, a simple cytoskeleton, and two small nuclei that are each surrounded by a membrane. The organism has neither mitochondria nor chloroplasts. This organism most probably is a
  a. radiozoan.
  b. diplomonad.
  c. ciliate.
  d. prokaryote.
  e. *Chlamydomonas*.

a
Conceptual
Understanding

26.43 The small size and simple construction of prokaryotes imposes limits on the
  a. number of simultaneous metabolic activities and the number of genes present.
  b. type of habitat they occupy and the frequency of reproduction that can occur.
  c. number of cells that can be associated in organized colonies.
  d. number of organelles present and the size of the nucleus.
  e. type of reproduction and the number of offspring that can be produced.

b
Application

26.44 The endoplasmic reticulum and Golgi apparatus are very similar among the groups of alga-like protists, but chloroplasts differ significantly and appear to be related to different prokaryotes. What do these facts imply about the evolution of the endomembrane organelle system of eukaryotic cells?
  a. The Golgi apparatus evolved before the endomembrane system.
  b. Endomembrane systems evolved before chloroplasts.
  c. Endomembrane systems evolved from symbiotic prokaryotes.
  d. Endomembrane systems evolved after chloroplasts.
  e. Chloroplasts evolved before the endoplasmic reticulum.

e
Factual Recall

26.45 All of the following statements about the alternation of generations in algal protists are true EXCEPT:
  a. Diploid sporophytes produce spores.
  b. Diploid stages are multicellular.
  c. Haploid gametophytes produce gametes.
  d. Spores produce gametophytes.
  e. Haploid stages are unicellular.

c
Application

26.46 Why is the filamentous body form of the slime and water molds considered a case of convergent evolution with the hyphae of fungi?
  a. Fungi are closely related to the slime and water molds.
  b. Body shape reflects ancestor–descendant relationships among organisms.
  c. Filamentous shape is an adaptation for a nutritional mode as a decomposer.
  d. Hyphae and filaments are necessary for locomotion in both groups.
  e. Filamentous body shape is evolutionarily primitive for all eukaryotes.

a
Conceptual
Understanding

26.47 Many biologists consider the kingdom Protista to be polyphyletic. Which of the following statements is consistent with this conclusion?
  a. Various combinations of prokaryote ancestors gave rise to different lineages of protists.
  b. Animals, plants, and fungi arose from different protistan ancestors.
  c. Multicellularity has evolved independently in different groups of protists.
  d. Chloroplasts in different eukaryotes are similar to different prokaryotes.
  e. Archezoans are intermediate and should not be considered part of the Protista.

b
Factual Recall

26.48 Which organelles do scientists believe originated by symbiotic relationships between primitive eukaryotes and certain prokaryotes?
  a. nuclei and ribosomes
  b. chloroplasts and mitochondria
  c. vacuoles and the Golgi apparatus
  d. storage vesicles and thylakoids
  e. cristae and the endoplasmic reticulum

Chapter 27

b
Conceptual
Understanding

27.1 Fruits have contributed to the success of angiosperms by
    a. nourishing the plants that make them.
    b. facilitating dispersal of seeds by wind and animals.
    c. attracting insects to the pollen inside.
    d. producing sperm and eggs inside a protective coat.
    e. producing triploid cells via double fertilization.

c
Factual Recall

27.2 Bryophytes have all of the following characteristics EXCEPT
    a. multicellularity.
    b. specialized cells and tissues.
    c. well-developed vascular tissue.
    d. a protected, stationary egg cell.
    e. a reduced, dependent sporophyte.

c
Conceptual
Understanding

27.3 Which of the following INCORRECTLY pairs a sporophyte embryo with its food source?
    a. pine embryo—female gametophyte tissue in nucellus
    b. grass embryo—$3n$ endosperm tissue in seed
    c. moss embryo—female sporophyte tissue
    d. fern embryo—photosynthetic gametophyte
    e. club moss embryo—subterranean, nonphotosynthetic gametophyte

c
Application

27.4 A botanist discovers a new species of plant in a tropical rainforest. After observing its anatomy and life cycle, the following characteristics are noted: flagellated sperm, xylem with tracheids, separate gametophyte and sporophyte phases, and no seeds. This plant is probably most closely related to
    a. mosses.
    b. conifers.
    c. ferns.
    d. liverworts.
    e. flowering plants.

e
Conceptual
Understanding

27.5 Angiosperms are the most successful terrestrial plants. This success is due to all the following EXCEPT
    a. animal pollination.
    b. reduced gametophytes.
    c. fruits enclosing seeds.
    d. xylem with vessels.
    e. sperm cells with flagella.

c
Conceptual
Understanding

27.6 All of the following plant structures are adaptations specifically for a terrestrial environment EXCEPT
  a. roots.
  b. xylem.
  c. cell walls.
  d. waxy cuticle.
  e. seeds.

c
Factual Recall

27.7 All of the following are characteristic of angiosperms EXCEPT
  a. coevolution with animal pollinators.
  b. double internal fertilization.
  c. free-living gametophytes.
  d. pistils.
  e. fruit.

c
Factual Recall

27.8 The ancestors of land plants were most likely similar to modern-day members of the
  a. Cyanobacteria (blue-green algae).
  b. Rhodophyta (red algae).
  c. Chlorophyta (green algae).
  d. Phaeophyta (brown algae).
  e. Chrysophyta (diatoms and golden-brown algae).

a
Conceptual
Understanding

27.9 Which of the following is true concerning the sporophyte and gametophyte generations in flowering plants?
  a. All of the below are true.
  b. The sporophyte generation is dominant.
  c. The sporophyte generation is what we see when observing a plant.
  d. Unlike ferns, the gametophyte generation is not photosynthetic.
  e. The gametophyte generation is relatively few cells in the flower.

d
Factual Recall

27.10 Along with the seed, the seed plants have evolved several additional adaptations to the land environment. Which one of the following is NOT such an adaptation?
  a. Flagellated gametes are not required for seed formation.
  b. The female gametophyte is protected from dessication by the surrounding tissues of the sporophyte.
  c. The seed and/or associated structures serve as a means of dispersal.
  d. Seed formation introduces a new type of genetic recombination.
  e. The seed contains nutrients for the enclosed embryo.

e
Conceptual
Understanding

27.11 What is one reason why the Chlorophyta are believed to be the ancestors of plants?
  a. Some of their members have developed holdfast, stipe, and blades—ancestral to root, stem, and leaves.
  b. They do not have flagellated gametes.
  c. They are the only multicellular algal protists.
  d. They exhibit an alternation of generations.
  e. They have similar chloroplasts and pigment composition.

c
Factual Recall

27.12 Which of the following organisms do not have a jacket of sterile cells that protect developing gametes and embryos?
a. mosses
b. vascular plants
c. brown algae
d. ferns
e. liverworts

b
Factual Recall

27.13 Which of the following is the dominant stage in the life cycle of a moss?
a. sporophyte
b. gametophyte
c. diploid
d. sporangium
e. flowering stage

a
Factual Recall

27.14 The term *living fossil* is sometimes used to describe a living member of a mostly extinct group. An example of a living fossil in the plant kingdom would be a
a. horsetail.
b. bristle-cone pine.
c. sunflower.
d. moss.
e. wind-pollinated angiosperm.

c
Application

27.15 A land plant produces flagellated sperm and has a dominant diploid generation. The plant is probably a
a. moss.
b. green alga.
c. fern.
d. conifer.
e. flowering plant.

d
Conceptual
Understanding

27.16 Which of the following is FALSE about the life cycle of mosses?
a. External water is required for fertilization.
b. Flagellated sperm are produced.
c. Antheridia and archegonia are produced by gametophytes.
d. Gametes are directly produced by meiosis.
e. Gametophytes arise from the protonema.

b
Factual Recall

27.17 In ferns, what does the spore become?
a. fiddlehead
b. gametophyte
c. rhizome
d. sporangium
e. sporophyte

Questions 27.18–27.21 refer to the generalized life cycle for plants shown in Figure 27.1. Each number within a circle or square represents a specific plant or plant part, and each number over an arrow represents either meiosis, mitosis, or fertilization.

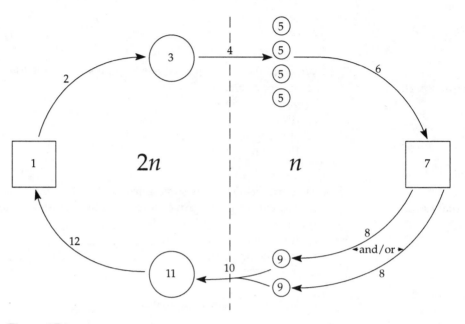

Figure 27.1

d
Conceptual
Understanding

27.18 A moss gametophyte is represented by
    a. 1
    b. 3
    c. 5
    d. 7
    e. 11

c
Conceptual
Understanding

27.19 Which number represents the embryo sac of an angiosperm flower?
    a. 1
    b. 3
    c. 7
    d. 9
    e. 11

c
Conceptual
Understanding

27.20 Meiosis is represented by
    a. 2 only.
    b. 3 only.
    c. 4 only.
    d. 8 only.
    e. both 4 and 8.

b
Conceptual
Understanding

27.21 Which number is a megaspore mother cell?
a. 1
b. 3
c. 5
d. 7
e. 11

c
Factual Recall

27.22 Plant spores give rise directly to
a. sporophytes.
b. gametes.
c. gametophytes.
d. zygotes.
e. seeds.

d
Conceptual
Understanding

27.23 In flowering plants, meiosis occurs specifically in the
a. megaspore mother cells.
b. microspore mother cells.
c. endosperm.
d. Only a and b are correct.
e. a, b, and c are correct.

d
Factual Recall

27.24 Danger of desiccation and the need for gas exchange are two conflicting
problems that were partially solved through the evolution of
a. phloem.
b. stomates.
c. cuticle.
d. Only b and c are correct.
e. a, b, and c are correct.

d
Conceptual
Understanding

27.25 Which of the following was not a problem for the first land plants?
a. sources of water
b. sperm transfer
c. desiccation
d. animal predation
e. gravity

c
Factual Recall

27.26 Which of the following represents the male gametophyte of an angiosperm?
a. ovule
b. microspore mother cell
c. pollen
d. embryo sac
e. fertilized egg

Use the following choices to identify the phrases for Questions 27.27–27.31.
a. *Bryophyta*
b. *Pterophyta*
c. *Coniferophyta*
d. *Anthophyta*
e. *Hepatophyta*

b
Factual Recall

27.27 Dominant sporophyte, small gametophyte, swimming sperm.

d
Factual Recall

27.28 Nonmotile sperm, both wind and insect pollinated.

d
Factual Recall

27.29 Endosperm, xylem vessels, and fruit.

e
Factual Recall

27.30 Flattened thallus, dominant gametophyte, motile sperm.

c
Factual Recall

27.31 Needle-like leaves, "naked" seeds, nonmotile sperm.

e
Application

27.32 A botanist discovers a new species of plant with a dominant sporophyte, chlorophyll *a* and *b*, and a cell wall made of cellulose. In assigning this plant to a division, all of the following would provide useful information EXCEPT whether or not the plant has
a. endosperm.
b. seeds.
c. flagellated sperm.
d. flowers.
e. starch.

c
Factual Recall

27.33 Plants with a dominant sporophyte are successful on land because
a. having no stomata, they lose less water.
b. they all disperse by means of seeds.
c. diploid plants are protected from the effects of mutation.
d. their gametophytes are all parasitic on the sporophytes.
e. eggs and sperm need not be produced.

a
Application

27.34 Larch trees are conifers that lose their leaves each fall; from this information, what can be concluded?
a. Not all conifers are evergreens.
b. Larch trees live where winters are dry.
c. Larch trees have been classified incorrectly.
d. Larch trees live where the growing season is long.
e. Larch trees are not as well-adapted as pines.

d
Factual Recall

27.35 Conifers are noted for all of the following EXCEPT
a. size.
b. longevity.
c. utility to humans.
d. great diversity of species.
e. success in cold climates.

e
Conceptual
Understanding

27.36 All of the following statements correctly describe portions of the pine life cycle
EXCEPT:
a. Female gametophytes have archegonia.
b. Seeds are produced in ovulate cones.
c. Meiosis occurs in sporangia.
d. Pollen grains are male gametophytes.
e. Pollination and fertilization are the same process.

d
Factual Recall

27.37 Gymnosperms differ from ferns in that gymnosperms
a. produce seeds.
b. have macrophylls.
c. have pollen.
d. Only a and c are correct.
e. a, b, and c are correct.

a
Factual Recall

27.38 All of the following are valid arguments for preserving tropical forests
EXCEPT:
a. People in the tropics do not need more agricultural land.
b. Many organisms are becoming extinct.
c. Plants that are possible sources of medicines are being lost.
d. Plants that could be developed into new crops are being lost.
e. Clearing land for agriculture results in soil destruction.

b
Application

27.39 Assume a botanist was visiting a tropical region for the purpose of
discovering plants with medicinal properties. All of the following might be
ways of identifying potentially useful plants EXCEPT
a. observing which plants sick animals seek out.
b. observing which plants are the most used food plants.
c. observing which plants animals do not eat.
d. collecting plants and subjecting them to chemical analysis.
e. asking local people which plants they use as medicine.

*Questions 27.40 and 27.41 refer to Figure 27.2, which shows the relative dominance of
gametophyte and sporophyte generations in an angiosperm, gymnosperm, fern, and
moss. Mentally arrange them in a sequence that shows increasing size and structural
complexity of the sporophyte.*

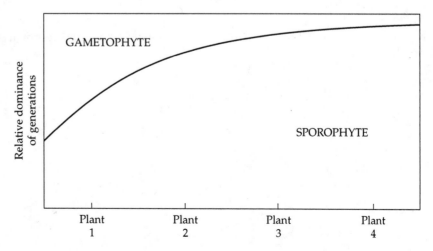

Figure 27.2

b
Application

27.40 A fern would be
   a. plant 1.
   b. plant 2.
   c. plant 3.
   d. plant 4.
   e. either plant 2 or 4.

d
Application

27.41 An angiosperm would be
   a. plant 1.
   b. plant 2.
   c. plant 3.
   d. plant 4.
   e. both plants 1 and 4.

e
Factual Recall

27.42 The term *Embryophyta* refers to which characteristic in which of the following groups?
   a. flagellated swimming sperm observed in the mosses, liverworts, and hornworts
   b. seed formation observed in the gymnosperms and angiosperms
   c. spore formation observed in the lycopods and ferns
   d. free-living embryos observed in the green algae, Chlorophyta
   e. retention of embryos in maternal tissues observed in the Plantae

d
Factual Recall

27.43 In addition to seeds, which of the following characteristics are unique to the seed-producing land plants?
   a. a haploid gametophyte retained within tissues of the diploid sporophyte
   b. lignin present in cell walls
   c. pollen
   d. Only a and c are correct.
   e. a, b, and c are correct.

d
Factual Recall

27.44 Why are Charophyte algae NOT considered to alternate generations during their life cycle?
   a. The haploid stage is not dependent on the diploid stage.
   b. The diploid stage is not dependent on the haploid stage.
   c. The zygote is diploid but is surrounded by nonreproductive cells.
   d. The diploid stage is only unicellular.
   e. The haploid stage is dominant.

b
Factual Recall

27.45 Of the following list, flagellated (swimming) sperm are present in which groups?
   1. Lycophyta
   2. Bryophyta
   3. Angiospermae
   4. Chlorophyta
   5. Pterophyta

a. 1, 2, 3
b. 1, 2, 4, 5
c. 1, 3, 4, 5
d. 2, 3, 5
e. 2, 3, 4, 5

a
Conceptual
Understanding

27.46 A number of characteristics are very similar between the Charophyta green algae and the kingdom Plantae. Of the following, which characteristic does NOT provide evidence for an evolutionarily close relationship beteween these two groups?
a. alternation of generations
b. chloroplast structure
c. cell plate formation during cytokinesis
d. sperm cell structure
e. ribosomal RNA base sequences

c
Conceptual
Understanding

27.47 A major change that occurred in the evolution of plants from their algal ancestors was the origin of a multicellular diploid stage. What advantage would multicellularity provide in this stage of the life cycle?
a. enhanced potential for independence of the diploid stage from the haploid stage
b. increased gamete production
c. increased spore production from each fertilization event
d. increased fertilization rate
e. increased size of the diploid stage

d
Factual Recall

27.48 One of the major functions of double fertilization in angiosperms is to
a. decrease the potential for mutation by insulating the embryo with other cells.
b. increase the number of fertilization events and offspring produced.
c. promote diversity in flower shape and color.
d. coordinate developmental timing between the embryo and its food stores.
e. emphasize embryonic survival by increasing embryo size.

e
Factual Recall

27.49 Heterospory refers to the condition of some plants in which
a. both male and female reproductive organs are found on the same plant.
b. a single individual exhibits two different types of plant growth.
c. spores are produced twice during a reproductive cycle.
d. different gametes are produced by the same individual.
e. two different spore types are produced.

d
Conceptual
Understanding

27.50 Agricultural modifications of plants have progressed to the point that a number of cultivated plant species probably could not survive in the wild. Why is this so?
a. Environmental conditions have changed since the plants evolved.
b. Seeds can be obtained only from seed banks in agricultural countries.
c. Cultivated plants are more vulnerable to human-caused pollution and disasters.
d. Special conditions not found in nature are needed for their growth and reproduction.
e. Their seeds cannot be dispersed without agricultural machinery.

b
Conceptual
Understanding

27.51 One of the major distinctions between plants and the algal protists is that
   a. only algal protists have flagellated, swimming sperm.
   b. embryos are not retained within parental tissues in protists.
   c. meiosis proceeds at a faster pace in protists than in plants.
   d. chlorophyll pigments in algal protists are different from those in plants.
   e. only plants form a cell plate during cytokinesis.

Chapter 28

d
Factual Recall

28.1 In fungi, karyogamy does not immediately follow plasmogamy, which
   a. means that sexual reproduction can occur in specialized structures.
   b. results in more genetic variation during sexual reproduction.
   c. allows fungi to reproduce asexually most of the time.
   d. creates dikaryotic cells.
   e. is necessary to create coenocytic hyphae.

e
Factual Recall

28.2 The division Deuteromycota
   a. includes members of all three divisions of fungi.
   b. is the division in which all the fungal components of lichens are classified.
   c. includes the imperfect fungi that have abnormal forms of sexual reproduction.
   d. is the classification of molds, yeasts, and lichens.
   e. is a practical classification that includes molds that usually reproduce asexually by conidia.

d
Factual Recall

28.3 Which of the following do all fungi have in common?
   a. meiosis in basidia
   b. coenocytic hyphae
   c. sexual life cycle
   d. absorption of nutrients
   e. symbioses with algae

b
Factual Recall

28.4 Fungi are of considerable ecological and/or economic importance in all of the following ways EXCEPT as
   a. symbiotic partners with plant roots.
   b. food for plants.
   c. decomposers of organic matter.
   d. human pathogens.
   e. producers of fuel.

e
Factual Recall

28.5 What is the threadlike basic structural element of a fungus?
   a. filament
   b. mold
   c. pseudoparenchyma
   d. mycelium
   e. hypha

c
Factual Recall

28.6 Gilled mushrooms typically available in supermarkets have meiotically produced spores located in or on _____ and belong to the division _____.
   a. asci—Basidiomycota
   b. hyphae—Phycomycota
   c. basidia—Basidiomycota
   d. asci—Ascomycota
   e. hyphae—Ascomycota

Questions 28.7–28.11 refer to the following divisions. Each term may be used once, more than once, or not at all.
  a. *Zygomycota*
  b. *Ascomycota*
  c. *Basidiomycota*
  d. *Deuteromycota*
  e. *Lichens*

a
Factual Recall

28.7 This division has the fewest species.

d
Factual Recall

28.8 This division is characterized by the lack of an observed sexual phase in its life cycle.

a
Factual Recall

28.9 This division is made up of coenocytic hyphae with asexual spores that develop in aerial sporangia.

b
Factual Recall

28.10 This division produces two kinds of haploid spores, one kind being asexually produced conidia.

c
Factual Recall

28.11 This division contains the mushrooms, shelf fungi, puffballs, and stinkhorns.

c
Factual Recall

28.12 All of the following are characteristic of fungi EXCEPT:
  a. They acquire their nutrients by absorption.
  b. Their body plan is a netlike mass of filaments called hyphae.
  c. Their cell walls consist mainly of cellulose microfibrils.
  d. They are specialized as saprobes, parasites, or mutualistic symbionts.
  e. The nuclei of the mycelia are haploid.

b
Application

28.13 You are given an organism to identify. It has a fruiting body that contains eight haploid spores lined up in a row. What kind of a fungus is it most likely to be?
  a. zygomycete
  b. ascomycete
  c. deuteromycete
  d. lichen
  e. basidiomycete

b
Conceptual
Understanding

28.14 A coenocytic structure implies being
  a. multicellular.
  b. multinucleate.
  c. commensalistic.
  d. saprobic.
  e. heterotrophic.

b
Factual Recall

28.15 What are mycorrhizae?
    a. the fruiting bodies of basidiomycetes
    b. mutualistic associations of plant roots and fungi
    c. the pores in fungi that allow ribosomes, mitochondria, and cell nuclei to flow from cell to cell
    d. the horizontal hyphae that spread out over food
    e. asexual structures formed by deuteromycetes

e
Factual Recall

28.16 The fungus that produces a compound responsible for the disease ergotism (St. Anthony's fire) is a(n)
    a. deuteromycete.
    b. zygomycete.
    c. lichen.
    d. basidiomycete.
    e. ascomycete.

c
Factual Recall

28.17 A mycelium is characteristic of most
    a. bacteria.
    b. protozoa.
    c. fungi.
    d. mosses.
    e. sponges.

d
Conceptual
Understanding

28.18 What do fungi and arthropods have in common?
    a. Both groups are commonly coenocytic.
    b. The haploid state is dominant in both groups.
    c. Both groups are predominantly saprobic in nutrition.
    d. Both groups use chitin for the construction of protective coats.
    e. Both groups have cell walls.

a
Factual Recall

28.19 Which characteristic is found in all fungal groups?
    a. heterotrophic nutrition
    b. saprobic nutrition
    c. multicellularity
    d. dikaryotic hyphae
    e. parasitism

e
Factual Recall

28.20 Parasitic fungi have specialized hyphae called
    a. aseptate hyphae.
    b. coenocytic hyphae.
    c. sporangia.
    d. dikaryotic hyphae.
    e. haustoria.

a
Factual Recall

28.21 The sporangia of bread molds are
    a. asexual structures that produce haploid spores.
    b. asexual structures that produce diploid spores.
    c. sexual structures that produce haploid spores.
    d. sexual structures that produce diploid spores.
    e. vegetative structures with no role in reproduction.

c
Factual Recall

28.22 Which of these fungal structures is associated with asexual reproduction?
  a. zygospore
  b. basidium
  c. conidium
  d. ascus
  e. antheridium

a
Factual Recall

28.23 Mushrooms and toadstools are classified as
  a. basidiomycetes.
  b. ascomycetes.
  c. deuteromycetes.
  d. zygomycetes.
  e. lichens.

e
Factual Recall

28.24 Lichens are symbiotic communities consisting of fungi and
  a. mosses only.
  b. cyanobacteria only.
  c. chlorophytes only.
  d. Only a and b are correct.
  e. Only b and c are correct.

b
Factual Recall

28.25 The symbiotic associations called mycorrhizae are considered to be
  a. parasitic.
  b. mutualistic.
  c. commensal.
  d. harmful to the plant partner.
  e. the beginning stages of the formation of lichens.

d
Factual Recall

28.26 What is the best definition of a fungus?
  a. eukaryotic heterotrophic plants
  b. eukaryotic, parasitic plants
  c. saprobic plants
  d. eukaryotic, multicellular heterotrophs
  e. saprobic heterotrophs

b
Factual Recall

28.27 The body of a fungus is made of a network of filaments called
  a. septa.
  b. mycelia.
  c. hyphae.
  d. haustoria.
  e. dikaryons.

e
Factual Recall

28.28 What does the word *dikaryon* mean?
  a. only two cells
  b. two hyphae
  c. only two mitochondria
  d. two chromosomes
  e. two nuclei

d
Factual Recall

28.29 About how many different kinds of fungi are there?
    a. 1,000
    b. 10,000
    c. 50,000
    d. 100,000
    e. 500,000

d
Factual Recall

28.30 A downy mildew is a member of which of the following groups?
    a. Deuteromycota
    b. Basidiomycota
    c. Ascomycota
    d. Oomycota
    e. Zygomycota

e
Factual Recall

28.31 Bread covered with a black fungus is most likely providing nutrition to what kind of organism?
    a. lichen
    b. ascomycete
    c. basidiomycete
    d. deuteromycete
    e. zygomycete

c
Factual Recall

28.32 Yeast is a member of which division?
    a. Deuteromycota
    b. Basidiomycota
    c. Ascomycota
    d. Zygomycota
    e. lichens

d
Conceptual
Understanding

28.33 If there were no mycorrhizae, which of the following would be true?
    a. There would be fewer infectious diseases.
    b. We wouldn't have antibodies like penicillin.
    c. There would be no mushrooms for pizza.
    d. A lot of trees would not grow well.
    e. Cheeses like blue cheese or Roquefort could not exist.

c
Conceptual
Understanding

28.34 The sac fungi get their name from which aspect of their life cycle?
    a. vegetative growth form
    b. asexual spore production
    c. sexual structures
    d. shape of the spore
    e. type of vegetative mycelium

e
Factual Recall

28.35 Both *Penicillium* and *Aspergillus* produce asexual spores at the tips of the
    a. asci.
    b. antheridia.
    c. rhizoids.
    d. gametangia.
    e. conidiophores.

a
Conceptual
Understanding

28.36 The primary role of the mushroom's subterranean mycelium is
    a. absorbing nutrients.
    b. anchoring.
    c. sexual reproduction.
    d. asexual reproduction.
    e. protection.

c
Factual Recall

28.37 Which aspects of mitotic division are unique to fungi?
    a. Cytokinesis involves formation of a cell plate after the nuclei have split.
    b. Chromosomes are few in number and the nuclei are small.
    c. The nuclear membrane remains intact around an internal spindle.
    d. An internal spindle does not form.
    e. Spores are always produced by mitotic division.

b
Conceptual
Understanding

28.38 In septate fungi, what structures allow cytoplasmic streaming to distribute needed nutrients, synthesized compounds, and organelles within the hyphae?
    a. chitinous layers in cell walls
    b. pores in septal walls
    c. complex microtubular cytoskeletons
    d. two nuclei
    e. tight junctions that form in septal walls between cells

a
Factual Recall

28.39 What accounts for the extremely fast growth of a fungal mycelium?
    a. a rapid distribution of synthesized proteins by cytoplasmic streaming
    b. their lack of motility that requires rapid spread of hyphae
    c. an increased surface area provided by long tubular body shape
    d. the readily available nutrients from their absorptive mode of nutrition
    e. a dikaryotic condition that supplies greater amounts of proteins and nutrients

e
Factual Recall

28.40 Lichens sometimes reproduce asexually by forming soredia, which are
    a. aseptate fungal hyphae located within algal cells.
    b. the fruiting bodies of fungi.
    c. flagellated, conjoined spores of both the fungus and alga.
    d. specialized conidiophores.
    e. clusters of fungal hyphae with imbedded algal cells.

c
Conceptual
Understanding

28.41 Why do biologists who study lichens sometimes refer to the symbiotic relationship between fungus and alga as "controlled parasitism"?
    a. Together, the fungus and alga may parasitize and kill other living organisms, such as plants.
    b. Each contributes to the maintenance of the other.
    c. Fungal haustoria may kill algal cells, but at a pace slow enough not to destroy all the algae present.
    d. Algal cells die at a faster rate than fungal cells.
    e. Fungal cells reproduce slower than the algae, thus becoming enclosed and unable to grow.

Questions 28.42 and 28.43 refer to the following information. A biologist is trying to classify a new fungal organism on the basis of the following characteristics: moldlike in appearance, reproduces asexually by conidia, and parasitizes woody plants.

a
Application

28.42 If asked for advice, to which group would you assign this new species?
    a. Deuteromycota
    b. Oomycota
    c. Ascomycota
    d. Basidiomycota
    e. Chytridiomycota

b
Application

28.43 Knowledge of which additional characteristic would enable you to refine or confirm your identification?
    a. presence of soredia
    b. form of sexual reproduction
    c. chemical composition of cell walls
    d. presence of flagellated sperm
    e. detection of algal cells imbedded in mycelium

d
Conceptual
Understanding

28.44 Fossil fungi date back to 440 million years ago, which coincides with the origin and early evolution of plants. What combination of environmental change and morphological change is similar in both fungi and plants?
    a. presence of "coal forests" and change in mode of nutrition
    b. periods of drought and presence of filamentous body shape
    c. predominance of swamps and presence of cellulose in cell walls
    d. colonization of land and loss of flagellated cells
    e. continental drift and mode of spore dispersal

b
Conceptual
Understanding

28.45 Which of the following characteristics do NOT provide evidence for a close evolutionary relationship between fungi and chytrids?
    a. presence of hyphae
    b. flagellated zoospores
    c. absorptive mode of nutrition
    d. chitinous cell walls
    e. amino acid base sequences of some enzymes

b
Factual Recall

28.46 Some biologists believe that the kingdoms Fungi and Animalia share a common protistan ancestor. Which protist group is currently considered the best candidate?
    a. cyanobacteria
    b. choanoflagellates
    c. chytridiomycotes
    d. deuteromycotes
    e. ciliates

Chapter 29

29.1 Which of the following is descriptive of protostomes?
a. spiral and indeterminate cleavage, coelom forms as split in solid mass of mesoderm
b. spiral and determinate cleavage, blastopore becomes mouth, schizocoelous development
c. spiral and determinate cleavage, enterocoelous development
d. radial and determinate cleavage, enterocoelous development, blastopore becomes anus
e. radial and determinate cleavage, blastopore becomes mouth, schizocoelous development

a
Factual Recall

29.2 Which of the following excretory structures is incorrectly matched with its class?
a. flame cells—Hydrozoa
b. malpighian tubules—Insecta
c. flame cells—Turbellaria
d. thin region of cuticle—Crustacea
e. metanephridia—Oligochaeta

b
Application

29.3 A new species of marine animal is discovered with the following characteristics: bilateral symmetry; pseudocoelom; complete digestive tract. Further examination of this organism would probably show that
a. in embryological development, its blastopore becomes an anus.
b. it has a blood vascular system.
c. it has cnidocytes containing nematocysts.
d. it has an external skeleton (shell).
e. it is segmented.

c
Factual Recall

29.4 All of the following animal groups have evolved terrestrial life forms EXCEPT
a. Mollusca.
b. Crustacea.
c. Echinodermata.
d. Arthropoda.
e. Vertebrata.

d
Application

29.5 A new species of terrestrial animal is discovered with the following characteristics: exoskeleton; tracheal system for gas exchange; modified segmentation. A knowledgeable zoologist would predict that it probably also would have
a. eight legs.
b. a water vascular system.
c. a sessile lifestyle.
d. wings.
e. parapodia.

For Questions 29.6–29.8, match the phrases with the choices below. Each choice may be used once, more than once, or not at all.

    a. Cnidaria
    b. Annelida
    c. Mollusca
    d. Arthropoda
    e. Echinodermata

d
Factual Recall

29.6 Protostomes that have an open circulatory system and an exoskeleton of chitin.

e
Factual Recall

29.7 Deuterostomes that have an internal skeleton.

b
Factual Recall

29.8 Protostomes that have a closed circulatory system and true segmentation.

b
Factual Recall

29.9 A radially symmetrical animal that has two embryonic germ layers belongs to which phylum?
    a. Porifera (Parazoa)
    b. Cnidaria
    c. Platyhelminthes
    d. Aschelminthes
    e. Echinodermata

c
Factual Recall

29.10 An arthropod has all the following characteristics EXCEPT
    a. protostome development.
    b. bilateral symmetry.
    c. pseudocoelom.
    d. three embryonic germ layers.
    e. true tissues.

*Use Figure 29.1 to answer Questions 29.11–29.14. The following chart of the animal king-
dom is set up as a phylogenetic tree. The dashed lines on the tree represent certain evolutionary changes. For example, above line x, all organisms are triploblastic (three germ
layers). Place each of the characteristics in the questions that follow on the appropriate line
in the chart.*

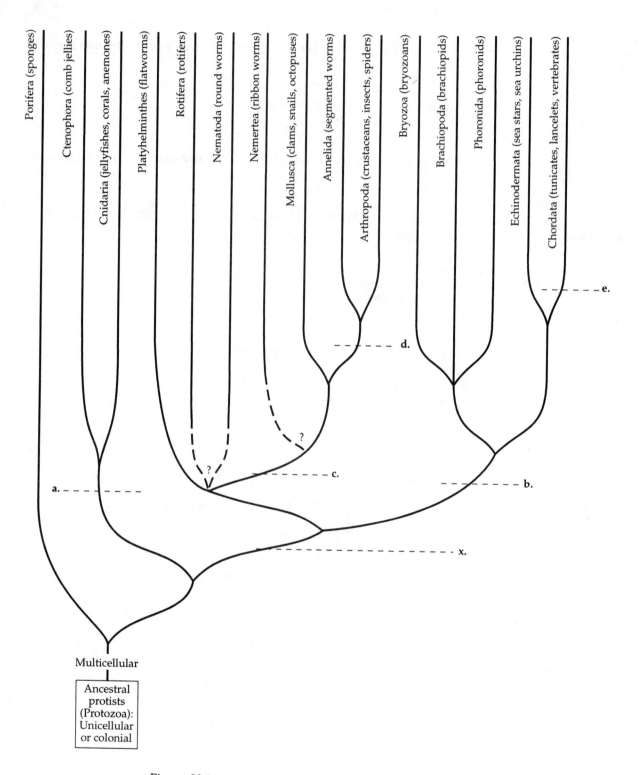

Figure 29.1

a
Application

29.11 Radial symmetry.

c
Application

29.12 Protostome development—schizocoelomate.

b
Application

29.13 Deuterostome development—enterocoelomate.

e
Factual Recall

29.14 Notochord.

b
Factual Recall

29.15 All of the following are protostomes EXCEPT
    a. mollusks.
    b. echinoderms.
    c. segmented worms.
    d. insects.
    e. spiders.

Figure 29.2 shows a chart of the Animal kingdom set up as a modified phylogenetic tree.
Use the diagram to answer Questions 29.16–29.18.

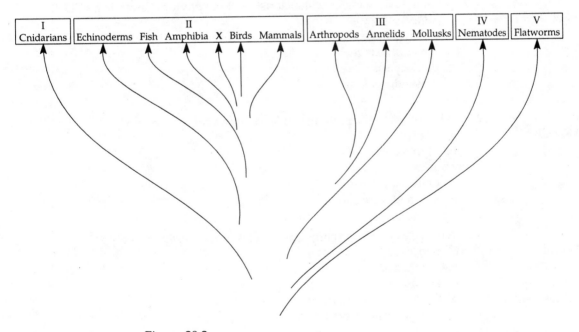

Figure 29.2

d
Factual Recall

29.16 A pseudocoelom is characteristic of which of the following groups?
    a. I
    b. II
    c. III
    d. IV
    e. V

e
Factual Recall

29.17 One opening to the digestive tract is characteristic of which of the following groups?
a. I only
b. III only
c. IV only
d. V only
e. I and V

b
Factual Recall

29.18 Which group includes organisms that are deuterostomes with radial symmetry?
a. I
b. II
c. III
d. IV
e. V

b
Factual Recall

29.19 Which of the following characteristics correctly applies to protosome development?
a. radial cleavage
b. determinate cleavage
c. enterocoelous
d. blastopore becomes the anus
e. archenteron absent

e
Factual Recall

29.20 All of the following are characteristics of the phylum Cnidaria EXCEPT
a. a gastrovascular cavity.
b. a polyp stage.
c. a medusa stage.
d. cnidocytes.
e. a pseudocoelom.

c
Factual Recall

29.21 Which class of the phylum Cnidaria occurs *only* as a polyp?
a. hydrozoa
b. scyphozoa
c. anthozoa
d. Only a and c are correct.
e. a, b, and c are correct.

d
Factual Recall

29.22 Which class of the phylum Cnidaria occurs *primarily* as a polyp?
a. hydrozoa
b. scyphozoa
c. anthozoa
d. Only a and c are correct.
e. a, b, and c are correct.

b
Factual Recall

29.23 Which class of the phylum Cnidaria includes the animals commonly called jellyfish?
a. Hydrozoa
b. Scyphozoa
c. Anthozoa
d. Only a and c are correct.
e. a, b, and c are correct.

c
Factual Recall

29.24 Which of the following are NOT found in sponges?
  a. oscula
  b. spongocoels
  c. cnidocytes
  d. spicules
  e. amoebocytes

a
Factual Recall

29.25 In the phylum Platyhelminthes, which of the following classes is mostly nonparasitic?
  a. Turbellaria
  b. Trematoda
  c. Cestoda
  d. Only b and c are correct.
  e. a, b, and c are correct.

d
Factual Recall

29.26 In the phylum Platyhelminthes, which of the following classes is mostly parasitic?
  a. Turbellaria
  b. Trematoda
  c. Cestoda
  d. Only b and c are correct.
  e. a, b, and c are correct.

b
Factual Recall

29.27 In the phylum Platyhelminthes, which of the following classes has life cycles that are typically an alternation of sexual and asexual phases and may require an intermediate host?
  a. Turbellaria
  b. Trematoda
  c. Cestoda
  d. Only b and c are correct.
  e. a, b, and c are correct.

b
Factual Recall

29.28 All of the following correctly characterize nematodes EXCEPT that
  a. they play an important role in decomposition.
  b. they have both circular and longitudinal muscles.
  c. they have a pseudocoelom.
  d. they have a complete digestive system.
  e. they are often parasitic.

d
Factual Recall

29.29 Which molluscan class includes animals that undergo embryonic torsion?
  a. Polyplacophora
  b. Bivalvia
  c. Cephalopoda
  d. Gastropoda
  e. All molluscan classes have this characteristic.

b
Factual Recall

29.30 Which molluscan class includes clams?
  a. Polyplacophora
  b. Bivalvia
  c. Cephalopoda
  d. Gastropoda
  e. None of the above; clams are not mollusks.

c
Factual Recall

29.31 Which molluscan class includes the most "intelligent" invertebrates?
  a. Polyplacophora
  b. Bivalvia
  c. Cephalopoda
  d. Gastropoda
  e. Both b and c are equally "intelligent."

e
Factual Recall

29.32 Annelids are abundant and successful organisms characterized accurately by all of the following EXCEPT
  a. a hydrostatic skeleton.
  b. segmentation.
  c. a complete digestive system.
  d. some parasitic forms.
  e. a cuticle made of chitin.

e
Factual Recall

29.33 All of the following are characteristics of arthropods EXCEPT
  a. an exoskeleton.
  b. numerous species.
  c. jointed appendages.
  d. a diversity of gas exchange structures.
  e. a dorsal nerve cord.

a
Factual Recall

29.34 Which of the following is a characteristic of echinoderms?
  a. radial symmetry
  b. spiral cleavage
  c. incomplete digestive system
  d. external skeleton
  e. a lophophore

d
Conceptual
Understanding

29.35 All of the following have contributed to hypotheses about the origins of multicellular animals EXCEPT
  a. embryological development.
  b. increased atmospheric oxygen during the Cambrian.
  c. the ediacaran fauna.
  d. the vertebrate fossil record.
  e. the diversity of simple invertebrates early in the fossil record.

d
Conceptual
Understanding

29.36 Sponges are limited to feeding on small food particles because
  a. they have no mouth.
  b. they have an incomplete digestive tract.
  c. their cell membranes are highly selective.
  d. their digestion is entirely intracellular.
  e. they lack a mechanism for bringing food into their bodies.

b
Factual Recall

29.37 Muscles and nerves in their simplest forms occur in the
    a. sponges.
    b. cnidarians.
    c. nematodes.
    d. flatworms.
    e. ribbon worms.

e
Conceptual
Understanding

29.38 The best way to describe the brain of a sea anemone would be as
    a. a thick ring around the mouth.
    b. a series of ganglia at the bases of the tentacles.
    c. a pair of ganglia at the anterior end.
    d. a single ganglion in the body wall.
    e. nonexistent.

d
Factual Recall

29.39 Which characteristic is shared by both cnidarians and flatworms?
    a. Both b and d below are correct.
    b. flame cells
    c. radial symmetry
    d. a gut with a single opening
    e. dorsoventrally flattened bodies

b
Factual Recall

29.40 One method of reducing the incidence of blood flukes in a human population
would be to
    a. reduce the mosquito population.
    b. reduce the freshwater snail population.
    c. purify all drinking water.
    d. ensure that all meat is properly cooked.
    e. carefully wash all raw fruits and vegetables.

b
Factual Recall

29.41 The larvae of many common human tapeworms are usually found
    a. encysted in human muscle.
    b. encysted in the muscle of an animal such as a cow or pig.
    c. in the abdominal blood vessels of humans.
    d. in the human intestine.
    e. in the intestines of cows and pigs.

b
Factual Recall

29.42 While snorkeling, a student observes an active marine animal that has a series
of muscular tentacles bearing suckers associated with its head. There is no
evidence of segmentation, but a pair of large, well-developed eyes are
evident. The student is observing an animal belonging to the class
    a. Gastropoda.
    b. Cephalopoda.
    c. Polyplacophora.
    d. Polychaeta.
    e. Bivalvia.

d
Factual Recall

29.43 Among the invertebrates, arthropods are unique in possessing
    a. a notochord.
    b. ventral nerve cords.
    c. open circulation.
    d. jointed appendages.
    e. segmented bodies.

d
Factual Recall

29.44 The presence or absence of mandibles can be used to distinguish between
a. insects and centipedes.
b. insects and crustaceans.
c. insects and millipedes.
d. insects and spiders.
e. centipedes and millipedes.

b
Factual Recall

29.45 The possession of two pairs of antennae will distinguish
a. spiders from insects.
b. crustaceans from insects.
c. millipedes from centipedes.
d. millipedes from insects.
e. insects from centipedes.

a
Application

29.46 While working in your garden, you uncover an animal with many legs, mostly as two pairs per segment. The animal must be a
a. millipede.
b. caterpillar.
c. centipede.
d. polychaete worm.
e. sow bug.

d
Factual Recall

29.47 The developmental process of which of the following animals involves a process called incomplete metamorphosis?
a. starfish
b. butterfly
c. spider
d. grasshopper
e. crayfish

b
Application

29.48 You find a small animal with eight legs crawling up your bedroom wall. Closer examination will reveal that this animal has
a. Both c and e below are correct.
b. Both d and e below are correct.
c. antennae.
d. no antennae.
e. chelicera.

b
Factual Recall

29.49 Both echinoderms and cnidarians
a. Both d and e below are correct.
b. are radially symmetrical.
c. are segmented.
d. have stinging cells.
e. have three embryonic tissue layers.

Answer Questions 29.50–29.54 with the choices below. Each choice may be used once, more than once, or not at all.
a. *class Crinoidea (sea lilies)*
b. *class Asteroidea (sea stars)*
c. *class Ophiuroidea (brittle stars)*
d. *class Echinoidea (sea urchins and sand dollars)*
e. *class Holothuroidea (sea cucumbers)*

b
Factual Recall

29.50 They can evert their stomach through their mouth to feed.

c
Factual Recall

29.51 They have distinct central disks and long, flexible arms.

e
Factual Recall

29.52 They are elongated in the oral–aboral axis.

a
Factual Recall

29.53 Their mouth is directed upward.

d
Factual Recall

29.54 They have long, movable spines.

a
Factual Recall

29.55 All of the major body plans we see today appeared in the fossil record 5 million to 10 million years ago at the beginning of the
   a. Cambrian period.
   b. Ediacaran period.
   c. Burgess period.
   d. Carboniferous period.
   e. Cretaceous period.

e
Conceptual
Understanding

29.56 Cephalization is primarily
   a. Both b and c below are correct.
   b. an adaptation to the method of feeding.
   c. due to the fate of the blastopore.
   d. the result of the type of digestive system.
   e. an adaptation to movement.

b
Conceptual
Understanding

29.57 Humans can acquire trichinosis by
   a. failing to practice safe sex.
   b. eating undercooked pork.
   c. inhaling the eggs of worms.
   d. eating undercooked beef.
   e. being bitten by tsetse flies.

c
Application

29.58 While sampling marine plankton, a student encounters large numbers of eggs in her samples. She rears some of the eggs in the laboratory for further study and finds that the blastopore becomes the mouth in a complete digestive system. The embryo develops into a trochophore larva and eventually has a coelom and open circulation. These eggs belonged to a(n)
   a. annelid.
   b. echinoderm.
   c. mollusk.
   d. nemertine.
   e. arthropod.

Chapter 30

c
Factual Recall

30.1 Which of the following is a shared characteristic of all chordates?
   a. hair
   b. skull
   c. notochord
   d. vertebral column
   e. four-chambered heart

b
Factual Recall

30.2 Pharyngeal gill slits appear to have functioned first as
   a. the opening to the digestive system or mouth.
   b. suspension-feeding devices.
   c. components of the jaw.
   d. gill slits for respiration.
   e. portions of the inner ear.

e
Factual Recall

30.3 Jaws first occurred in which of the following classes?
   a. Agnatha
   b. Chondrichthyes
   c. Osteichthyes
   d. Ostracodermi
   e. Placodermi

a
Factual Recall

30.4 Which chordate group is postulated to be most like the earliest chordates?
   a. Cephalochordata
   b. adult Urochordata
   c. Amphibia
   d. Reptilia
   e. Chondrichthyes

b
Application

30.5 A new species of aquatic chordate is discovered that closely resembles an ancient form. It has the following characteristics: external "armor" of bony plates; no paired fins; and a suspension-feeding mode of nutrition. In addition to these characteristics, it will probably have which of the following characteristics?
   a. legs
   b. no jaws
   c. an amniotic egg
   d. metamorphosis
   e. endothermy

d
Conceptual
Understanding

30.6 The most primitive hominid
   a. may have hunted dinosaurs.
   b. lived 1.2 million years ago.
   c. closely resembled a chimpanzee.
   d. walked on two legs.
   e. had a relatively large brain.

b
Factual Recall

30.7 What is one characteristic that separates chordates from all other animals?
   a. true coelom
   b. hollow dorsal nerve cord
   c. blastopore, which becomes the anus
   d. bilateral symmetry
   e. segmentation

d
Factual Recall

30.8 Which of the following is characteristic of all chordates?
   a. a jointed endoskeleton
   b. a ventral nerve cord
   c. gills
   d. a notochord
   e. two pairs of appendages

e
Factual Recall

30.9 Which of the following is a feature that sets primates apart from all other mammals?
   a. placental embryonic development
   b. hairy bodies
   c. naked faces
   d. ability to produce milk
   e. opposable thumbs in many species

a
Factual Recall

30.10 With which of the following statements would a biologist be most inclined to agree?
   a. Humans and apes probably represent divergent lines of evolution from common ancestors.
   b. Humans evolved from New World monkeys.
   c. Humans have stopped evolving and now represent the pinnacle of evolution.
   d. Apes evolved from humans.
   e. Humans and apes are the result of disruptive selection in a species of *Gorilla*.

For Questions 30.11–30.15, match the characteristic or description with the class. Each choice may be used once, more than once, or not at all.
   a. *Amphibia*
   b. *Aves*
   c. *Chondrichthyes*
   d. *Mammalia*
   e. *Reptilia*

c
Factual Recall

30.11 Some members have a cartilaginous endoskeleton.

e
Factual Recall

30.12 Internal fertilization, amniotic egg, skin that resists drying, evolved in late Carboniferous.

d
Factual Recall

30.13 Three major groups: egg-laying, pouched, and placental.

a
Factual Recall

30.14 Includes salamanders, frogs, and toads.

e
Factual Recall

30.15 Includes snakes, turtles, and lizards.

a
Factual Recall

30.16 Which one of the following has a two-chambered heart?
    a. Osteichthyes
    b. Amphibia
    c. Reptilia
    d. Aves
    e. Mammalia

d
Conceptual
Understanding

30.17 Why is the term *cold-blooded* not very appropriate for reptiles?
    a. The keratinized skin of reptiles serves to insulate and conserve heat.
    b. The metabolism of reptiles can generate internal heat for temperature control.
    c. The scales of reptiles serve to rid excess body heat by reradiation to the environment.
    d. Reptiles regulate body temperature by using various mechanisms such as behavioral adaptations.
    e. Reptiles swallow large prey whole to provide enough food to generate body heat.

e
Conceptual
Understanding

30.18 Structures that are made of keratin include which of the following?
    I. avian feathers
    II. reptilian scales
    III. mammalian hair

    a. I only
    b. III only
    c. I and II
    d. II and III
    e. I, II, and III

c
Factual Recall

30.19 From which of the following groups are snakes most likely descended?
    a. dinosaurs
    b. plesiosaurs
    c. lizards
    d. crocodiles
    e. synapsids

e
Conceptual
Understanding

30.20 Which of the following statements about human evolution is CORRECT?
    a. Modern humans are the only human species to have evolved on Earth.
    b. Human ancestors were virtually identical to chimpanzees.
    c. Human evolution occurred by phyletic change within an unbranched lineage.
    d. The upright posture and enlarged brain of humans evolved simultaneously.
    e. Fossil evidence indicates that early anthropoids were arboreal, were cat-sized, and lived about 35 million years ago.

c
Conceptual
Understanding

30.21 Which sequence of evolutionary relationships is consistent with the fossil
      record?
      a. reptiles → amphibians → birds → fishes
      b. reptiles → birds → fishes → amphibians
      c. fishes → amphibians → reptiles → birds
      d. fishes → birds → reptiles → amphibians
      e. reptiles → birds → amphibians → fishes

d
Factual Recall

30.22 When did the first amphibians appear on Earth?
      a. Eocene epoch
      b. Paleocene epoch
      c. Precambrian era
      d. Devonian period
      e. Permian period

b
Factual Recall

30.23 All of the following are reptilian characteristics EXCEPT
      a. ectothermy.
      b. brachiation.
      c. amniote egg.
      d. keratinized skin.
      e. conical teeth that are relatively uniform in size.

a
Factual Recall

30.24 When did dinosaurs and pterosaurs become extinct?
      a. Cretaceous "crisis"
      b. Permian extinctions
      c. Devonian "disaster"
      d. Phanerozoic eon
      e. Hadeon eon

b
Factual Recall

30.25 Which of the following is the era known as the "age of reptiles"?
      a. Cenozoic
      b. Mesozoic
      c. Paleozoic
      d. Precambrian
      e. Cambrian

b
Factual Recall

30.26 Which of the following structures is or are characteristic of vertebrates?
      a. c, d, and e below are all correct.
      b. d and e below are both correct.
      c. open circulation
      d. pharyngeal slits
      e. dorsal, hollow nerve cord

a
Conceptual
Understanding

30.27 Which of the following statements about mammalian evolution is correct?
      a. Mammals evolved from reptilian stocks even earlier than birds.
      b. The first mammals were large predators like the saber-tooth tigers.
      c. Mammals did not coexist with the dominant dinosaurs.
      d. The early mammals were most similar to small, bipedal, ratite birds.
      e. Mammals evolved from the marsupials during the Pleistocene epoch.

a
Conceptual
Understanding

30.28 Why is it thought that the Neanderthals contributed little to the gene pool of modern humanity?
   a. Recent studies of the DNA found in mitochondria of modern humans have shown incredible similarities among modern populations.
   b. The fossils found in the Neander Valley were a hoax and the "Neanderthals" never really existed.
   c. Neanderthals had degenerated brain capacity and thus could not have contributed to human ancestry.
   d. There is no evidence that Neanderthals were capable of walking upright or using tools.
   e. Humans, the "naked apes," have nothing in common with the Neanderthals, the "hairy apes."

c
Factual Recall

30.29 The amniote egg first evolved in which of the following groups?
   a. fish
   b. birds
   c. reptiles
   d. amphibians
   e. egg-laying mammals (monotremes)

d
Factual Recall

30.30 Which of the following are the only modern animals that descended directly from dinosaurs?
   a. lizards
   b. crocodiles
   c. snakes
   d. birds
   e. mammals

b
Factual Recall

30.31 Which of the following classifications would not apply to both dogs and humans?
   a. class Mammalia
   b. order Primata
   c. phylum Chordata
   d. kingdom Animalia
   e. subphylum Vertebrata

b
Conceptual
Understanding

30.32 Humans and apes are presently classified in the same category at all of the following levels EXCEPT
   a. class.
   b. genus.
   c. kingdom.
   d. order.
   e. phylum.

c
Factual Recall

30.33 Which of the following are not considered apes?
   a. gibbon
   b. gorilla
   c. lemur
   d. orangutan
   e. chimpanzee

c
Factual Recall

30.34 Which of the following structures are possessed by birds only?
  a. enlarged pectoral muscles and a four-chambered heart
  b. light bones and a four-chambered heart
  c. feathers and carinate sternum
  d. a short tail and mammary glands
  e. a large brain and endothermy

Questions 30.35–30.39 refer to the phylogenetic tree shown in Figure 30.1.

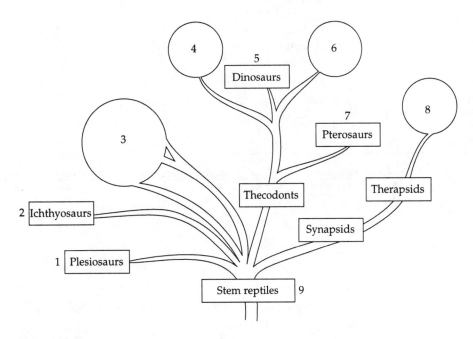

Figure 30.1

b
Conceptual
Understanding

30.35 The organisms represented by 8 most likely are
  a. birds.
  b. all mammals.
  c. flying reptiles.
  d. modern reptiles.
  e. all mammals except humans.

a
Conceptual
Understanding

30.36 Which organisms are represented by 6?
  a. birds
  b. all mammals
  c. flying reptiles
  d. modern reptiles
  e. all mammals except humans

e
Conceptual
Understanding

30.37 Which number represents Cotylosaurs?
  a. 3
  b. 4
  c. 6
  d. 8
  e. 9

a
Conceptual
Understanding

30.38 Which pair of numbers represents extinct reptiles that had returned to an aquatic life?
   a. 1 and 2
   b. 3 and 4
   c. 5 and 7
   d. 6 and 8
   e. 7 and 9

b
Conceptual
Understanding

30.39 Which pair of numbers most likely represents modern reptiles?
   a. 1 and 2
   b. 3 and 4
   c. 5 and 7
   d. 6 and 8
   e. 7 and 9

Questions 30.40 and 30.41 refer to Figure 30.2.

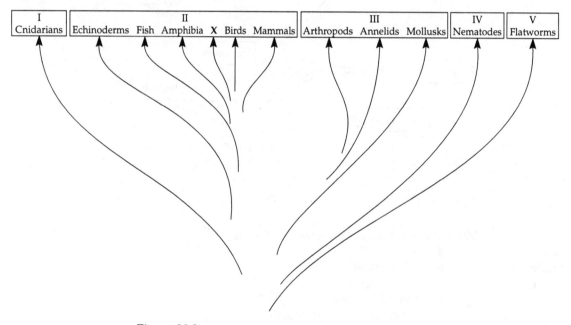

Figure 30.2

c
Application

30.40 Which of the following is a correct description of the class represented by X?
   a. endotherms; skin covered with feathers; embryo enclosed in protective membranes and a shell
   b. endotherms; skin usually covered with hair; young nourished with milk secreted by the mother
   c. ectotherms; skin usually covered with scales; embryo enclosed in protective membranes and a shell
   d. ectotherms; naked skin; embryo not protected by membranes or a shell
   e. ectotherms; bony skeleton; breathe by means of gills; usually have an air bladder

c
Factual Recall

30.41 Excluding the echinoderms, all the animals in group II have which of the following characteristics in common?
    a. protostome embryo; coelom; radial symmetry; dorsal nerve cord
    b. protostome embryo; no coelom; bilateral symmetry; ventral nerve cord
    c. deuterostome embryo; coelom; bilateral symmetry; dorsal nerve cord
    d. deuterostome embryo; pseudocoelom; asymmetry; ventral nerve cord
    e. deuterostome embryo; no coelom; radial symmetry; dorsal nerve cord

d
Conceptual
Understanding

30.42 Which of the following statements would be LEAST acceptable to most zoologists?
    a. Modern cephalochordates are contemporaries of vertebrates, not their ancestors.
    b. The first fossils resembling cephalochordates appeared in the fossil record at least 550 million years ago.
    c. Recent work in molecular systematics supports the hypothesis that cephalochordates are the closest relatives of the vertebrates.
    d. The modern cephalochordates are the immediate ancestors of the vertebrates.
    e. Cephalochordates display the same method of swimming as do fishes.

b
Factual Recall

30.43 The lobe-finned fishes appear to have given rise directly to
    a sharks.
    b. amphibians.
    c. stem reptiles.
    d. freshwater ray-finned fishes.
    e. placoderms.

b
Factual Recall

30.44 The jaws of vertebrates were derived by the modification of
    a. scales of the lower lip.
    b. one or more gill arches.
    c. one or more gill slits.
    d. one or more of the bones of the cranium.
    e. one or more of the vertebrae.

b
Factual Recall

30.45 The swim bladder of modern bony fishes
    a. both c and e below are correct.
    b. was modified from simple lungs.
    c. developed into lungs in some fishes.
    d. first appeared in sharks.
    e. provides buoyancy but at a high energy cost.

e
Factual Recall

30.46 Bony fishes (Osteichthyes) originally evolved
    a. in a marine environment.
    b. directly from lampreys and hagfish.
    c. early in the Cambrian.
    d. directly from cephalochordates.
    e. in freshwater environments.

d
Factual Recall

30.47 In which classes of vertebrates is fertilization exclusively internal?
   a. Chondricthyes, Osteichthyes, and Mammalia
   b. Amphibia, Mammalia, and Aves
   c. Chondrichthyes, Osteichthyes, and Reptilia
   d. Reptilia, Aves, and Mammalia
   e. Mammalia, Aves, and Amphibia

d
Factual Recall

30.48 Examination of the fossils of *Archaeopteryx* reveals that in common with modern birds it had
   a. both c and d are correct.
   b. both d and e are correct.
   c. a long tail containing vertebrae.
   d. feathers.
   e. teeth.

e
Factual Recall

30.49 Differentiation of teeth is greatest in
   a. sharks.
   b. bony fishes.
   c. amphibians.
   d. reptiles.
   e. mammals.

d
Factual Recall

30.50 A sheet of muscle called the diaphragm is found in
   a. both c and d below are correct.
   b. both d and e below are correct.
   c. birds.
   d. mammals.
   e. reptiles.

b
Factual Recall

30.51 Which is NOT characteristic of all mammals?
   a. a four-chambered heart that prevents mixing of oxygenated and deoxygenated blood
   b. give birth to live young (viviparous)
   c. have hair during at least some period of their life
   d. have glands to produce milk to nourish their offspring
   e. have a diaphragm to assist in ventilating the lungs

b
Factual Recall

30.52 All of the following are trends in primate evolution EXCEPT
   a. enhanced depth perception.
   b. well-developed claws for clinging to trees.
   c. a shoulder joint adapted to brachiation.
   d. reduction of the number of young to a single birth at one time.
   e. a long period of parental care of the offspring.

e
Conceptual
Understanding

30.53 The adaptation to arboreal life by early human ancestors can explain, at least in part, all of the following human characteristics EXCEPT
a. limber shoulder joints.
b. dextrous hands with opposable thumbs.
c. excellent eye–hand coordination.
d. enhanced depth perception.
e. reduced body hair.

a
Factual Recall

30.54 The major and dramatic alteration of hominid anatomy was primarily the result of
a. an upright stance.
b. enlargement of the brain.
c. protracted postnatal development of offspring.
d. the development of speech.
e. the adoption of tool use.

a
Factual Recall

30.55 *Australopithecus africanus* and *Homo habilis* differed in that
a. Both d and e below are correct.
b. *Homo habilis* had a smaller brain.
c. *Homo habilis* used tools while *Australopithecus africanus* did not.
d. *Homo habilis* was bipedal while *Australopithecus africanus* was not.
e. *Homo habilis* had an opposable thumb while *Australopithecus africanus* did not.

d
Factual Recall

30.56 Based on current evidence, which of the following statements best describes the evolution of humans?
a. Humans evolved from the chimpanzee.
b. Humans evolved in a single, orderly series of stages in which each stage became more advanced than its predecessor.
c. The various characteristics that we associate with humans evolved in unison over long periods of time.
d. Humans and apes diverged from a common ancestor about 6–8 million years ago.
e. Humans are more closely related to gorillas than to chimpanzees.

Chapter 31

e
Factual Recall

31.1 Vascular plant tissue includes all of the following cell types EXCEPT
   a. vessels.
   b. sieve cells.
   c. tracheids.
   d. companion cells.
   e. cambium cells.

c
Factual Recall

31.2 Which of the following is incorrectly paired with its structure and function?
   a. sclerenchyma—supporting cells with thick secondary walls
   b. periderm—protective coat of woody stems and roots
   c. pericycle—waterproof ring of cells surrounding central stele in roots
   d. mesophyll—parenchyma cells functioning in photosynthesis in leaves
   e. ground meristem—primary meristem that produces ground tissue system

b
Factual Recall

31.3 Which of the following is a true statement about growth in plants?
   a. Only primary growth is localized at meristems.
   b. Some plants lack secondary growth.
   c. Only stems have secondary growth.
   d. Only secondary growth produces reproductive structures.
   e. Monocots have only primary growth and dicots have only secondary growth.

a
Factual Recall

31.4 Land plants are composed of all of the following tissue types EXCEPT
   a. mesoderm.
   b. epidermal.
   c. meristematic.
   d. vascular.
   e. ground tissue.

e
Factual Recall

31.5 Which of the following are primary meristems?
   a. procambium
   b. protoderm
   c. ground meristem
   d. Only a and c are correct.
   e. a, b, and c are correct.

d
Conceptual
Understanding

31.6 What effect does "pinching back" a house plant have on the plant?
   a. increases apical dominance
   b. inhibits the growth of lateral buds
   c. produces a plant that will grow taller
   d. produces a plant that will grow fuller
   e. increases the flow of auxin down the shoot

b
Factual Recall

31.7 Most of the water and minerals taken up from the soil by a plant are absorbed by
  a. taproots.
  b. root hairs.
  c. the thick parts of the roots near the base of the stem.
  d. storage roots.
  e. sections of the root that have secondary xylem.

a
Factual Recall

31.8 The photosynthetic cells in the interior of a leaf are what kind of cells?
  a. parenchyma
  b. collenchyma
  c. sclerenchyma
  d. phloem
  e. endodermis

e
Factual Recall

31.9 Which functional plant cells lack a nucleus?
  a. xylem only
  b. sieve tube cells only
  c. companion cells only
  d. both companion and parenchyma cells
  e. both xylem and sieve tube cells

c
Factual Recall

31.10 The largest organelle in most mature living plant cells is the
  a. chloroplast.
  b. nucleus.
  c. central vacuole.
  d. dictyosome (Golgi apparatus).
  e. mitochondrion.

b
Factual Recall

31.11 The fiber cells of plants are a type of
  a. parenchyma.
  b. sclerenchyma.
  c. collenchyma.
  d. meristematic cell.
  e. xylem cell.

c
Conceptual
Understanding

31.12 One important difference between the anatomy of roots and the anatomy of leaves is that
  a. only leaves have phloem and only roots have xylem.
  b. the cells of roots have cell walls that are lacking in leaf cells.
  c. a waxy cuticle covers leaves but is absent in roots.
  d. vascular tissue is found in roots but is absent from leaves.
  e. leaves have surface tissue (epidermis), while such tissue is absent from roots.

c
Factual Recall

31.13 What does primary phloem in the root develop from?
  a. protoderm
  b. endoderm
  c. procambium
  d. ground tissue
  e. vascular cambium

b
Factual Recall

31.14 What tissue makes up most of the wood of a tree?
    a. primary xylem
    b. secondary xylem
    c. secondary phloem
    d. mesophyll cells
    e. vascular cambium

e
Conceptual
Understanding

31.15 All of the following cell types are correctly matched with their functions EXCEPT
    a. mesophyll—photosynthesis.
    b. guard cell—regulation of transpiration.
    c. sieve-tube member—translocation.
    d. vessel element—water transport.
    e. companion cell—formation of secondary xylem and phloem.

e
Factual Recall

31.16 Which of the following root tissues gives rise to secondary roots?
    a. endodermis
    b. phloem
    c. cortex
    d. epidermis
    e. pericycle

a
Conceptual
Understanding

31.17 Additional vascular tissue produced as secondary growth in a root originates from which cells?
    a. vascular cambium
    b. apical meristem
    c. endodermis
    d. phloem
    e. xylem

e
Conceptual
Understanding

31.18 The vascular system of a three-year-old dicot stem consists of
    a. 3 rings of xylem and 3 of phloem.
    b. 2 rings of xylem and 2 of phloem.
    c. 2 rings of xylem and 1 of phloem.
    d. 2 rings of xylem and 3 of phloem.
    e. 3 rings of xylem and 1 of phloem.

a
Factual Recall

31.19 All of the following are primary meristems of a plant EXCEPT
    a. epidermis only.
    b. protoderm only.
    c. ground meristem only.
    d. procambium only.
    e. both procambium and protoderm.

e
Factual Recall

31.20 All of the following are derived from ground meristem EXCEPT
    a. collenchyma.
    b. sclerenchyma.
    c. parenchyma.
    d. sclereids.
    e. phloem.

Questions 31.21–31.24 are based on the drawing of root or stem cross sections shown in Figure 31.1.

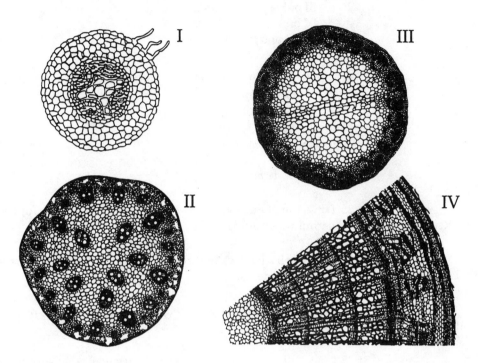

Figure 31.1

a
Conceptual
Understanding

31.21 Endodermis is present in
  a. I only.
  b. II only.
  c. III only.
  d. IV only.
  e. both I and III.

d
Conceptual
Understanding

31.22 A woody dicot is represented by
  a. I only.
  b. II only.
  c. III only.
  d. IV only.
  e. both I and III.

b
Conceptual
Understanding

31.23 A monocot stem is represented by
  a. I only.
  b. II only.
  c. III only.
  d. IV only.
  e. both I and III.

d
Conceptual
Understanding

31.24 A plant that is at least three years old is represented by
    a. I only.
    b. II only.
    c. III only.
    d. IV only.
    e. both I and III.

Questions 31.25–31.29 use the following answers. Each answer may be used once, more than once, or not at all.
    a. *parenchyma*
    b. *collenchyma*
    c. *sclerenchyma*
    d. *tracheids*
    e. *sieve-tube member*s

d
Factual Recall

31.25 Long, thin, tapered cells with lignified cell walls that function in support and permit water flow through pits.

e
Factual Recall

31.26 Living cells that lack nuclei and ribosomes; they transport sucrose and other organic nutrients.

a
Factual Recall

31.27 The least specialized plant cells, which serve general metabolic, synthetic, and storage functions.

b
Factual Recall

31.28 Cells with unevenly thickened primary walls that support young parts of the plant.

c
Factual Recall

31.29 Mature cells without protoplasts but with thick, lignified secondary walls that may form fibers or sclereids.

c
Conceptual
Understanding

31.30 In a root, the ground meristem differentiates and forms the
    a. epidermis only.
    b. cork cambium only.
    c. cortex only.
    d. procambium only.
    e. epidermis and procambium.

d
Conceptual
Understanding

31.31 In which of the following is a vascular cambium most likely to be found?
    a. herbaceous dicots only
    b. woody dicots only
    c. monocots only
    d. herbaceous and woody dicots
    e. monocots and herbaceous dicots

a
Application

31.32 A student examining leaf cross sections under a microscope finds many loosely packed cells with relatively thin cell walls. The cells have numerous chloroplasts. What cells are these?
    a. parenchyma
    b. xylem
    c. endodermis
    d. collenchyma
    e. sclerenchyma

a
Factual Recall

31.33 Which of the following tissues is incorrectly matched with its characteristics?
    a. collenchyma—uniformly thick-walled supportive tissue
    b. epidermis—protective outer covering of plant body
    c. sclerenchyma—heavily lignified secondary walls
    d. meristematic tissue—undifferentiated tissue capable of cell division
    e. parenchyma—thin-walled, loosely packed, unspecialized cells

a
Factual Recall

31.34 Which of the following is NOT a normal function of roots?
    a. photosynthesis
    b. anchoring the plant
    c. absorbing minerals
    d. storing food
    e. conducting water

d
Factual Recall

31.35 Pores on the leaf surface that function in gas exchange are called
    a. hairs.
    b. xylem cells.
    c. phloem cells.
    d. stomata.
    e. sclereids.

b
Application

31.36 You are studying a plant from the arid southwestern United States. Which of the following adaptations is LEAST likely to have evolved in response to water shortages?
    a. closing the stomata during the hottest time of the day
    b. development of large leaf surfaces to absorb water
    c. formation of a fibrous root system spread over a large area
    d. mycorrhizae associated with the root system
    e. a thick waxy cuticle on the epidermis

e
Factual Recall

31.37 Plant physiologists often incubate plant tissue in an extract of termite gut to dissolve the cell wall. After this incubation treatment, the structure left is called
    a. the parenchyma.
    b. a guard cell.
    c. a companion cell.
    d. the endodermis.
    e. a protoplast.

e
Factual Recall

31.38 Which of the following is NOT a characteristic of parenchyma cells?
    a. thin primary walls
    b. flexible primary walls
    c. unspecialized
    d. lack secondary walls
    e. little metabolism and synthesis

c
Factual Recall

31.39 An evolutionary adaptation that increases exposure of a plant to light in a dense forest is
    a. closing the stomates.
    b. lateral buds.
    c. apical dominance.
    d. absence of petioles.
    e. intercalary meristems.

a
Factual Recall

31.40 Bark
    a. is composed of phloem plus periderm.
    b. is associated with annuals but not perennials.
    c. is formed by apical meristems.
    d. has no identifiable function in trees.
    e. forms annual rings.

a
Application

31.41 If you were able to walk into an opening cut into the center of a large redwood tree, when you exit from the middle of the trunk (stem) outward, you would cross, in order,
    a. the annual rings, phloem, and bark.
    b. the newest xylem, oldest phloem, and periderm.
    c. the vascular cambium, oldest xylem, and newest xylem.
    d. the secondary xylem, secondary phloem, and vascular cambium.
    e. the summer wood, bark, and phloem.

b
Application

31.42 As a youngster, you drive a nail in the trunk of a young tree that is 3 meters tall. The nail is about 1.5 meters from the ground. Fifteen years later, you return and discover the tree has grown to a height of 30 meters. The nail is now _____ meters above the ground.
    a. 0.5
    b. 1.5
    c. 3.0
    d. 15.0
    e. 28.5

c
Application

31.43 A friend has discovered a new plant and brings it to you to describe. The plant has the following characteristics: a fibrous root system; no petioles; parallel leaf veins; thick, lignified cell walls; and a vascular cambium. Which of the following best describes the new plant?
    a. herbaceous dicot
    b. woody dicot
    c. woody monocot
    d. herbaceous monocot
    e. woody annual

a
Application

31.44 The gardener on your campus regularly removes apical dominance by
    a. pruning.
    b. deep watering of the roots.
    c. fertilizing.
    d. transplanting.
    e. feeding the plants nutrients.

e
Factual Recall

31.45 Which of the following is NOT a fundamental difference between monocot and dicot morphology and anatomy? Monocots have _____, while dicots have _____.
    a. one cotyledon; two cotyledons
    b. parallel veins; net veins
    c. fibrous roots; tap roots
    d. vascular bundles in a ring; vascular bundles scattered throughout the stem
    e. long petioles; no petioles

Chapter 32

b
Factual Recall

32.1 Which of the following is true concerning the water potential of a plant cell?
a. It is higher than that of air.
b. It is equal to zero when the cell is in pure water and is turgid.
c. It is equal to 0.23 MPa.
d. It becomes higher when K^+ ions are actively moved into the cell.
e. It becomes lower after the uptake of water by osmosis.

d
Factual Recall

32.2 Hydrophytes, plants that are adapted to live in aquatic habitats, are most likely to show which of the following morphologies?
a. no vascular tissue
b. no roots
c. leaves reduced to spines
d. stomata located on the top of leaves
e. stomata located in pits

d
Application

32.3 If the guard cells and surrounding epidermal cells in a plant are deficient in potassium ions, all of the following would occur EXCEPT
a. photosynthesis would decrease.
b. roots would take up less water.
c. phloem transport rates would decrease.
d. leaf temperatures would decrease.
e. wilting would become more likely.

a
Factual Recall

32.4 The opening of stomates is thought to involve
a. an increase in the osmotic concentration of the guard cells.
b. a decrease in the osmotic concentration of the stoma.
c. active transport of water out of the guard cells.
d. decreased turgor pressure in guard cells.
e. movement of K^+ out of guard cells.

d
Factual Recall

32.5 The following factors may sometimes play a role in the movement of sap through xylem. Which one depends upon the direct expenditure of ATP by the plant?
a. capillarity of water within the xylem
b. evaporation of water from leaves
c. cohesion among water molecules
d. concentration of ions in the symplast
e. bulk flow of water in the root apoplast

b
Factual Recall

32.6 Which one of the following statements about transport of nutrients in phloem is FALSE?
    a. Solute particles can be actively transported into phloem at the source.
    b. Companion cells control the rate and direction of movement of phloem sap.
    c. Differences in osmotic concentration at the source and sink cause a hydrostatic pressure gradient to be formed.
    d. A sink is that part of the plant where a particular solute is consumed or stored.
    e. A source may be located anywhere in the plant.

c
Conceptual
Understanding

32.7 Which of the following is responsible for the cohesion of water molecules?
    a. hydrogen bonds between the oxygen atoms of two adjacent water molecules
    b. covalent bonds between the hydrogen atoms of two adjacent water molecules
    c. hydrogen bonds between the oxygen atom of one water molecule and a hydrogen atom of another water molecule
    d. covalent bonds between the oxygen atom of one water molecule and a hydrogen atom of another water molecule
    e. Cohesion has nothing to do with the bonding but is the result of the tight packing of the water molecules in the xylem column.

c
Factual Recall

32.8 The water within xylem vessels moves toward the top of a tree as a result of
    a. active transport of ions into the stele.
    b. atmospheric pressure on roots.
    c. evaporation of water through stoma.
    d. the force of root pressure.
    e. osmosis in the root.

a
Factual Recall

32.9 Guttation in plants is a consequence of
    a. root pressure.
    b. transpiration.
    c. pressure flow in phloem.
    d. plant injury.
    e. condensation of atmospheric water.

a
Application

32.10 According to the pressure-flow hypothesis of phloem transport,
    a. solute moves from a high concentration in the "source" to a lower concentration in the "sink."
    b. water is actively transported into the "source" region of the phloem to create the turgor pressure needed.
    c. the combination of a high turgor pressure in the "source" and transpiration water loss from the "sink" moves solutes through phloem conduits.
    d. the formation of starch from sugar in the "sink" increases the osmotic concentration.
    e. the pressure in the phloem of a root is normally greater than the pressure in the phloem of a leaf.

e
Conceptual
Understanding

32.11 According to the pressure-flow model of translocation in plants, sieve tubes near photosynthetic cells are sites of
a. relatively high hydrostatic pressure.
b. relatively low hydrostatic pressure.
c. relatively high concentrations of organic nutrients.
d. active pumping of sucrose out of the sieve tube.
e. Both a and c are correct.

c
Conceptual
Understanding

32.12 Water entering the stele of the root from the cortex must pass through the
a. Casparian strip.
b. phloem.
c. endodermal cytoplasm.
d. epidermis.
e. xylem.

b
Application

32.13 George Washington completely removed the bark from around a cherry tree but was stopped by his father before cutting the tree down. It was noticed that the leaves retained their normal appearance for several weeks, but that the tree eventually died. The tissue(s) that George left functional was (were) the
a. phloem.
b. xylem.
c. cork cambium.
d. cortex.
e. companion and sieve cells.

b
Conceptual
Understanding

32.14 All of the following are factors in the movement of water through a terrestrial plant EXCEPT
a. hydrogen bonds linking water molecules.
b. the influence of plant hormones on cell expansion.
c. capillary action.
d. root pressure.
e. evaporation of water from the leaves.

c
Conceptual
Understanding

32.15 Arrange the following five events in an order that explains the mass flow of materials in the phloem.
1. Water diffuses into the sieve elements.
2. Leaf cells produce sugar by photosynthesis.
3. Solutes are actively transported into sieve elements.
4. Sugar is transported from cell to cell in the leaf.
5. Sugar moves down the stem.

a. 2, 1, 4, 3, 5
b. 1, 2, 3, 4, 5
c. 2, 4, 3, 1, 5
d. 4, 2, 1, 3, 5
e. 2, 4, 1, 3, 5

b
Application

32.16 Which of the following experimental procedures would most likely reduce transpiration while allowing normal growth of a plant?
  a. subjecting the leaves of the plant to a partial vacuum
  b. increasing the level of carbon dioxide around the plant
  c. putting the plant in drier soil
  d. decreasing the relative humidity around the plant
  e. injecting potassium ions into the guard cells of the plant

a
Conceptual
Understanding

32.17 Which of the following would be the best analogy for describing water movement in the xylem of a tree trunk?
  a. drinking through a soda straw
  b. turning on a faucet
  c. opening the flood gates of a dam
  d. pushing water with an oar
  e. pumping blood with a heart

b
Conceptual
Understanding

32.18 Active transport would be least important in the normal functioning of which of the following plant tissue types?
  a. leaf transfer cells
  b. stem xylem
  c. root endodermis
  d. leaf mesophyll
  e. root phloem

c
Application

32.19 Assume a particular chemical interferes with the establishment and maintenance of proton gradients across the membranes of plant cells. All of the following processes would be directly affected by this chemical EXCEPT
  a. photosynthesis.
  b. phloem loading.
  c. xylem transport.
  d. cellular respiration.
  e. stomatal opening.

d
Conceptual
Understanding

32.20 A water molecule could move all the way through a plant from soil to root to leaf to air and pass through a living cell only once. This living cell would be a part of which structure?
  a. the Casparian strip
  b. a guard cell
  c. the root epidermis
  d. the endodermis
  e. the root cortex

e
Application

32.21 Plants do not have a circulatory system like that of some animals. If a given water molecule did "circulate" (that is, go from one point in a plant to another and back), it would require the activity of
  a. only the xylem.
  b. only the phloem.
  c. only the endodermis.
  d. both the xylem and the endodermis.
  e. both the xylem and the phloem.

d
Conceptual
Understanding

32.22 Phloem transport of sucrose can be described as going from "source to sink." Which of the following would not normally function as a sink?
   a. growing leaf
   b. growing root
   c. storage organ in summer
   d. mature leaf
   e. shoot tip

d
Factual Recall

32.23 The water lost during transpiration is an unfortunate side effect of the plant's exchange of gases. However, the plant derives some benefit from this water loss in the form of
   a. evaporative cooling.
   b. mineral transport.
   c. increased turgor.
   d. Only a and b are correct.
   e. a, b, and c are correct.

b
Conceptual
Understanding

32.24 In which plant tissue would the pressure component of water potential often be negative?
   a. leaf mesophyll
   b. stem xylem
   c. stem phloem
   d. root cortex
   e. root epidermis

b
Factual Recall

32.25 One is most likely to see guttation in small plants when the
   a. transpiration rates are high.
   b. root pressure exceeds transpiration pull.
   c. preceding evening was hot, windy, and dry.
   d. stomata are partially closed.
   e. roots are not absorbing minerals from the soil.

a
Conceptual
Understanding

32.26 Water potential is generally most negative in which of the following parts of a plant?
   a. mesophyll cells
   b. xylem vessels in leaves
   c. xylem vessels in roots
   d. cells of the root cortex
   e. root hairs

d
Application

32.27 Photosynthesis begins to decline when leaves wilt because
   a. flaccid cells are incapable of photosynthesis.
   b. CO_2 accumulates in the leaves and inhibits photosynthesis.
   c. there is insufficient water for photolysis during light reactions.
   d. stomata close, preventing CO_2 entry into the leaf.
   e. the chlorophyll of flaccid cells cannot absorb light.

b
Conceptual
Understanding

32.28 Ignoring all other factors, what kind of day would result in the fastest delivery of water and minerals to the leaves of a tree?
   a. cool, dry day
   b. warm, dry day
   c. warm, humid day
   d. cool, humid day
   e. very hot, dry, windy day

b
Factual Recall

32.29 Which of the following is a CORRECT statement about sugar movement in phloem?
   a. Diffusion can account for the observed rates of transport.
   b. Movement can occur both upward and downward in the plant.
   c. Sugar is translocated from sinks to sources.
   d. Only phloem cells with nuclei can perform sugar movement.
   e. Sugar transport does not require energy.

b
Conceptual
Understanding

32.30 In the pressure-flow hypothesis of translocation, what causes the pressure?
   a. root pressure
   b. the osmotic uptake of water by sieve tubes at the source
   c. the accumulation of minerals and water by the stele in the root
   d. the osmotic uptake of water by the sieve tubes of the sink
   e. hydrostatic pressure in xylem vessels

e
Conceptual
Understanding

32.31 Which of the following statements about transport in plants is FALSE?
   a. Weak bonding between water molecules and the walls of xylem vessels or tracheids helps support the columns of water in the xylem.
   b. Hydrogen bonding between water molecules, which results in the high cohesion of the water, is essential for the rise of water in tall trees.
   c. Although some angiosperm plants develop considerable root pressure, this is not sufficient to raise water to the tops of tall trees.
   d. Most plant physiologists now agree that the pull from the top of the plant resulting from transpiration is sufficient, when combined with the cohesion of water, to explain the rise of water in the xylem in even the tallest trees.
   e. Gymnosperms can sometimes develop especially high root pressure, which may account for the rise of water in tall pine trees without transpiration pull.

b
Conceptual
Understanding

32.32 Water flows into the source end of a sieve tube because
   a. sucrose has diffused into the sieve tube, making it hypertonic.
   b. sucrose has been actively transported into the sieve tube, making it hypertonic.
   c. water pressure outside the sieve tube forces in water.
   d. the companion cell of a sieve tube actively pumps in water.
   e. sucrose has been dumped from the sieve tube by active transport.

d
Conceptual
Understanding

32.33 Which of the following statements is FALSE concerning the xylem?
    a. Xylem tracheids and vessels fulfill their vital function only after their death.
    b. The cell walls of the tracheids are greatly strengthened with cellulose fibrils forming thickened rings or spirals.
    c. Water molecules are transpired from the cells of the leaves and replaced by water molecules in the xylem pulled up from the roots due to the cohesion of water molecules.
    d. Movement of materials is by mass flow; materials move owing to a turgor pressure gradient from "source" to "sink."
    e. In the morning, sap in the xylem begins to move first in the twigs of the upper portion of the tree, and later in the lower trunk.

e
Conceptual
Understanding

32.34 Which of the following has the lowest (most negative) water potential?
    a. soil
    b. root xylem
    c. trunk xylem
    d. leaf cell walls
    e. leaf air spaces

a
Conceptual
Understanding

32.35 Which of the following factors that sometimes play a role in the rise of sap in the xylem depends on the expenditure of energy by the plant?
    a. secretion of ions into the stele of the root
    b. cohesion of water molecules due to hydrogen bonding
    c. transpiration of water from the leaves
    d. capillarity of water in xylem tracheids and vessels
    e. diffusion of water into the root hairs

d
Conceptual
Understanding

32.36 Transpiration in plants requires all of the following EXCEPT
    a. adhesion of water molecules to cellulose.
    b. cohesion between water molecules.
    c. evaporation of water molecules.
    d. active transport through xylem cells.
    e. transport through tracheids.

c
Factual Recall

32.37 Which of the following statements about xylem is INCORRECT?
    a. Xylem conducts material upward.
    b. Xylem conduction occurs within dead cells.
    c. Xylem carries mainly sugars and amino acids.
    d. Xylem has a lower water potential than the soil does.
    e. No energy input from the plant is required for xylem transport.

c
Factual Recall

32.38 Root hairs are most important to a plant because they
    a. anchor a plant in the soil.
    b. store starches.
    c. increase the surface area for absorption.
    d. provide a habitat for nitrogen-fixing bacteria.
    e. contain xylem tissue.

a
Factual Recall

32.39 All of the following are adaptations that help reduce water loss from a plant EXCEPT
   a. transpiration.
   b. sunken stomates.
   c. C_4 photosynthesis.
   d. small, thick leaves.
   e. Crassulacean acid metabolism.

d
Conceptual
Understanding

32.40 In plant roots, the Casparian strip is correctly described by which of the following?
   a. It is located in the walls between endodermal cells and cortex cells.
   b. It provides energy for the active transport of minerals into the stele from the cortex.
   c. It ensures that all minerals are absorbed from the soil in equal amounts.
   d. It ensures that all water and dissolved substances must pass through a cell before entering the stele.
   e. It provides increased surface area for the absorption of mineral nutrients.

a
Conceptual
Understanding

32.41 All of the following normally enter the plant through the roots EXCEPT
   a. carbon dioxide.
   b. nitrogen.
   c. potassium.
   d. water.
   e. calcium.

a
Conceptual
Understanding

32.42 Active transport involves all of the following EXCEPT
   a. slow movement through the lipid bilayer of a membrane.
   b. pumping of solutes across the membrane.
   c. hydrolysis of ATP.
   d. transport of solute against a concentration gradient.
   e. a specific transport protein in the membrane.

d
Conceptual
Understanding

32.43 Which of the following is NOT a characteristic of the plasma membrane proton pump?
   a. hydrolyzes ATP
   b. produces a proton gradient
   c. generates a membrane potential
   d. equalizes the charge on each side of a membrane
   e. stores potential energy on one side of a membrane

b
Factual Recall

32.44 The unifying principle of cellular energetics is
   a. active transport.
   b. chemiosmosis.
   c. ATP hydrolysis.
   d. water potential.
   e. source–sink relationships.

e
Conceptual
Understanding

32.45 Which of the following is an example of osmosis?
    a. flow of water out of a cell
    b. pumping of water into a cell
    c. flow of water between cells
    d. Both a and b are true of osmosis.
    e. Both a and c are true of osmosis.

e
Conceptual
Understanding

32.46 Your laboratory partner has an open beaker of pure water. By definition, the Ψ of this water is
    a. not meaningful, because it is an open beaker and not plant tissue.
    b. a negative number set by the volume of the beaker.
    c. a positive number set by the volume of the beaker.
    d. equal to the atmospheric pressure.
    e. zero.

c
Application

32.47 If $\Psi_P = 0.3$ MPa and $\Psi_S = -0.45$ MPa, the resulting Ψ is
    a. +0.75 MPa.
    b. −0.75 MPa.
    c. −0.15 MPa.
    d. +0.15 MPa.
    e. impossible to calculate with this information.

a
Application

32.48 The value for Ψ in Question 32.47 was measured for root tissue. If you take the root tissue and place it in a 0.1 M solution of sucrose ($\Psi = -0.23$), net water flow would
    a. be from the tissue into the sucrose solution.
    b. be from the sucrose solution into the tissue.
    c. be in both directions and the concentrations would remain equal.
    d. occur only as ATP was hydrolyzed in the tissue.
    e. be impossible to determine from the values given here.

a
Conceptual
Understanding

32.49 The main mechanisms determining the direction of short-distance transport within a potato tuber are
    a. diffusion due to concentration differences, and bulk flow due to pressure differences.
    b. diffusion due to pressure and concentration differences.
    c. active transport due to the hydrolysis of ATP and ion transport into the tuber cells.
    d. determined by the stucture and function of the tonoplast of the tuber cells.
    e. not affected by temperature and pressure.

b
Conceptual
Understanding

32.50 Which of the following would likely NOT contribute to the surface area available for water absorption from the soil by a plant root system?
    a. root hairs
    b. endodermis
    c. mycorrhizae
    d. fungi associated with the roots
    e. fibrous arrangement of the roots

c
Application

32.51 An astronaut brings you a plant from Mars. The morphology of the plant looks normal enough, but a plant anatomist discovers that the roots lack a Casparian strip in the endodermis. Most likely this plant would
   a. grow fine on Earth if given plenty of water.
   b. be able to accumulate ions by passive transport in the symplast.
   c. not be able to effectively concentrate K^+ and other ions inside the stele.
   d. be able to accumulate minerals by active transport in the apoplast.
   e. not grow on Earth because of a lack of sucrose production.

a
Application

32.52 Which of the following best explains why no tall trees seem to be CAM plants?
   a. They would be unable to move water and minerals to the top of the plant during the day.
   b. They would be unable to supply sufficient sucrose for active transport of minerals into the roots during the day or night.
   c. Transpiration occurs only at night, and this would cause a highly negative Ψ in the roots of a tall plant during the day.
   d. Since the stomates are closed in the leaves, the Casparian strip is closed in the endodermis of the root.
   e. With the stomates open at night, the transpiration rate would limit plant height.

e
Factual Recall

32.53 Guard cells
   a. protect the endodermis.
   b. accumulate K^+ ions and close the stomates.
   c. contain chloroplasts that directly import K^+ ions into the cells.
   d. guard against mineral loss through the stomates.
   e. help balance the photosynthesis–transpiration compromise.

e
Application

32.54 As the biologist for the New Mexico state agriculture department, it is your job to look for plants that have evolved structures with a selective advantage in dry, hot conditions. Which of the following adaptations would be LEAST likely to meet your objective?
   a. CAM plants that grow rapidly
   b. small, thick leaves with stomates on the lower surface
   c. a thick cuticle on fleshy leaves
   d. large, fleshy stems with the ability to carry out photosynthesis
   e. plants that do not produce abscisic acid and have a short, thick tap root

Chapter 33

c
Factual Recall

33.1 What is meant by the term *chlorosis*?
   a. the uptake of the micronutrient chlorine by a plant
   b. the formation of chlorophyll within the thylakoid membranes of a plant
   c. the yellowing of leaves due to decreased chlorophyll production
   d. a contamination of glassware in hydroponic culture
   e. release of negatively charged minerals such as chloride from clay particles in soil

d
Factual Recall

33.2 What are epiphytes?
   a. aerial vines common in tropical regions
   b. haustoria for anchoring to host plants and obtaining xylem sap
   c. plants that live in poor soil and digest insects to obtain nitrogen
   d. plants that grow on other plants but do not obtain nutrients from their hosts
   e. plants that have a symbiotic relationship with roots

For Questions 33.3–33.6, match the element with its source and use in typical terrestrial plants.

Source	*Use*
a. *soil solution*	*nucleic acids and proteins*
b. *soil solution*	*most or all organic molecules*
c. *air*	*nucleic acids and proteins*
d. *air*	*most or all organic molecules*
e. *air and soil solution*	*most or all organic molecules*

d
Conceptual
Understanding

33.3 Carbon.

b
Conceptual
Understanding

33.4 Hydrogen.

e
Conceptual
Understanding

33.5 Oxygen.

a
Conceptual
Understanding

33.6 Nitrogen.

c
Factual Recall

33.7 There are several properties of a soil in which typical plants would grow well. Of the following, which would be the LEAST conducive to plant growth?
a. abundant humus
b. numerous soil organisms
c. high clay content
d. high porosity
e. high cation exchange capacity

c
Factual Recall

33.8 Which of the following describes the fate of most of the water taken up by a plant?
a. It is used as a solvent.
b. It is used as a hydrogen source in photosynthesis.
c. It is lost during transpiration.
d. It makes cell elongation possible.
e. It is used to keep cells turgid.

a
Factual Recall

33.9 Which of the following best describes the general role of micronutrients in plants?
a. They are cofactors in enzyme reactions.
b. They are necessary for essential regulatory functions.
c. They prevent chlorosis.
d. They are components of nucleic acids.
e. They are necessary for the formation of cell walls.

e
Factual Recall

33.10 A soil well suited for the growth of most plants would have all of the following properties EXCEPT
a. abundant humus.
b. air spaces.
c. good drainage.
d. high cation exchange capacity.
e. a high pH.

d
Application

33.11 A young farmer purchases some land in a relatively arid area and is interested in earning a reasonable profit for many years. Which of the following strategies would best allow such a goal to be achieved?
a. establishing an extensive irrigation system
b. using plenty of the best fertilizers
c. finding a way to sell all parts of crop plants
d. selecting crops adapted to arid areas
e. converting hillsides into fields

b
Factual Recall

33.12 In what way do nitrogen compounds differ from other minerals needed by plants?
a. Only nitrogen can be lost from the soil.
b. Only nitrogen is not derived from the breakdown of parent rock.
c. Only nitrogen is needed for protein synthesis.
d. Only nitrogen is held by cation exchange capacity in the soil.
e. Only nitrogen can be absorbed by root hairs.

b
Conceptual
Understanding

33.13 Why is nitrogen fixation such an important process?
a. Nitrogen fixation can only be done by certain prokaryotes.
b. Fixed nitrogen is most often the limiting factor in plant growth.
c. Nitrogen fixation is very expensive in terms of metabolic energy.
d. Nitrogen fixers are sometimes symbiotic with legumes.
e. Nitrogen fixing capacity can be genetically engineered.

a
Factual Recall

33.14 The enzyme nitrogenase is inhibited by oxygen. Which one of the following adaptations has evolved in response to this inhibition?
a. leghemoglobin
b. root nodules
c. water ferns
d. carbohydrate transfer from hosts
e. bacteroids

d
Factual Recall

33.15 Carnivorous plants have evolved mechanisms that trap and digest small animals. The products of this digestion are used to supplement the plant's supply of
a. energy.
b. carbohydrates.
c. lipids and steroids.
d. minerals.
e. water.

c
Application

33.16 A farmer who was allergic to fungal spores sprayed a woodlot with a fungicide. What would be the most serious result of such spraying?
a. a decrease in food for animals that eat mushrooms
b. a decrease in rates of wood decay
c. a decrease in tree growth due to the death of mycorrhizae
d. an increase in the number of decomposing bacteria
e. There would be no serious results.

c
Conceptual
Understanding

33.17 A plant that has started growth exhibits chlorosis of the leaves of the entire plant. The chlorosis is probably due to a deficiency of which of the following macronutrients?
a. carbon
b. oxygen
c. nitrogen
d. calcium
e. hydrogen

e
Application

33.18 Which of the following elements is incorrectly paired with its function in a plant?
a. nitrogen—component of nucleic acids, proteins, hormones, coenzymes
b. magnesium—component of chlorophyll; activates many enzymes
c. phosphorus—component of nucleic acids, phospholipids, ATP, several coenzymes
d. potassium—cofactor functional in protein synthesis; osmosis; operation of stomata
e. sulfur—component of DNA; activates some enzymes

d
Conceptual
Understanding

33.19 Which two elements make up more than 90% of the dry weight of plants?
    a. carbon and nitrogen
    b. oxygen and hydrogen
    c. nitrogen and oxygen
    d. oxygen and carbon
    e. carbon and potassium

a
Conceptual
Understanding

33.20 All of the following are elements that plants need in very small amounts (micronutrients) EXCEPT
    a. hydrogen.
    b. iron.
    c. chlorine.
    d. copper.
    e. zinc.

b
Factual Recall

33.21 The greatest proportion of the water taken up by plants is
    a. split during photosynthesis.
    b. lost through stomata during transpiration.
    c. absorbed by central vacuoles during cell elongation.
    d. returned to the soil by roots.
    e. stored in the xylem.

b
Conceptual
Understanding

33.22 The bulk of a plant's dry weight is derived from
    a. soil minerals.
    b. CO_2.
    c. the hydrogen from H_2O.
    d. the oxygen from H_2O.
    e. the uptake of organic nutrients from the soil.

c
Factual Recall

33.23 In the nutrition of a plant, which element is classified as a macronutrient?
    a. zinc
    b. chlorine
    c. calcium
    d. molybdenum
    e. manganese

d
Conceptual
Understanding

33.24 Which soil mineral is most likely to be leached away due to a hard rain?
    a. Na^+
    b. K^+
    c. CA^{++}
    d. NO_3^-
    e. H^+

d
Factual Recall

33.25 The N-P-K percentages on a package of fertilizer refer to the
    a. total protein content of the three major ingredients of the fertilizer.
    b. percentages of manure collected from different types of animals.
    c. relative percentages of organic and inorganic nutrients in the fertilizer.
    d. percentages of three important mineral nutrients.
    e. proportions of three different nitrogen sources.

d
Conceptual
Understanding

33.26 Most crop plants acquire their nitrogen mainly in the form of
    a. NH_3.
    b. N_2.
    c. CN_2H_2.
    d. NO_3^-.
    e. amino acids absorbed from the soil.

c
Conceptual
Understanding

33.27 Which of the following is a true statement about nitrogen fixation in root
    nodules?
    a. The plant contributes the nitrogenase enzyme.
    b. The process is relatively inexpensive in terms of ATP costs.
    c. Leghemoglobin helps maintain a low O_2 concentration within the nodule.
    d. The process tends to deplete nitrogen compounds in the soil.
    e. The bacteria of the nodule are autotrophic.

b
Factual Recall

33.28 Among important crop plants, nitrogen-fixing root nodules are most
    commonly an attribute of
    a. corn.
    b. legumes.
    c. wheat.
    d. members of the potato family.
    e. cabbage and other members of the brassica family.

e
Factual Recall

33.29 The fraction of the dry weight of a plant that is organic material is closest to
    a. 6%.
    b. 17%.
    c. 67%.
    d. 81%.
    e. 96%.

a
Conceptual
Understanding

33.30 Which of the following is the major role of potassium in plants?
    a. osmotic regulation
    b. photosynthesis
    c. ATP synthesis
    d. reproduction
    e. lipid metabolism

e
Application

33.31 If an African violet has chlorosis, which of the following elements would be a
    useful addition to the soil?
    a. chlorine
    b. chlorophyll
    c. copper
    d. iodine
    e. magnesium

e
Factual Recall

33.32 What soil(s) is (are) the most fertile?
    a. humus only
    b. loam only
    c. silt only
    d. clay only
    e. both humus and loam

a
Factual Recall

33.33 What should be added to soil to prevent minerals from leaching away?
    a. humus
    b. sand
    c. mycorrhizae
    d. nitrogen
    e. silt

b
Conceptual
Understanding

33.34 The enzyme nitrogenase reduces atmospheric nitrogen to form
    a. N_2.
    b. NH_3.
    c. NO_2.
    d. NO^-.
    e. NH^-.

d
Conceptual
Understanding

33.35 If a legume is infected with *Rhizobium*, what is the probable effect on the plant?
    a. It gets chlorosis.
    b. It dies.
    c. It desiccates.
    d. It obtains nitrogen from nitrogen fixation.
    e. It contributes water to the soil.

c
Factual Recall

33.36 What is the mutualistic association between roots and fungi called?
    a. nitrogen fixation
    b. *Rhizobium* infection
    c. mycorrhizae
    d. parasitism
    e. root hair enhancement

a
Application

33.37 You are conducting an experiment on plant growth. You take a plant fresh from the soil and it weighs 5 kg. Then you dry the plant overnight and determine the dry weight to be 1 kg. Of this dry weight, how much would you expect to be made up of inorganic minerals?
    a. 50 grams
    b. 500 grams
    c. 1 kg
    d. 4 kg
    e. 5 kg

d
Application

33.38 Which of the following is NOT true of micronutrients in plants?
    a. They are the elements required in relatively small amounts.
    b. They are required for a plant to grow from a seed and complete its life cycle.
    c. They generally help in catalytic functions in the plant.
    d. They are the essential elements of small size and molecular weight.
    e. Overdoses of them can be toxic.

Figure 33.1 shows the results of a study to determine the effect of soil air spaces on plant growth. Use these data to answer Questions 33.39 and 33.40.

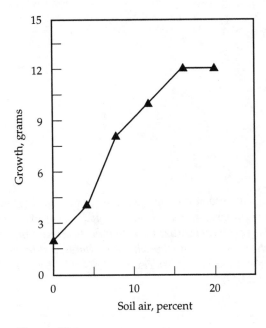

Figure 33.1

c
Application

33.39 The best conclusion from the data in Figure 33.1 is that the plant
 a. grows best without air in the soil.
 b. grows fastest in 5 to 10 percent air.
 c. grows best in air levels above 15 percent.
 d. does not respond differently to different levels of air in the soil.
 e. would grow to 24 grams in 40 percent air.

e
Conceptual
Understanding

33.40 The best explanation for the shape of this growth response curve is that
 a. the plant requires air in the soil for photosynthesis.
 b. the roots are able to absorb more nitrogen (N_2) in high levels of air.
 c. most of the decrease in weight at low air levels is due to transpiration from the leaves.
 d. increased soil air produces more root mass in the soil but does not affect the top stems and leaves.
 e. the roots require oxygen for respiration and growth.

For Questions 33.41–33.45, match the element to its major function in plants. Use each choice only once.

   Function
 a. *component of lignin-biosynthetic enzymes*
 b. *component of DNA and RNA*
 c. *active in chlorophyll formation*
 d. *active in amino acid formation*
 e. *formation and stability of cell walls*

c
Factual Recall

33.41 Zinc.

b
Factual Recall

33.42 Nitrogen.

a
Factual Recall

33.43 Copper.

e
Factual Recall

33.44 Calcium.

d
Factual Recall

33.45 Manganese.

In west Texas, cotton has become an important crop in the last three decades. However, in this hot, dry part of the country there is little rainfall, so farmers irrigate their cotton fields. They must also regularly fertilize the cotton fields because the soil is very sandy. Figure 33.2 shows the record of annual productivity (measured in kilograms of cotton per hectare of land) since 1960 in a west Texas cotton field. Use these data to answer Questions 33.46 and 33.47.

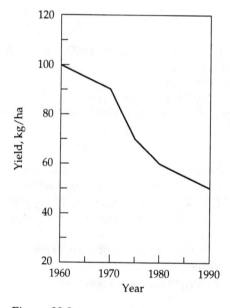

Figure 33.2

d
Application

33.46 Based on the information provided here, what is the most likely cause of the decline in productivity?
   a. The farmer used the wrong kind of fertilizer.
   b. The cotton is developing a restistance to the fertilizer and to irrigation water.
   c. Water has accumulated in the soil due to irrigation.
   d. The soil has become hypertonic to the roots.
   e. The rate of photosynthesis has declined due to irrigation.

a
Application

33.47 If you were the county agriculture agent, what would be the best advice you could give the farmer who owns the field under study in Figure 33.2?
   a. Plant a variety of cotton that requires less water and can tolerate salinity.
   b. Continue to fertilize, but stop irrigating and rely on rainfall.
   c. Continue to irrigate, but stop fertilizing and rely on organic nutrients in the soil.
   d. Continue to fertilize and irrigate, but add the nitrogen-fixing bacteria, *Rhizobium*, to the irrigation water until the productivity increases.
   e. Add acid to the soil and increase its cation exchange capabilities so more nutrients are retained in the soil.

e
Application

33.48 Dwarf mistletoe grows on many of the pine trees in the Rockies. Although the mistletoe is green, it is probably not sufficiently active in photosynthesis to produce all the sugar it needs. The mistletoe also produces haustoria. Thus, dwarf mistletoe growing on pine trees is best classified as
   a. an epiphyte.
   b. a nitrogen-fixing legume.
   c. a carnivorous plant.
   d. a symbiotic plant.
   e. a parasite.

e
Factual Recall

33.49 A farming commitment embracing a variety of methods that are conservation-minded, environmentally safe, and profitable is called
   a. hydroponics.
   b. nitrogen fixation.
   c. responsible irrigation.
   d. genetic engineering.
   e. sustainable agriculture.

e
Factual Recall

33.50 In a nodule, the gene coding for nitrogenase
   a. is inactivated by leghemoglobin.
   b. is absent in active bacteroids.
   c. is found in the cells of the pericycle.
   d. protects the nodule from nitrogen.
   e. is part of the *Rhizobium* chromosome.

Chapter 34

34.1 A unique feature of fertilization in angiosperms is that
    a. All of the below are unique to angiosperms.
    b. a chemical attractant guides the sperm nucleus toward the egg.
    c. cross fertilization is common.
    d. sperm do not swim.
    e. it is a double fertilization.

34.2 Which of the following statements is true of protoplast fusion?
    a. It occurs when the second sperm nucleus fuses with the polar nuclei in the embryo sac.
    b. It can be used to form new plant varieties by combining genomes from two plants.
    c. It is used to develop gene banks to preserve genetic variability.
    d. It is the method of test-tube cloning that produces whole plants from explants.
    e. It occurs within a callus that is developing in tissue culture.

34.3 Morphogenesis in plants is largely a result of
    a. genetic differences among the cell lineages involved.
    b. the plane of cell divisions and the direction of cell expansion.
    c. morphogens that create gradients in developing leaves.
    d. Only b and c are correct.
    e. a, b, and c are correct.

34.4 Which of the following "vegetables" is technically a fruit?
    a. potato
    b. lettuce
    c. broccoli
    d. celery
    e. green beans

34.5 As flowers develop, all of the following transitions occur EXCEPT that
    a. the microspores become pollen grains.
    b. the ovary becomes a fruit.
    c. the petals are discarded.
    d. the tube nucleus becomes a sperm nucleus.
    e. the ovules become seeds.

34.6 In plants, which of the following could be an advantage of sexual reproduction as opposed to asexual reproduction?
    a. genetic variation
    b. mitosis
    c. stable populations
    d. rapid population increase
    e. greater longevity

b
Factual Recall

34.7 Which of the following occurs in an angiosperm ovule?
   a. An archegonium forms from the megasporophyte.
   b. A megaspore mother cell undergoes meiosis.
   c. The egg nucleus is usually diploid.
   d. The fusion nucleus develops into the embryo.
   e. The endosperm surrounds the megaspore mother cell.

c
Factual Recall

34.8 Where and by which process are sperm produced in plants?
   a. meiosis in pollen grains
   b. meiosis in anthers
   c. mitosis in male gametophytes
   d. mitosis in the micropyle
   e. mitosis in the embryo sac

a
Factual Recall

34.9 The product of meiosis in plants is always which of the following?
   a. spores
   b. eggs
   c. sperm
   d. seeds
   e. Both b and c are correct.

e
Factual Recall

34.10 Based on studies of plant evolution, which flower part is not a modified leaf?
   a. stamen
   b. carpel
   c. petal
   d. sepal
   e. receptacle

b
Conceptual
Understanding

34.11 Meiosis occurs within all of the following flower parts EXCEPT the
   a. ovule.
   b. style.
   c. megasporangium.
   d. anther.
   e. ovary.

d
Factual Recall

34.12 All of the following are features responsible for the evolutionary success of angiosperms EXCEPT
   a. a triploid endosperm.
   b. an ovary that becomes a fruit.
   c. animal pollination.
   d. a reduced sporophyte phase.
   e. double fertilization.

a
Conceptual
Understanding

34.13 When seeds germinate, the radicle emerges before the shoot. This allows the seedling to
   a. obtain a dependable water supply.
   b. mobilize stored carbohydrates.
   c. protect the emerging coleoptile.
   d. avoid etiolation.
   e. initiate photosynthesis.

e
Conceptual
Understanding

34.14 All of the following could be considered advantages of asexual reproduction in plants EXCEPT
   a. success in a stable environment.
   b. increased agricultural productivity.
   c. cloning an exceptional plant.
   d. production of artificial seeds.
   e. adaptation to change.

e
Factual Recall

34.15 Which developmental process transforms a fertilized egg into a plant?
   a. growth
   b. morphogenesis
   c. cellular differentiation
   d. Only a and c are correct.
   e. a, b, and c are correct.

e
Factual Recall

34.16 A flower that lacks stamens is said to be
   a. incomplete only.
   b. imperfect only.
   c. staminate only.
   d. monoecious only.
   e. both imperfect and incomplete.

a
Factual Recall

34.17 In flowering plants, pollen is released from the
   a. anther.
   b. stigma.
   c. carpel.
   d. sepal.
   e. pollen tube.

d
Conceptual
Understanding

34.18 In the life cycle of an angiosperm, which of the following stages is diploid?
   a. megaspore only
   b. generative nucleus of a pollen grain only
   c. polar nuclei of the embryo sac only
   d. microsporocyte only
   e. both megaspore and microsporocyte

b
Conceptual
Understanding

34.19 In which of the following pairs are the two terms truly equivalent?
   a. ovule—egg
   b. embryo sac—female gametophyte
   c. endosperm—male gametophyte
   d. seed—zygote
   e. microspore—pollen grain

c
Conceptual
Understanding

34.20 In the life cycle of an angiosperm, double fertilization refers to
 a. fertilization of the egg by sperm nuclei from two different pollen grains.
 b. fertilization of the egg by two sperm nuclei released from a single pollen tube.
 c. a pollen tube releasing two sperm nuclei, one fertilizing the egg and the other combining with the two polar nuclei.
 d. fertilization of two different eggs within the embryo sac.
 e. fertilization of two eggs within the archegonium.

c
Factual Recall

34.21 The radicle of a plant embryo gives rise to which structure(s)?
 a. leaves
 b. cotyledons
 c. root
 d. stem
 e. shoot

c
Factual Recall

34.22 Which of these structures is unique to the seed of a monocot?
 a. cotyledon
 b. endosperm
 c. coleoptile
 d. radicle
 e. seed coat

c
Factual Recall

34.23 The first step in the germination of a seed is usually
 a. pollination.
 b. fertilization.
 c. imbibition of water.
 d. hydrolysis of starch and other food reserves.
 e. emergence of the radicle.

e
Conceptual
Understanding

34.24 In flowering plants, meiosis occurs specifically in the
 a. megasporocyte only.
 b. microsporocyte only.
 c. endosperm only.
 d. pollen tube only.
 e. megasporocyte and microsporocyte.

c
Factual Recall

34.25 All of the following floral parts are directly involved in pollination or fertilization EXCEPT the
 a. stigma.
 b. anther.
 c. sepal.
 d. carpel.
 e. style.

d
Conceptual
Understanding

34.26 The largest cell(s) of the angiosperm embryo sac is (are) the
   a. egg cell.
   b. antipodals.
   c. synergids.
   d. polar nuclei cell.
   e. microgametophye.

c
Factual Recall

34.27 The male gametophyte of flowering plants is the
   a. ovule.
   b. microsporocyte.
   c. pollen.
   d. embryo sac.
   e. stamen.

c
Application

34.28 Which of the following is a correct sequence of processes that take place when
   a flowering plant reproduces?
   a. meiosis—fertilization—ovulation—germination
   b. fertilization—meiosis—nuclear fusion—formation of embryo and endosperm
   c. meiosis—pollination—nuclear fusion—formation of embryo and endosperm
   d. growth of pollen tube—pollination—germination—fertilization
   e. meiosis—mitosis—nuclear fusion—pollen

*Questions 34.29–34.31 refer to the diagram of an embryo sac of an angiosperm
(Figure 34.1).*

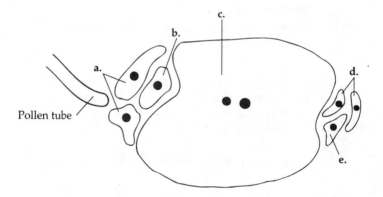

Figure 34.1

b
Factual Recall

34.29 Which cell(s) after fertilization, give(s) rise to the embryo plant?

c
Factual Recall

34.30 Which cell(s) become(s) the triploid endosperm?

b
Factual Recall

34.31 Which cell(s) is (are) the egg cell(s)?

Questions 34.32–34.36 refer to the following answers. Each answer may be used once, more than once, or not at all.

    a. stamen
    b. stigma
    c. embryo sac
    d. sepal
    e. coleoptile

e
Factual Recall

34.32 The sheath covering the embryonic shoot in a monocot.

a
Factual Recall

34.33 Flower part modified as a male reproductive structure.

d
Factual Recall

34.34 The outer modified leaf of a flower.

c
Factual Recall

34.35 The female gametophyte of angiosperms.

b
Factual Recall

34.36 The sticky top of a carpel.

b
Factual Recall

34.37 How many sperm nuclei are needed for the double fertilization of angiosperms?
    a. 1 (for the egg)
    b. 2 (1 for the egg, 1 for the polar nuclei)
    c. 4 (2 for each embryo sac)
    d. 3 (1 for the egg, 2 for each of the polar nuclei)
    e. 3 (2 for the egg, 1 for the polar nuclei, thus making a $3n$ nucleus)

e
Conceptual
Understanding

34.38 In angiosperms, the terminal cell divides to become the
    a. suspensor only.
    b. proembryo only.
    c. cotyledons only.
    d. suspensor and the proembryo.
    e. proembryo and the cotyledons.

d
Factual Recall

34.39 The embryonic root is the
    a. plumule.
    b. hypocotyl.
    c. epicotyl.
    d. radicle.
    e. plumule.

e
Factual Recall

34.40 A perfect flower is correctly described as a flower that
    a. has no sepals.
    b. has fused carpels.
    c. is on a dioecious plant.
    d. has no endosperm.
    e. has both stamens and carpels.

a
Factual Recall

34.41 Which of these is incorrectly paired with its life-cycle generation?
    a. anther—gametophyte
    b. pollen—gametophyte
    c. embryo sac—gametophyte
    d. pistil—sporophyte
    e. embryo—sporophyte

b
Factual Recall

34.42 Which of the following statements is correct about the basal cell in a zygote?
    a. It develops into the root of the embryo.
    b. It forms the suspensor that anchors the embryo.
    c. It results from the fertilization of the polar nuclei by a sperm nucleus.
    d. It divides to form the cotyledons.
    e. It forms the proembryo.

a
Conceptual
Understanding

34.43 A unique feature of fertilization in angiosperms is that
    a. one sperm fertilizes the egg; another combines with the polar nuclei.
    b. the sperm may be carried by the wind to the female organ.
    c. a pollen tube carries a sperm nucleus into the female gametophyte.
    d. a chemical attractant guides the sperm toward the egg.
    e. the sperm cells have flagella for locomotion.

d
Conceptual
Understanding

34.44 All of the following statements regarding the endosperm are true EXCEPT:
    a. It may be absorbed by the cotyledons in the seeds of dicots.
    b. It is a triploid tissue.
    c. It is digested by enzymes in monocot seeds following hydration.
    d. It develops from the fertilized egg.
    e. It is rich in nutrients, which it provides to the embryo.

a
Factual Recall

34.45 All of the following are true of the hypocotyl hook EXCEPT:
    a. It is the first structure to emerge from a dicot seed.
    b. It pulls the cotyledons up through the soil.
    c. It straightens when exposed to light.
    d. It becomes very long in an etiolated seedling.
    e. It is the region just below the cotyledons.

b
Conceptual
Understanding

34.46 A disadvantage of monoculture is that
    a. the whole crop ripens at one time.
    b. genetic uniformity makes a crop vulnerable to a new pest or disease.
    c. it predominantly uses vegetative propagation.
    d. most grain crops self-pollinate.
    e. it allows for the cultivation of large areas of land.

c
Factual Recall

34.47 Which of the following statements is correct about protoplast fusion?
    a. It is used to develop gene banks to maintain genetic variability.
    b. It is the method of test-tube cloning thousands of copies.
    c. It can be used to form new plant species.
    d. It occurs within a callus.
    e. It requires that the cell wall remains intact during the fusion process.

b
Conceptual
Understanding

34.48 What substance is released from the barley seed embryo and initiates enzyme production in the aleurone layer?
    a. alpha amylase
    b. gibberellic acid
    c. water
    d. sugar
    e. starch

a
Conceptual
Understanding

34.49 Which of the following is the correct sequence during alternation of generations in a flowering plant?
    a. sporophyte—meiosis—gametophyte—gametes—fertilization—diploid zygote
    b. sporophyte—mitosis—gametophyte—meiosis—sporophyte
    c. haploid gametophyte—gametes—meiosis—fertilization—diploid sporophyte
    d. sporophyte—spores—meiosis—gametophyte—gametes
    e. haploid sporophyte—spores—fertilization—diploid gametophyte

d
Conceptual
Understanding

34.50 Which of the following is the correct order of floral organs from the outside to the inside of a complete flower?
    a. petals—sepals—stamens—carpels
    b. sepals—stamens—petals—carpels
    c. spores—gametes—zygote—embryo
    d. sepals—petals—stamens—carpels
    d. male gametophyte—female gametophyte—sepals—petals

b
Factual Recall

34.51 Which of the following is NOT a primary function of flowers?
    a. pollen production
    b. photosynthesis
    c. meiosis
    d. egg production
    e. sexual reproduction

e
Conceptual
Understanding

34.52 Carpellate flowers
    a. are perfect.
    b. are complete.
    c. produce pollen.
    d. are found only on dioecious plants.
    e. develop into fruits.

e
Conceptual
Understanding

34.53 Which of the following is the correct sequence of events in a pollen sac?
   a. sporangia—meiosis—two haploid cells—meiosis—two pollen grains per cell
   b. pollen grain—meiosis—two generative cells—two tube cells per pollen grain
   c. two haploid cells—meiosis—generative cell—tube cell—fertilization—pollen grain
   d. pollen grain—mitosis—microspores—meiosis—generative cell plus tube cell
   e. microsporocyte—meiosis—microspores—mitosis—two haploid cells per pollen grain

c
Factual Recall

34.54 In flowering plants, a mature male gametophyte contains
   a. two haploid gametes and a diploid pollen grain.
   b. a generative cell and a tube cell.
   c. two sperm nuclei and one tube cell nucleus.
   d. two haploid microspores per gametophyte.
   e. a haploid nucleus and a diploid pollen wall.

a
Factual Recall

34.55 Within the female gametophyte, three mitotic divisions of the megaspore produce
   a. three antipodal cells, two polar nuclei, one egg, and two synergids.
   b. the triple fusion nucleus.
   c. three pollen grains.
   d. two antipodals, two polar nuclei, two eggs, and two synergids.
   e. a tube nucleus, a generative cell, and a sperm cell.

a
Factual Recall

34.56 Which of the following types of plants is NOT able to self-pollinate?
   a. dioecious
   b. monoecious
   c. complete
   d. wind-pollinated
   e. insect-pollinated

b
Conceptual
Understanding

34.57 Recent research has shown that pollination requires that carpels recognize pollen grains as "self or nonself." For self-incompatibility, the system requires
   a. rejection of nonself cells.
   b. the rejection of self.
   c. carpel incompatibility with the egg cells.
   d. that the flowers be incomplete.
   e. the union of genetically identical sperm and egg cells.

e
Conceptual
Understanding

34.58 Pollen and stigmas with a common allele at the S-locus
   a. are always found in dioecious flowers.
   b. are always found in wind-pollinated, complete flowers.
   c. usually produce viable seeds following fertilization.
   d. do not produce sperm and egg by meiosis.
   e. interact abnormally.

a
Application

34.59 You are studying a plant from the Amazon that shows strong self-incompatibility. To characterize this reproductive mechanism you would look for
a. ribonuclease activity in stigma cells.
b. RNA in the plants.
c. pollen grains with very thick walls.
d. carpels that cannot produce eggs by meiosis.
e. systems of wind, but not insect pollination.

e
Factual Recall

34.60 One possible practical application of the studies on self-incompatibility is
a. to improve the nutritional quality of the grain.
b. the elimination of wind pollination in incompatible plants.
c. the production of crosses between nitrogen-fixing crops.
d. to develop self-incompatible plants that are drought tolerant.
e. the transfer of key regions of the S-locus to self-compatible plants.

e
Application

34.61 You are working for a company that grows and distributes avocados. One problem your company wishes to overcome is the rapid ripening of the avocado after it is picked. As the head of the avocado research effort, you would most likely try to
a. develop a plant that produces fruit without the seed inside.
b. alter the genes that produce the soft pericarp.
c. alter the S-locus to slow ripening.
d. slow germination and prevent rapid ripening.
e. genetically alter the cell walls to resist digestion after the harvest.

e
Factual Recall

34.62 Vegetative reproduction
a. involves both meiosis and mitosis to produce haploid and diploid cells.
b. produces vegetables.
c. involves meiosis only.
d. can lead to genetically altered forms of the species.
e. produces clones.

b
Factual Recall

34.63 Which of the following is thought to be important in the control of differential enlargement as a plant organ takes shape?
a. S-locus
b. cytoskeleton
c. acid accumulation in the cells
d. meiosis
e. synthesis of protein-rich cytoplasm

Chapter 35

b
Factual Recall

35.1 All of the following may function in signal transduction in plants EXCEPT
   a. calcium ions.
   b. nonrandom mutations.
   c. receptor proteins.
   d. phytochrome.
   e. second messengers.

d
Conceptual
Understanding

35.2 In order to flower, what does a short-day plant need?
   a. a burst of red light in the middle of the night
   b. a burst of far red light in the middle of the night
   c. a day that is longer than a certain length
   d. a night that is longer than a certain length
   e. a higher ratio of $P_r : P_{fr}$

d
Factual Recall

35.3 What is the function of calmodulin in a signal-transduction pathway?
   a. to receive the stimulus and activate the second messenger in the transduction step
   b. to induce the selective activation of genes
   c. to be a membrane-bound hormone receptor that causes an influx of Ca^{2+}
   d. to form a complex with Ca^{2+} and activate specific molecules
   e. to induce rapid responses such as stomatal closing or cell elongation

e
Conceptual
Understanding

35.4 The effects of a plant hormone are dependent on which of the following?
   a. concentrations of other hormones
   b. concentration of the hormone
   c. developmental stage of the plant
   d. Only b and c are correct.
   e. a, b, and c are correct.

b
Application

35.5 Vines in tropical rain forests must grow toward large trees before being able to grow toward the sun. To reach a large tree, the most useful kind of growth movement for a tropical vine would be approximately the OPPOSITE of
   a. positive thigmotropism.
   b. positive phototropism.
   c. positive gravitropism.
   d. sleep movements.
   e. circadian rhythms.

a
Application

35.6 Plants often use changes in daylength (photoperiod) to trigger events such as dormancy and flowering. It is logical that plants have evolved this mechanism because photoperiod changes
   a. are more predictable than air temperature changes.
   b. alter the amount of energy available to the plant.
   c. are modified by soil temperature changes.
   d. are independent of the biological clock.
   e. are correlated with moisture availability.

d
Application

35.7 In nature, poinsettias bloom in early March. Research has shown that these plants are triggered to flower three months before they actually bloom. The trigger is the length of the light-dark cycle. In order to get poinsettias to bloom in December, florists change the length of the light-dark cycle in September. Given the information and clues above, which of the following is FALSE?
a. Poinsettias are short-day plants.
b. Poinsettias require a light period shorter than some maximum.
c. Poinsettias require a longer dark period than is available in September.
d. The dark period can be interrupted without affecting flowering.
e. Poinsettias will flower even if there are brief periods of dark during the day-time.

b
Conceptual
Understanding

35.8 What do results of research on gravitropic responses of roots and stems show?
a. Different tissues have the same response to auxin.
b. The effect of a plant hormone can depend on the tissue.
c. Some responses of plants require no hormones at all.
d. Light is required for the gravitropic response.
e. Cytokinin can only function in the presence of auxin.

d
Factual Recall

35.9 While responses to plant hormones are normally slow, one hormone has been shown to be involved in the rapid opening and closing of stomata. Which of the following is that hormone?
a. auxin
b. cytokinin
c. ethylene
d. abscisic acid
e. gibberellin

e
Application

35.10 A botanist exposed two groups of plants (of the same species) to two photoperiods, one with fourteen hours of light and ten hours of dark and the other with ten hours of light and fourteen hours of dark. Under the first set of conditions, the plants flowered, but they failed to flower under the second set of conditions. Which of the following conclusions would be consistent with these results?
a. The critical night length is fourteen hours.
b. The plants are short-day plants.
c. The critical day length is ten hours.
d. The plants can convert phytochrome to florigen.
e. The plants flower in the spring.

c
Conceptual
Understanding

35.11 Which of these conclusions is supported by the research of both Went and the Darwins on shoot responses to light?
a. When shoots are exposed to light, a chemical substance migrates toward the light.
b. Agar contains a chemical substance that mimics a plant hormone.
c. A chemical substance involved in shoot bending is produced in shoot tips.
d. Once shoot tips have been cut, normal growth cannot be induced.
e. Light stimulates the synthesis of a plant hormone that responds to light.

a
Conceptual
Understanding

35.12 The stimulation of rooting in stem cuttings and the rapid and lethal stem growth of broad-leaved dicot weeds can be accomplished by the use of a molecule which is a synthetic
  a. auxin.
  b. cytokinin.
  c. oligosaccharin.
  d. gibberellin.
  e. ethylene.

For Questions 35.14–35.17, match the hormone with the classic description of that hormone. Each choice may be used once, more than once, or not at all.
  a. *auxin*
  b. *cytokinin*
  c. *gibberellin*
  d. *ethylene*
  e. *abscisic acid*

e
Factual Recall

35.13 Inhibits growth; closes stomata during water stress.

b
Factual Recall

35.14 Stimulates cell division by influencing the synthesis or activation of proteins required for mitosis.

a
Factual Recall

35.15 Acts by increasing the plasticity of the cell wall.

d
Factual Recall

35.16 A gas that hastens fruit ripening.

c
Factual Recall

35.17 Promotes internode elongation; promotes germination of certain seeds.

c
Factual Recall

35.18 Plant hormones can be characterized by all of the following EXCEPT that they
  a. may act by altering gene expression.
  b. have a multiplicity of effects.
  c. function independently of other hormones.
  d. control plant growth and development.
  e. affect division, elongation, and differentiation of cells.

Use the following answers in your responses to Questions 35.19–35.21. Each one may be used once, more than once, or not at all.
  a. *auxin*
  b. *cytokinins*
  c. *gibberellins*
  d. *abscisic acid*
  e. *ethylene*

a
Application

35.19 Growing a plant in space under conditions of microgravity is most likely to affect the activity of which hormone?

d
Factual Recall

35.20 Which hormone has been shown to trigger a rapid plant response to a sudden decrease in humidity?

c
Factual Recall

35.21 A botanist is unable to germinate the seeds of a wild plant and decides to treat them with a hormone to promote germination. Based on classical studies, which hormone is the logical one to try first?

d
Factual Recall

35.22 If a plant is mechanically stimulated, it will grow shorter, thicker stems. This response is
a. the result of ethylene production.
b. caused by an increase in turgor.
c. an adaptation to windy environments.
d. Only a and c are correct.
e. a, b, and c are correct.

b
Application

35.23 If a population of plants is transplanted to a higher latitude, which of the following processes is the most likely to be modified by natural selection?
a. circadian rhythm
b. photoperiodic response
c. phototropic response
d. biological clock
e. thigmomorphogenesis

a
Application

35.24 If atmospheric carbon dioxide levels continue to increase, which of the following is the most likely response of plants?
a. increased growth
b. greater water loss
c. more rapid senescence
d. slower signal transduction
e. altered photoperiodic response

a
Factual Recall

35.25 Charles and Francis Darwin concluded from their experiments on phototropism by oat seedlings that the part of the seedling that detects the direction of light is
a. the tip of the coleoptile.
b. the part of the coleoptile that bends during the response.
c. the root tip.
d. the cotyledon.
e. phytochrome.

d
Conceptual
Understanding

35.26 If a short-day plant has a critical night length of 15 hours, then which of the following 24-hour cycles will prevent flowering?
a. 8 hours light/16 hours dark
b. 4 hours light/20 hours dark
c. 6 hours light/2 hours dark/light flash/16 hours dark
d. 8 hours light/8 hours dark/light flash/8 hours dark
e. 2 hours light/20 hours dark/2 hours light

d
Factual Recall

35.27 A long-day plant will flower if
   a. the duration of continuous light exceeds a critical length.
   b. the duration of continuous light is less than a critical length.
   c. the duration of continuous darkness exceeds a critical length.
   d. the duration of continuous darkness is less than a critical length.
   e. it is kept in continuous far-red light.

b
Factual Recall

35.28 There is some experimental evidence that a hypothetical flowering hormone may be produced by
   a. flowers.
   b. leaves.
   c. roots.
   d. seeds.
   e. floral buds.

e
Factual Recall

35.29 One effect of gibberellins is to stimulate the aleurone layer of certain seeds to produce
   a. RuBP carboxylase.
   b. lipids.
   c. abscisic acid.
   d. starch.
   e. amylase.

b
Conceptual
Understanding

35.30 Which of the following has NOT been established as an aspect of auxin's role in cell elongation?
   a. Auxin instigates a loosening of cell wall fibers.
   b. Auxin increases the quantity of cytoplasm in the cell.
   c. Through auxin activity, vacuoles increase in size.
   d. Auxin activity permits an increase in turgor pressure.
   e. Auxin stimulates proton pumps.

d
Conceptual
Understanding

35.31 Plants that have their flowering inhibited by floodlights turned on at night are
   a. day-neutral plants.
   b. short-night plants.
   c. devoid of phytochrome.
   d. short-day plants.
   e. long-day plants.

c
Conceptual
Understanding

35.32 We know from the experiments of the past that plants bend toward light because
   a. they need sunlight energy for photosynthesis.
   b. the sun stimulates stem growth.
   c. cell expansion is greater on the dark side of the stem.
   d. auxin is inactive on the dark side of the stem.
   e. phytochrome stimulates florigen formation.

c
Factual Recall

35.33 The application of which of the following hormones would be a logical first choice in an attempt to produce normal growth in mutant dwarf plants?
a. indoleacetic acid
b. cytokinin
c. gibberellin
d. abscisic acid
e. ethylene

d
Factual Recall

35.34 In attempting to make a seed break dormancy, one could logically treat it with
a. cytokinins.
b. 2, 4-D.
c. CO_2.
d. gibberellins.
e. abscisic acid.

d
Conceptual
Understanding

35.35 According to the acid growth hypothesis, auxin works by
a. dissolving sieve plates, permitting more rapid transport of nutrients.
b. dissolving the cell membranes temporarily, permitting cells that were on the verge of dividing to divide more rapidly.
c. changing the pH within the cell, which would permit the electron transport chain to operate more efficiently.
d. allowing the affected cell walls to stretch.
e. greatly increasing the rate of deposition of cell wall material.

a
Conceptual
Understanding

35.36 If you take a short-day plant and put it in a lab under conditions where it will flower (long nights and short days), but interrupt its day period with a few minutes of darkness, what will happen?
a. It will flower.
b. It will not flower.
c. It will die.
d. It will lose its ability to photosynthesize.
e. It will form new shoots from the axillary buds.

d
Conceptual
Understanding

35.37 Plant hormones produce their effects by
a. altering the expression of genes only.
b. modifying the permeability of the plasma membrane only.
c. modifying the structure of the nuclear envelope membrane only.
d. Both a and b are correct.
e. Both b and c are correct.

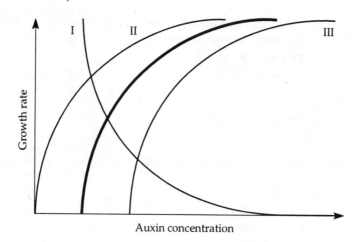

Figure 35.1

a
Application

35.38 The heavy line in Figure 35.1 illustrates the relationship between auxin concentration and cell growth in stem tissues. If the same range of concentrations was applied to lateral buds, what curve would probably be produced?
   a. I only
   b. II only
   c. III only
   d. II and III
   e. either I or III

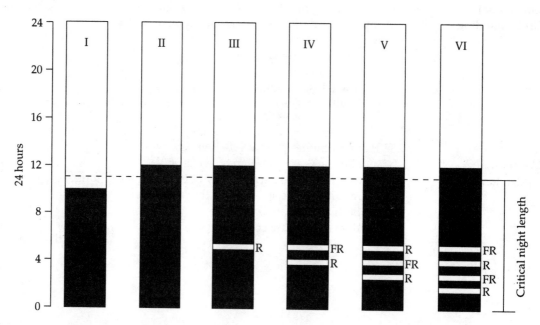

Figure 35.2

Key
white bars = daylight
black bars = night
R = intense flash of red light
FR = intense flash of far-red light

d
Application

35.39 A certain long-day plant has a critical night length of 11 hours for flowering. Under which of the following light regimens will this plant flower? (Refer to Figure 35.2.)
a. I only
b. II only
c. III only
d. I, III, and V
e. II, V, and VI

a
Factual Recall

35.40 Which of the following is NOT presently considered to be a major mechanism whereby hormones control plant development?
a. affecting cell respiration via regulation of the Krebs cycle
b. affecting cell division via the cell cycle
c. affecting cell elongation through acid growth
d. affecting cell differentiation through altered gene activity
e. mediating short-term physiological responses to environmental stimuli

d
Conceptual
Understanding

35.41 Ethylene, as an example of a plant hormone, may have multiple effects on a plant, depending on all of the following EXCEPT the
a. site of action within the plant.
b. developmental stage of the plant.
c. concentration of ethylene.
d. altered chemical structure of ethylene from a gas to a liquid.
e. readiness of cell membrane receptors for the ethylene.

b
Factual Recall

35.42 The only known naturally occurring auxin is
a 2, 4-D.
b. IAA.
c. GA.
d. ABA.
e. TCA.

c
Application

35.43 Suppose there is a large oak tree on your campus and the city places a very bright street light right next to it on a tall pole. A botanist on the faculty complains to the city council and asks them to remove the light. Most likely the botanist is concerned because the light
a. will alter the photosynthetic rate of the tree and keep it growing at night.
b. may cause the stomates to close because of increased ABA synthesis. This could starve the tree for CO_2 and it could die.
c. will change the photoperiod and cause the tree to retain its leaves during the winter. This could cause dehydration and loss of the tree.
d. will cause the tree to bend toward the light on the pole and it could fall.
e. will stimulate ethylene production, premature senescence, and early death of the tree.

a
Factual Recall

35.44 Evidence for phototropism due to the asymmetric distribution of auxin moving down the stem
a. has not been found in dicots such as sunflower and radish.
b. has been found in all monocots and most dicots.
c. has been shown to involve only IAA stimulation of cell elongation on the dark side of the stem.
d. can be demonstrated with unilateral red light but not blue light.
e. is now thought by most plant scientists NOT to involve the shoot tip.

a
Conceptual
Understanding

35.45 A botanist discovers a plant that lacks the ability to form starch grains in root cells, yet the roots still grow downward. This evidence refutes the long-standing hypothesis that
a. falling statoliths trigger gravitropism.
b. starch accumulation triggers the negative phototropic response of roots.
c. starch grains block the acid growth response in roots.
d. starch is converted to auxin, which causes the downward bending in roots.
e. starch and downward movement are necessary for thigmotropism.

b
Factual Recall

35.46 In legumes, it has been shown that sleep movements are correlated with
a. positive thigmotropisms.
b. rhythmic opening and closing of K^+ channels in motor cell membranes.
c. senescence (the aging process in plants).
d. flowering and fruit development.
e. ABA-stimulated closing of guard cells caused by loss of K^+.

c
Factual Recall

35.47 The transduction pathway that activates systemic acquired resistance in plants is initially signaled by
a. antisense RNA.
b. P_{fr} phytochrome.
c. salicylic acid.
d. abscisic acid.
e. red, but not far-red, light.

a
Factual Recall

35.48 You are part of a desert plant research team trying to discover crops that will be productive in arid climates. You discover a plant that produces a guard-cell hormone under water deficit conditions. Most likely the hormone is
a. ABA.
b. GA.
c. IAA.
d. 2, 4-D.
e. salicylic acid.

e
Conceptual
Understanding

35.49 The best analogy to the way phytochrome functions in plants is
a. a computer with no monitor but considerable memory.
b. a dam and electrical power plant on a large river.
c. a rock falling down a metabolic waterfall.
d. a red gene connected to a far-red enzyme.
e. a molecular light switch.

e
Factual Recall

35.50 Plant cells begin synthesizing large quantities of heat-shock proteins
  a. after the induction of chaperone proteins.
  b. in response to the lack of CO_2 following the closing of stomates by ethylene.
  c. when desert plants are quickly removed from high temperatures.
  d. when they are subjected to moist heat (steam) followed by electrical shock.
  e. when the air is above 49°C around temperate region species.

e
Factual Recall

35.51 In extremely cold regions, woody species may survive freezing temperatures by
  a. emptying water from the vacuoles to prevent freezing.
  b. decreasing the numbers of phospholipids in cell membranes.
  c. decreasing the fluidity of all cellular membranes.
  d. producing canavanine as a natural antifreeze.
  e. changing solute concentrations inside cells to allow supercooling.

b
Factual Recall

35.52 A pathogenic fungus invades a plant. The infected plant produces _____ in response to the attack.
  a. antisense RNA
  b. phytoalexins
  c. phytochrome
  d. statoliths
  e. thickened cellulose microfibrils in the cell wall

Refer to Figure 35.3 to answer Questions 35.53 and 35.54.

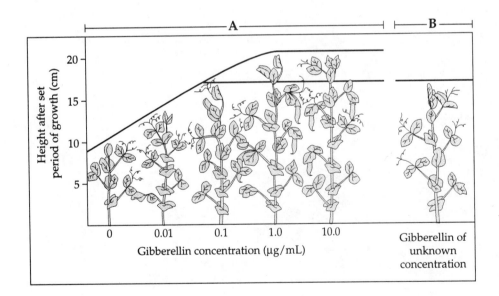

Figure 35.3

d
Conceptual
Understanding

35.53 The results of this experiment shown on the left of the graph (area A) may be used
   a. to show that these plants can live without gibberellin.
   b. to show that gibberellin is necessary in positive gravitropism.
   c. to show that taller plants with more gibberellin produce fruit (pods).
   d. as a bioassay for gibberellin.
   e. to study phytoalexins in plants.

c
Application

35.54 The amount of GA in the experimental plant (B) is approximately
   a. zero.
   b. 0.01 µg/mL.
   c. 0.1 µg/mL.
   d. 1.0 µg/mL.
   e. equal to the amount of gibberellin in the shortest plant.

Chapter 36

a
Factual Recall

36.1 What is stratified cuboidal epithelium composed of?
    a. several layers of boxlike cells
    b. a hierarchical arrangement of flat cells
    c. a tight layer of square cells attached to a basement membrane
    d. an irregularly arranged layer of pillarlike cells
    e. a layer of ciliated, mucus-secreting cells

d
Factual Recall

36.2 Interstitial fluid
    a. forms the extracellular matrix of connective tissue.
    b. is the internal environment found in animal cells.
    c. is composed of blood.
    d. provides for the exchange of materials between blood and cells.
    e. is found inside the small intestine

b
Conceptual
Understanding

36.3 In a typical multicellular animal, the circulatory system interacts with specialized surfaces in order to exchange materials with the exterior environment. Which of the following is NOT an example of such an exchange surface?
    a. lung
    b. muscle
    c. skin
    d. intestine
    e. kidney

d
Factual Recall

36.4 "Stratified columnar" is a description that might apply to what type of animal tissue?
    a. connective
    b. striated muscle
    c. nerve
    d. epithelial
    e. bone

e
Conceptual
Understanding

36.5 The epithelium best adapted for a body surface subject to abrasion is
    a. simple squamous.
    b. simple cuboidal.
    c. simple columnar.
    d. stratified columnar.
    e. stratified squamous.

b
Factual Recall

36.6 Muscles are joined to bones by
    a. ligaments.
    b. tendons.
    c. loose connective tissue.
    d. Haversian systems.
    e. positive feedback.

e
Factual Recall

36.7 The fibroblasts secrete
  a. fats.
  b. chondrin.
  c. interstitial fluids.
  d. calcium phosphate for bone.
  e. proteins for connective fibers.

b
Factual Recall

36.8 Which type of muscle is responsible for peristalsis along the digestive tract?
  a. cardiac
  b. smooth
  c. voluntary
  d. striated
  e. skeletal

c
Conceptual
Understanding

36.9 In mammals, the diaphragm separates the abdominal cavity from the
  a. coelom.
  b. pharynx.
  c. thoracic cavity.
  d. gastrovascular cavity.
  e. oral cavity.

a
Conceptual
Understanding

36.10 Which of the following is a problem that had to be solved as animals increased in size?
  I.   decreasing surface-to-volume ratio
  II. reproducing in aqueous environments
  III. increasing tendency for larger bodies to be more variable

  a. I only
  b. II only
  c. III only
  d. I and III only
  e. I, II, and III

a
Factual Recall

36.11 Cardiac muscle is which of the following?
  a. striated and branched
  b. striated and unbranched
  c. smooth and voluntary
  d. striated and voluntary
  e. smooth and involuntary

d
Factual Recall

36.12 Which of the following tissues lines kidney ducts?
  a. connective
  b. smooth muscle
  c. nervous
  d. epithelial
  e. adipose

c
Factual Recall

36.13 Which of the following apply to skeletal muscle?
  a. smooth and involuntary
  b. smooth and unbranched
  c. striated and voluntary
  d. smooth and voluntary
  e. striated and branched

a
Factual Recall

36.14 Cartilage is described as which of the following types of tissue?
    a. connective
    b. reproductive
    c. nervous
    d. epithelial
    e. adipose

e
Factual Recall

36.15 Bones are held together at joints by
    a. negative feedback.
    b. Haversian systems.
    c. loose connective tissue.
    d. tendons.
    e. ligaments.

c
Conceptual
Understanding

36.16 Which of the following fibers has the greatest tensile strength?
    a. elastin fibers
    b. fibrin fibers
    c. collagenous fibers
    d. reticular fibers
    e. spindle fibers

c
Conceptual
Understanding

36.17 Which of the following is an example of positive feedback?
    a. An increase in blood sugar concentration increases the amount of the hormone that stores sugar as glycogen.
    b. A decrease in blood sugar concentration increases the amount of the hormone that converts glycogen to glucose.
    c. An infant's suckling at the mother's breast increases the amount of the hormone that induces the release of milk from the mammary glands.
    d. An increase in calcium concentration increases the amount of the hormone that stores calcium in bone.
    e. A decrease in calcium concentration increases the amount of the hormone that releases calcium from bone.

b
Factual Recall

36.18 Why must multicellular organisms keep their cells awash in an "internal pond"?
    a. Negative feedback will only operate in interstitial fluids.
    b. All cells need an aqueous medium for the exchange of food, gases, and wastes.
    c. The cells of multicellular organisms tend to lose water because of osmosis.
    d. The cells of multicellular organisms tend to accumulate wastes, a consequence of diffusion.
    e. This phenomenon only occurs in aquatic organisms because terrestrial organisms have adapted to life in a dry situation.

Questions 36.19–36.23 refer to the diagrams of tissues shown in Figure 36.1.

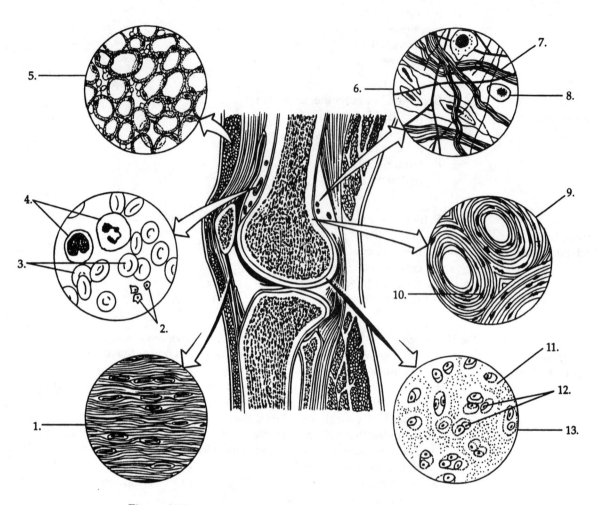

Figure 36.1

a
Conceptual
Understanding

36.19 Which of the following numbers represents a tissue found in tendons?
    a. 1
    b. 5
    c. 6
    d. 9
    e. 13

d
Conceptual
Understanding

36.20 Which of the following numbers represents a tissue rich in hydroxyapatite?
    a. 1
    b. 5
    c. 6
    d. 9
    e. 13

b
Conceptual
Understanding

36.21 Which of the following numbers represents a tissue rich in fat?
a. 1
b. 5
c. 6
d. 9
e. 13

e
Conceptual
Understanding

36.22 Which of the following numbers represents chondrocytes?
a. 3
b. 4
c. 8
d. 10
e. 12

d
Conceptual
Understanding

36.23 Which of the following numbers represents the location of osteocytes?
a. 2
b. 3
c. 8
d. 10
e. 12

Questions 36.24–36.28 refer to the structures listed below. From the list, select the term that best matches each of the following statements or descriptions. Each term may be used once, more than once, or not at all.
a. *dendrites*
b. *mesenteries*
c. *platelets*
d. *intercalated discs*
e. *sarcomeres*

e
Factual Recall

36.24 The basic contractile units of muscles.

c
Factual Recall

36.25 Participate in the blood clotting process.

d
Factual Recall

36.26 Help to relay impulses within cardiac muscle.

a
Factual Recall

36.27 Neuron processes that conduct an impulse toward the cell body.

b
Factual Recall

36.28 Functions to suspend organs within the abdominal cavity.

d
Conceptual
Understanding

36.29 Which of the following ideas is NOT consistent with our understanding of animal structure?
   a. The environment imposes similar problems on all animals.
   b. The evolution of structure in an animal is influenced by its environment.
   c. All but the simplest animals demonstrate the same hierarchical levels of organization.
   d. Different animals contain fundamentally different categories of tissues.
   e. Short-term adjustments to environmental changes are mediated by physiological organ systems.

c
Factual Recall

36.30 Which of the following statements best describes a basement membrane?
   a. made up of a single layer of cells
   b. made up of several layers of cells
   d. made up of secretions from cells
   d. associated with the cells of all four categories of tissue
   e. Its function remains a mystery to biologists.

a
Conceptual
Understanding

36.31 Which of the following characteristics of blood best explains its classification as connective tissue?
   a. Its cells are widely dispersed and surrounded by a fluid.
   b. It contains more than one type of blood cell.
   c. It is contained in vessels that "connect" different parts of an organism's body.
   d. Its cells can move from place to place.
   e. It is found within all the organs of the body.

e
Factual Recall

36.32 Which of the following traits is characteristic of ALL types of muscle tissue?
   a. intercalated discs that allow cells to communicate
   b. striated banding pattern seen under the microscope
   c. cells that lengthen when appropriately stimulated
   d. response that can be consciously controlled
   e. cells that contain actin and myosin

b
Conceptual
Understanding

36.33 What is the common functional significance of the many cells making up such seemingly different structures as the lining of the air sacs in the lungs and the wavy lining of the human intestine?
   a. increased oxygen demand from their metabolic activity
   b. increased exchange surface provided by their membranes
   c. greater numbers of cell organelles contained within their cytoplasm
   d. greater protection due to increased cellular mass
   e. lowered basal metabolic rate due to cooperation between cells

c
Factual Recall

36.34 The major role of most homeostatic control systems in animals is to maintain the constancy of fluid
   a. within cells.
   b. within blood vessels.
   c. around cells.
   d. within body cavities.
   e. within the intestine.

a
Conceptual
Understanding

36.35 Which of the following is the best example of an effector's response in negative feedback?
    a. an increase in body temperature resulting from shivering
    b. an increase in body temperature resulting from exercise
    c. an increase in body temperature resulting from exposure to the sun
    d. an increase in body temperature resulting from fever
    e. a decrease in body temperature resulting from shock

d
Application

36.36 After having measured the resting metabolic rate of a particular animal at 25° C, a researcher discovers that the organism's resting metabolic rate is significantly lower at 20° C. Which of the following organisms might this researcher have been studying?
    a. mouse
    b. elephant
    c. human
    d. snake
    e. bird

e
Conceptual
Understanding

36.37 Which of the following measurements would be the LEAST reliable indicator of an animal's metabolic rate?
    a. the amount of ATP produced within its cells
    b. the amount of heat it generates
    c. the amount of oxygen it inspires
    d. the amount of carbon dioxide it expires
    e. the amount of water it drinks

d
Conceptual
Understanding

36.38 Which of the following is an important distinction between the measurement of basal metabolic rate (BMR) and standard metabolic rate (SMR)?
    a. An animal must be fasting for the measurement of SMR.
    b. BMRs are performed only on ectothermic animals.
    c. An organism must be actively exercising for the measurement of BMR.
    d. SMRs must be determined at a specific temperature.
    e. The BMR for a particular animal is usually lower than that animal's SMR.

b
Conceptual
Understanding

36.39 Which of the following statements concerning the relationship of energy expenditure to body size in animals is TRUE?
    a. Large endotherms have higher metabolic rates than small endotherms.
    b. Large ectotherms have lower metabolic rates than small ectotherms.
    c. There is an inverse relationship between the total energy used by an animal and its body size.
    d. There is a direct relationship between metabolic rate and body size.
    e. Only endotherms show an inverse relationship between metabolic rate and body size.

a
Conceptual
Understanding

36.40 An increase in which of the following parameters is most important in the
evolution of specialized exchange surfaces such as the linings of the lungs or
intestines?
a. surface area
b. thickness
c. number of cell layers
d. metabolic rate of its component cells
e. volume of its component cells

b
Conceptual
Understanding

36.41 How does positive feedback differ from negative feedback?
a. Positive feedback benefits the organism, whereas negative feedback is
detrimental.
b. In positive feedback, the effector's response is in the same direction as the
initiating stimulus rather than opposite to it.
c. In positive feedback, the effector increases some parameter
(i.e., temperature), whereas in negative feedback it decreases.
d. Positive feedback systems have effectors, whereas negative feedback
systems utilize receptors.
e. Positive feedback systems have control centers that are lacking in negative
feedback systems.

Chapter 37

b
Conceptual
Understanding

37.1 Which of the following is an advantage of a complete digestive system over a gastrovascular cavity?
    a. Food items are retained longer.
    b. Specialized regions are possible.
    c. Digestive enzymes can be more specific.
    d. Extensive branching is possible.
    e. Intracellular digestion is easier.

d
Conceptual
Understanding

37.2 All of the following are adaptations to an herbivorous diet EXCEPT
    a. broad, flat teeth.
    b. a rumen.
    c. ingestion of feces.
    d. bile salts.
    e. amylase.

c
Factual Recall

37.3 Some nutrients are considered "essential" in the diets of certain animals because
    a. only those animals use the nutrients.
    b. they are subunits of important polymers.
    c. they cannot be manufactured by the organism.
    d. they are necessary coenzymes.
    e. only some foods contain them.

c
Factual Recall

37.4 Which of the following is TRUE of the mammalian digestive system?
    a. All foods begin their enzymatic digestion in the mouth.
    b. After leaving the oral cavity, the bolus enters the larynx.
    c. The epiglottis prevents food from entering the trachea.
    d. Enzymatic digestion continues in the esophagus.
    e. The trachea leads to the esophagus and then to the stomach.

b
Factual Recall

37.5 Which of these animals has a gastrovascular cavity?
    a. bird
    b. hydra
    c. mammal
    d. insect
    e. annelid

c
Factual Recall

37.6 In general, herbivorous mammals have teeth modified for
    a. cutting.
    b. ripping.
    c. grinding.
    d. splitting.
    e. piercing.

e
Conceptual
Understanding

37.7 Cows are able to survive on a diet consisting almost entirely of cellulose because
    a. cows are autotrophic.
    b. the cow, like the rabbit, reingests its feces.
    c. cows can manufacture all 20 amino acids out of sugars in the liver.
    d. unlike humans, the cow's saliva has enzymes capable of digesting cellulose.
    e. cows have cellulose-digesting, symbiotic microorganisms in their rumens.

b
Conceptual
Understanding

37.8 A digestive juice with a pH of 2 probably came from the
    a. mouth.
    b. stomach.
    c. pancreas.
    d. esophagus.
    e. small intestine.

d
Conceptual
Understanding

37.9 All of the following statements about digestion are correct EXCEPT:
    a. Digestion is catalyzed by enzymes.
    b. Digestion cleaves nucleic acids into nucleotides.
    c. Digestion cleaves fats into glycerol and fatty acids.
    d. During digestion the essential macromolecules are directly absorbed.
    e. During digestion polysaccharides and disaccharides are split into simple sugars.

e
Factual Recall

37.10 What are essential amino acids?
    a. those that are absent in fruits and vegetables
    b. the only amino acids found in human proteins
    c. those amino acids that are generally more abundant in vegetables than in meat
    d. one class of vitamins that is indispensable for neurological development
    e. molecules that can't be synthesized by most animals

b
Factual Recall

37.11 What is the substrate of salivary amylase?
    a. protein
    b. starch
    c. sucrose
    d. glucose
    e. maltose

c
Application

37.12 Which of the following would have the lowest metabolic rate (mm O_2/g/hr)?
    a. cows
    b. owls
    c. frogs
    d. humans
    e. pigeons

e
Factual Recall

37.13 What is peristalsis?
    a. a process of fat emulsification in the small intestine
    b. voluntary control of the rectal sphincters regulating defecation
    c. the transport of nutrients to the liver through the hepatic portal vein
    d. loss of appetite, fatigue, dehydration, and nervous disorders
    e. smooth muscle contractions that move food through the alimentary canal

b
Conceptual
Understanding

37.14 Which of the following enzymes has the lowest pH optimum?
  a. amylase
  b. pepsin
  c. lipase
  d. trypsin
  e. sucrase

a
Factual Recall

37.15 The source of trypsinogen and chymotrypsinogen is the
  a. pancreas.
  b. appendix.
  c. gallbladder.
  d. mouth.
  e. liver.

c
Factual Recall

37.16 Which of the following is a correct statement about bile salts?
  a. They are enzymes.
  b. They are manufactured by the pancreas.
  c. They help stabilize fat–water emulsions.
  d. They increase the efficiency of pepsin action.
  e. They are normally an ingredient of gastric juice.

Questions 37.17–37.21 refer to the digestive system structures in Figure 37.1.

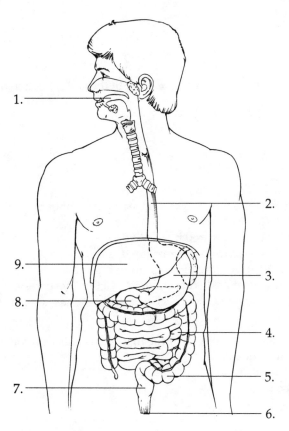

Figure 37.1

d
Factual Recall

37.17 The bicarbonate that neutralizes the acidity of chyme is produced by
  a. 1.
  b. 2.
  c. 3.
  d. 8.
  e. 9.

b
Factual Recall

37.18 Where does the digestion of carbohydrates occur?
  a. 1 and 3
  b. 1 and 4
  c. 3 and 4
  d. 4 and 5
  e. 5 and 7

c
Factual Recall

37.19 Where are the enzymes maltase, sucrase, and lactase produced?
  a. 1
  b. 3
  c. 4
  d. 8
  e. 10

b
Factual Recall

37.20 Where does the digestion of fats occur?
  a. 3 only
  b. 4 only
  c. 1 and 4
  d. 3 and 4
  e. 1, 3, and 4

a
Factual Recall

37.21 Where does the reabsorption of most of the water used in digestion occur?
  a. 4
  b. 5
  c. 6
  d. 7
  e. 8

e
Factual Recall

37.22 Which of the following is a correct statement about trypsin, chymotrypsin, and carboxypeptidase?
  a. They are manufactured by the liver.
  b. They are all forms of the enzyme lipase.
  c. They hydrolyze starch into disaccharides.
  d. They are denatured and rendered inactive by sucrase.
  e. They are activated by the action of enterokinase on zymogens.

a
Factual Recall

37.23 Most enzymatic hydrolysis of the macromolecules in food occurs in the
  a. small intestine.
  b. large intestine.
  c. stomach.
  d. liver.
  e. mouth.

d
Factual Recall

37.24 Most nutrients are absorbed across the epithelium of the
    a. colon.
    b. stomach.
    c. esophagus.
    d. small intestine.
    e. large intestine.

a
Factual Recall

37.25 All of the following statements about nutritional disorders are correct EXCEPT:
    a. Rickets is caused by a vitamin C deficiency.
    b. Weak bones are caused by a calcium deficiency.
    c. Obesity is caused by overnourishment.
    d. Kwashiorkor is caused by a protein deficiency.
    e. Beriberi is caused by a vitamin B_1 deficiency.

d
Factual Recall

37.26 Which of the following is a correct statement about pepsin?
    a. It is manufactured by the pancreas.
    b. It helps stabilize fat–water emulsions.
    c. It splits maltose into monosaccharides.
    d. It is activated by the action of HCl on pepsinogen.
    e. It is denatured and rendered inactive in solutions with low pH.

a
Factual Recall

37.27 An example of a fat-soluble vitamin is
    a. vitamin A.
    b. vitamin B_{12}.
    c. vitamin C.
    d. iodine.
    e. biotin.

d
Conceptual
Understanding

37.28 Blood sugar concentration is likely to vary most in which of these blood vessels?
    a. the abdominal artery
    b. the coronary arteries
    c. the pulmonary veins
    d. the hepatic portal vein
    e. the hepatic vein, which drains the liver

Questions 37.29–37.32 refer to the substances below. From the list, match the term that best fits each of the following descriptions. Each term may be used once, more than once, or not at all.
    a. *zymogen*
    b. *secretin*
    c. *gastrin*
    d. *enterogastrone*
    e. *cholecystokinin*

c
Factual Recall

37.29 Produced by the stomach lining.

e
Factual Recall

37.30 Stimulates the gallbladder to release bile.

b
Factual Recall

37.31 Released by the duodenum in response to the acidic pH of chyme.

d
Factual Recall

37.32 Peristalsis inhibitor that is released in response to a chyme rich in fats.

Questions 37.33–37.35 refer to the substances below. From the list, match the term that best fits each of the following descriptions. Each term may be used once, more than once, or not at all.
> a. *nucleases*
> b. *chylomicrons*
> c. *lipoproteins*
> d. *lysine and leucine*
> e. *thiamine and niacin*

d
Factual Recall

37.33 Essential amino acids.

e
Factual Recall

37.34 Water-soluble vitamins.

b
Factual Recall

37.35 Fat globules transported by exocytosis into the lacteals of the microvilli.

a
Conceptual
Understanding

37.36 Which of the following terms could be applied to any organism with a digestive system?
a. heterotroph
b. autotroph
c. herbivore
d. omnivore
e. bulk-feeder

c
Application

37.37 In a variation of the game "twenty questions," you are asked whether the animal being thought of is a suspension-feeder or not. Which of the following questions could NOT provide you with any useful information?
a. Is it aquatic?
b. Does it have teeth?
c. Is it bigger than your biology book?
d. Does it have claws?
e. Does it have tentacles?

e
Application

37.38 You are "designing" an animal but can provide it with only one protein-digesting enzyme. Which of the following would you choose so that your animal could absorb the maximum amount of amino acids?
a. trypsin
b. pepsin
c. a dipeptidase
d. enteropeptidase
e. carboxypeptidase

d
Factual Recall

37.39 A structure that produces no digestive secretions of any kind is the
   a. duodenum.
   b. pancreas.
   c. salivary gland.
   d. gallbladder.
   e. liver.

b
Conceptual
Understanding

37.40 The process of intracellular digestion is usually preceded by
   a. hydrolysis.
   b. endocytosis.
   c. absorption.
   d. elimination.
   e. secretion.

c
Conceptual
Understanding

37.41 During the process of digestion, fats are broken down when fatty acids are detached from glycerol, and proteins are degraded when amino acids are separated from each other. What do these two processes have in common?
   a. Both processes can be catalyzed by the same enzyme.
   b. Both processes occur intracellularly in most organisms.
   c. Both involve the addition of a water molecule to break bonds.
   d. Both require the presence of hydrochloric acid to lower pH.
   e. Both require ATP as an energy source.

a
Factual Recall

37.42 Because they accumulate in the body, excess ingestion of which of the following can have toxic effects?
   a. fat-soluble vitamins
   b. water-soluble vitamins
   c. calcium and phosphorus
   d. proteins
   e. sugars

d
Conceptual
Understanding

37.43 To actually enter the body, a substance must cross a cell membrane. During which stage of food processing does this first happen?
   a. ingestion
   b. digestion
   c. hydrolysis
   d. absorption
   e. elimination

c
Conceptual
Understanding

37.44 The body is capable of catabolizing many substances as sources of energy. Which of the following could be used as a source of energy but would be the last utilized for this purpose?
   a. fat in adipose tissue
   b. glucose in the blood
   c. protein in muscle cells
   d. glycogen in muscle cells
   e. calcium phosphate in bone

e
Factual Recall

37.45 A hormone produced by the epithelial lining of the stomach is
a. secretin.
b. chymotrypsin.
c. cholecystokinin.
d. enterogastrone.
e. gastrin.

d
Conceptual
Understanding

37.46 How does the digestion and absorption of fat differ from that of carbohydrates?
a. Processing of fat does not require any digestive enzymes, whereas the processing of carbohydrates does.
b. Fat absorption occurs in the stomach, whereas carbohydrates are absorbed from the small intestine.
c. Carbohydrates need to be emulsified before they can be digested, whereas fats do not.
d. Most absorbed fat first enters the lymphatic system, whereas carbohydrates directly enter the blood.
e. Fat must be worked on by bacteria in the large intestine before it can be absorbed, which is not the case for carbohydrates.

b
Conceptual
Understanding

37.47 Animals require 20 basic amino acids. An amino acid that is referred to as nonessential would be best described as one that
a. is not one of the 20.
b. can be made from other substances.
c. is not used by the organism in biosynthesis.
d. must be ingested in the diet.
e. is less important than an essential amino acid.

a
Conceptual
Understanding

37.48 Which of the following glandular secretions involved in digestion would be most likely to be released initially as zymogens?
a. protein-digesting enzymes
b. fat-solubilizing bile salts
c. acid-neutralizing bicarbonate
d. carbohydrate-digesting enzymes
e. hormones such as gastrin

e
Factual Recall

37.49 Which of the following is an enzyme produced by two entirely different accessory glands?
a. pepsin
b. trypsin
c. aminopeptidase
d. lactase
e. amylase

c
Conceptual
Understanding

37.50 Which of the following terms could be used accurately to describe absorbed nutrients?
a. macromolecules
b. polymers
c. monomers
d. enzymes
e. peptides

Chapter 38

d
Factual Recall

38.1 Where do air-breathing insects carry out gas exchange?
    a. in specialized external gills
    b. in specialized internal gills
    c. in the alveoli of their lungs
    d. across the membranes of cells
    e. across the thin cuticular exoskeleton

e
Conceptual
Understanding

38.2 Which one of these statements about lungs is FALSE?
    a. Gas exchange takes place across moist membranes.
    b. The gases move across the exchange membranes by diffusion.
    c. The total exchange surface area is large.
    d. The lining of the alveoli is only one cell thick.
    e. The concentration of CO_2 is higher in the air than in the alveolar capillaries.

d
Conceptual
Understanding

38.3 Which one of the following statements about gills operating in water is FALSE?
    a. Water can support the delicate gill features.
    b. Most fish actively pump water over their gills.
    c. Keeping membranes moist is no problem.
    d. Water carries more oxygen than air and therefore makes gills more efficient than lungs.
    e. Gills have evolved many times in aquatic animals.

a
Conceptual
Understanding

38.4 Which of the following features do all gas exchange systems have in common?
    a. The exchange surfaces are moist.
    b. They are enclosed within ribs.
    c. They are maintained at a constant temperature.
    d. They are exposed to air.
    e. They are found only in animals.

d
Factual Recall

38.5 In which animal does blood flow from the respiratory organ to the heart before circulating through the rest of the body?
    a. annelid
    b. mollusk
    c. fish
    d. frog
    e. insect

c
Conceptual
Understanding

38.6 Which of the following blood components would interfere with the functioning of an open circulatory system but not a closed one?
    a. electrolytes
    b. water
    c. red blood cells
    d. amino acids
    e. antibodies

c
Conceptual
Understanding

38.7 All of the following are reasons why gas exchange is more difficult for aquatic animals than it is for terrestrial animals EXCEPT:
a. Water is denser than air.
b. Water contains much less O_2 than air per unit volume.
c. Gills have less surface area than lungs.
d. Water is harder to pump than air.
e. Exchanging gases with water causes substantial heat loss.

d
Conceptual
Understanding

38.8 At sea level, atmospheric pressure is 760 mm Hg. Oxygen gas is approximately 20% of the total gases in the atmosphere. What is the partial pressure of oxygen?
a. 0.2 mm Hg
b. 20.0 mm Hg
c. 76.0 mm Hg
d. 152.0 mm Hg
e. 508.0 mm Hg

b
Application

38.9 A red blood cell is in an artery in the left arm of a human. How many capillary beds must this cell pass through before it is returned to the left ventricle of the heart?
a. one
b. two
c. three
d. four
e. five

a
Conceptual
Understanding

38.10 Which one of the following animals would have the highest heart rate?
a. rat
b. cat
c. human
d. horse
e. elephant

a
Conceptual
Understanding

38.11 Most of the carbon dioxide carried by the blood in humans is carried as
a. bicarbonate ions in the plasma.
b. CO_2 attached to hemoglobin.
c. carbonic acid in the erythrocytes.
d. CO_2 dissolved in the plasma.
e. bicarbonate attached to hemoglobin.

c
Conceptual
Understanding

38.12 What is the reason why fluid is forced out of systemic capillaries at the arteriolar end?
a. The osmotic pressure of the interstitial fluid is greater than that of the blood.
b. The hydrostatic pressure of the blood is less than that of the interstitial fluid.
c. The hydrostatic pressure of the blood is greater than the osmotic pressure of the interstitial fluid.
d. The osmotic pressure of the interstitial fluid is greater than the hydrostatic pressure of the blood.
e. The osmotic pressure of the blood is greater than the hydrostatic pressure of the interstitial fluid.

b
Application

38.13 If a molecule of CO_2 released into the blood in your left toe travels out of your nose, it must pass through all of the following structures EXCEPT the
    a. right atrium.
    b. pulmonary vein.
    c. alveolus.
    d. trachea.
    e. right ventricle.

e
Factual Recall

38.14 Which of the following is FALSE concerning the hemoglobin molecule?
    a. It contains amino acids.
    b. It contains iron.
    c. It is composed of four polypeptide chains.
    d. It can bind four O_2 molecules.
    e. It is found in humans only.

d
Factual Recall

38.15 Select the formula below that correctly represents the behavior of CO_2 in the blood.
    a. $CO_2 + H_2O \leftrightarrow H_2O_3 + CO$
    b. $CO_2 + H^+ \leftrightarrow HCO_3^+ \leftrightarrow H_2O + H_2CO_3$
    c. $CO_2 + HCO_3^- \leftrightarrow H^+ \leftrightarrow H_2O + H_2CO_3$
    d. $CO_2 + H_2O \leftrightarrow H_2CO_3 \leftrightarrow HCO_3^- + H^+$
    e. $CO_2 + H_2CO_3 + O_2 \leftrightarrow H_2O + CO_2 + H_2CO_3$

b
Conceptual
Understanding

38.16 What is necessary in order for a mother to supply her fetus with oxygen?
    a. The affinity of her hemoglobin for oxygen must increase during early pregnancy.
    b. Fetal hemoglobin must have a greater affinity for oxygen than does maternal hemoglobin.
    c. The fetus must synthesize myoglobin during early development.
    d. The pH of maternal blood must be kept as alkaline as possible.
    e. The blood types of mother and fetus must be compatible.

b
Factual Recall

38.17 In comparing myoglobin to adult hemoglobin, it would be correct to say that myoglobin
    a. has approximately the same molecular mass.
    b. has a greater affinity for oxygen.
    c. contains magnesium instead of iron.
    d. contains a greater amount of iron.
    e. Both a and d are correct.

Refer to Figure 38.1 to answer Questions 38.18–38.20.

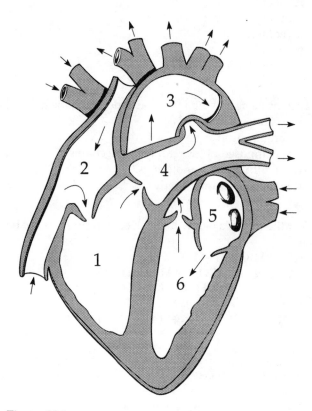

Figure 38.1

e
Factual Recall

38.18 Chambers or vessels that carry oxygenated blood include which of the
following?
   a. 1 and 2 only
   b. 3 and 4 only
   c. 5 and 6 only
   d. 1, 2, and 4
   e. 3, 5, and 6

c
Factual Recall

38.19 Blood is carried directly to the lungs from which of the following?
   a. 2
   b. 3
   c. 4
   d. 5
   e. 6

c
Factual Recall

38.20 The correct sequence of blood flow beginning at the pulmonary artery and
passing through the systematic circulation is
   a. 2—1—4—systematic—3—6—5.
   b. 3—6—5—systematic—4—1—2.
   c. 4—5—6—3—systematic—2—1.
   d. 4—6—3—systematic—2—1—5.
   e. 5—6—3—systematic—2—1—4—3.

e
Factual Recall

38.21 A person may be predisposed to a heart attack or stroke if he or she has which of the following conditions?
a. thrombosis
b. high blood pressure (hypertension)
c. atherosclerosis
d. Only a and b are correct.
e. a, b, and c are correct.

b
Factual Recall

38.22 Which of the following sequences does NOT indicate a direct pathway that blood might follow in mammalian circulation?
a. left ventricle → aorta
b. right ventricle → pulmonary vein
c. pulmonary vein → left atrium
d. vena cava → right atrium
e. right ventricle → pulmonary artery

b
Conceptual
Understanding

38.23 Through how many capillary beds must a red blood cell of a human travel if it takes the shortest possible route from the right ventricle to the right atrium?
a. 1
b. 2
c. 3
d. 4
e. 5

b
Factual Recall

38.24 What does it mean if the blood pressure of a human is 110/80?
a. The systolic pressure is 80.
b. The diastolic pressure is 80.
c. The pulse rate is 80 beats per minute.
d. The blood pressure during contraction of the heart is 80.
e. Both b and c are correct.

c
Conceptual
Understanding

38.25 Which of the following are the only vertebrates in which blood flows directly from respiratory organs to body tissues without first returning to the heart?
a. amphibians
b. birds
c. fishes
d. mammals
e. reptiles

c
Factual Recall

38.26 Tracheal systems for gas exchange are found in
a. crustaceans.
b. earthworms.
c. insects.
d. jellyfish.
e. vertebrates.

c
Conceptual
Understanding

38.27 Which of the following is correct for a blood pressure of 130/80?
    I. The systolic pressure is 130.
    II. The diastolic pressure is 80.
    III. The blood pressure during heart contraction is 80.

    a. I only
    b. III only
    c. I and II only
    d. II and III only
    e. I, II, and III

b
Factual Recall

38.28 The meshwork that forms the fabric of a blood clot mostly consists of which protein?
    a. fibrinogen
    b. fibrin
    c. thrombin
    d. prothrombin
    e. collagen

b
Application

38.29 If, during protein starvation, the osmotic pressure on the venous side of capillary beds drops below the hydrostatic pressure, then
    a. hemoglobin will not release oxygen.
    b. fluids will tend to accumulate in tissues.
    c. the pH of the interstitial fluids will increase.
    d. most carbon dioxide will be bound to hemoglobin and carried away from tissues.
    e. plasma proteins will escape through the endothelium of the capillaries.

c
Factual Recall

38.30 Human plasma proteins include which of the following?
    I. fibrinogen
    II. hemoglobin
    III. immunoglobulin

    a. I only
    b. II only
    c. I and III only
    d. II and III only
    e. I, II, and III

a
Factual Recall

38.31 The phenomenon that increases the gas exchange efficiency of fish gills is the
    a. countercurrent exchange mechanism.
    b. largest tidal volume of all vertebrates.
    c. high degree of oxygen saturation in water.
    d. back-and-forth movement of water that maximizes oxygen uptake.
    e. large blood flow velocity and pressure found in some fish species.

c
Conceptual
Understanding

38.32 All of the following respiratory surfaces are associated with capillary beds
EXCEPT the
a. gills of fishes.
b. alveoli of lungs.
c. tracheae of insects.
d. skin of earthworms.
e. skin of frogs.

a
Conceptual
Understanding

38.33 Which of the following occurs with the exhalation of air from human lungs?
a. The volume of the thoracic cavity decreases.
b. The residual volume of the lungs decreases.
c. The diaphragm contracts.
d. The epiglottis closes.
e. The rib cage expands.

b
Conceptual
Understanding

38.34 Air flows continuously in one direction through the lungs of which animals?
a. frogs
b. birds
c. mammals
d. crocodiles
e. flying insects

d
Conceptual
Understanding

38.35 Breathing is usually regulated by
a. erythropoietin levels in the blood.
b. the concentration of red blood cells.
c. hemoglobin levels in the blood.
d. CO_2 and O_2 concentration and pH-level sensors.
e. the lungs and the larynx.

*Questions 38.36–38.40 refer to the following conditions. Each term may be used once, more
than once, or not at all.*
a. *atherosclerosis*
b. *arteriosclerosis*
c. *hypertension*
d. *heart murmur*
e. *hemophilia*

c
Factual Recall

38.36 High blood pressure.

e
Factual Recall

38.37 An inherited defect in the clotting process.

d
Factual Recall

38.38 Defect in one or more of the valves of the heart.

b
Factual Recall

38.39 Calcification of plaques lining the inner walls of arteries.

354   **Chapter 38**

a
Factual Recall

38.40 Plaque formation by infiltration of lipids into arterial smooth muscles of arteries.

Questions 38.41–38.44 refer to the following answers. Each answer can be used once, more than once, or not at all.
   a. *low-density lipoproteins*
   b. *immunoglobulins*
   c. *erythropoietin*
   d. *epinephrine*
   e. *platelets*

d
Factual Recall

38.41 Speeds up heart rate.

e
Factual Recall

38.42 Part of the formed elements of the blood.

c
Factual Recall

38.43 Stimulates the production of red blood cells.

e
Factual Recall

38.44 An early participant in the clotting process.

Questions 38.45 and 38.46 refer to the data shown below. Blood entering a vertebrate's capillary bed was measured for the pressures exerted by various factors.

	Arterial End of Capillary Bed	Venous End of Capillary Bed
Hydrostatic pressure	8 mm Hg	14 mm Hg
Osmotic pressure	26 mm Hg	26 mm Hg
P_{O_2}	100 mm Hg	42 mm Hg
P_{CO_2}	40 mm Hg	46 mm Hg

e
Application

38.45 For this capillary bed, which of the following statements is CORRECT?
   a. The pH is lower on the arterial side than on the venous side.
   b. Oxygen is taken up by the erythrocytes within the capillaries.
   c. The osmotic pressure remains constant due to carbon dioxide compensation.
   d. The hydrostatic pressure declines from the arterial to the venous sides because oxygen is lost.
   e. Fluids will leave the capillaries on the arterial side of the bed and reenter on the venous side.

d
Conceptual
Understanding

38.46 The site of this capillary bed could be all of the following EXCEPT the
   a. pancreas.
   b. muscle tissue.
   c. medulla.
   d. alveoli.
   e. kidneys.

a
Conceptual
Understanding

38.47 Which of the following is an example of countercurrent flow?
   a. the flow of water across the gills of a fish and the flow of blood within those gills
   b. the flow of blood in the dorsal vessel of an insect and the flow of air within its tracheae
   c. the flow of air within the primary bronchi of a human and the flow of blood within the pulmonary veins
   d. the flow of water across the skin of a frog and the flow of blood within the ventricle of its heart
   e. the flow of fluid out of the arterial end of a capillary and the flow of fluid back into the venous end of that same capillary

b
Application

38.48 At the summit of a high mountain, the atmospheric pressure is only half of its normal value of 760 mm Hg. If the atmosphere is still composed of 21% oxygen, what is the partial pressure of oxygen at this altitude?
   a. 0 mm Hg
   b. 80 mm Hg
   c. 160 mm Hg
   d. 380 mm Hg
   e. 760 mm Hg

d
Factual Recall

38.49 Which of the following are characteristics of both hemoglobin and hemocyanin?
   a. found within blood cells
   b. red in color
   c. contain the element iron
   d. transport oxygen
   e. occur in mammals

e
Conceptual
Understanding

38.50 If the atrioventricular node could be surgically removed from the heart without disrupting signal transmission to the Purkinje fibers,
   a. no apparent effect on heart activity would be observed.
   b. the heart rate would be decreased.
   c. only the ventricles would contract.
   d. only the atria would contract.
   e. atria and ventricles would contract at about the same time.

d
Factual Recall

38.51 The velocity of blood flow is lowest in
   a. the aorta.
   b. arteries.
   c. arterioles.
   d. capillaries.
   e. veins.

e
Factual Recall

38.52 Average blood pressure is lowest in
    a. the aorta.
    b. arteries.
    c. arterioles.
    d. capillaries.
    e. veins.

a
Conceptual
Understanding

38.53 Air rushes into the lungs of humans during inspiration because
    a. the volume of the thoracic cavity increases.
    b. pressure in the alveoli increases.
    c. the diaphragm contracts and pushes upward on the chest cavity.
    d. pulmonary muscles contract and pull on the outer surface of the lungs.
    e. smooth muscle lining the trachea, bronchi, and bronchioles contracts and causes their volume to increase.

c
Conceptual
Understanding

38.54 If your erythrocytes contained NO hemoglobin, but your respiratory system was functioning normally, which of the following statements would be correct?
    a. Your blood would contain no oxygen.
    b. Your blood would contain its normal amount of oxygen.
    c. The partial pressure of oxygen in your blood would be normal.
    d. The partial pressure of oxygen in your blood would be zero.
    e. Your body cells would receive no oxygen.

e
Factual Recall

38.55 Plasma proteins in humans have many functions. Which of the following is NOT one of them?
    a. maintenance of blood osmotic pressure
    b. transport of water-insoluble lipids
    c. blood clotting
    d. immune responses
    e. oxygen transport

d
Factual Recall

38.56 The interstitial fluid of humans has the same composition as
    a. blood.
    b. plasma.
    c. intracellular fluid.
    d. lymph.
    e. hemolymph.

b
Factual Recall

38.57 Most of the carbon dioxide in the blood of humans is transported
    a. as dissolved CO_2 in plasma.
    b. as bicarbonate in plasma.
    c. attached to hemoglobin in red blood cells.
    d. attached to hemocyanin in plasma.
    e. as carbonic acid.

a
Factual Recall

38.58 Which of the following is NOT a normal event in the process of blood clotting?
    a. production of erythropoietin
    b. conversion of fibrinogen to fibrin
    c. activation of prothrombin to thrombin
    d. adhesion of platelets
    e. clotting factor release by clumped platelets

d
Conceptual
Understanding

38.59 If a person were suffering from edema, which of the following conditions would reduce it?
a. decreased plasma protein production by the liver
b. a prolonged starvation diet
c. an obstruction in the lymphatic system
d. lower blood pressure
e. enlarged clefts between capillary endothelial cells due to damage or inflammation

Chapter 39

39.1 What are antigens?
   a. proteins found in the blood that cause foreign blood cells to clump
   b. proteins embedded in B-cell membranes
   c. proteins that consist of two light and two heavy polypeptide chains
   d. antibody-generating foreign macromolecules
   e. Both a and c are correct.

39.2 A transfusion of type A blood given to a person who has type O blood would result in
   a. the recipient's B antigens reacting with the donated anti-B antibodies.
   b. the recipient's anti-A antibodies clumping the donated red blood cells.
   c. the recipient's anti-A and anti-O antibodies reacting with the donated red blood cells if the donor was a heterozygote (Ai) for blood type.
   d. no reaction because type O is a universal donor.
   e. no reaction because the O-type individual does not have antibodies.

39.3 The following events occur when a mammalian immune system first encounters a pathogen. Place them in correct sequence, and then choose the answer that indicates that sequence.
   I. Pathogen is destroyed.
   II. Lymphocytes secrete antibodies.
   III. Antigenic determinants from pathogen bind to antigen receptors on lymphocytes.
   IV. Lymphocytes specific to antigenic determinants from pathogen become numerous.
   V. Only memory cells remain.

   a. I, III, II, IV, V
   b. III, II, I, V, IV
   c. II, I, IV, III, V
   d. IV, II, III, I, V
   e. III, IV, II, I, V

39.4 Physical barriers to invasion by other organisms
   a. include the skin and the mucous membranes.
   b. are difficult for bacteria and viruses to penetrate.
   c. may work in conjunction with secretions like tears, perspiration, and mucus.
   d. Only a and c are correct.
   e. a, b, and c are correct.

d
Application

39.5 Jenner successfully used cowpox virus as a vaccine against a different virus that causes smallpox. Why was he successful even though he used viruses of different kinds?
a. All of the below are true.
b. The immune system responds nonspecifically to antigens.
c. The cowpox virus made antibodies in response to the presence of smallpox.
d. Cowpox and smallpox are antibodies with similar immunizing properties.
e. There are some antigenic determinants common to both pox viruses.

c
Conceptual
Understanding

39.6 The clonal selection theory is an explanation for
a. how a single type of stem cell can produce both red blood cells and white blood cells.
b. how antibody proteins can be molded to fit antigens after the antigen interacts with an antibody-producing type of cell.
c. how an antigen can provoke development of very few cells to result in production of high levels of specific antibodies.
d. how HIV can disrupt the immune system.
e. how macrophages can recognize specific T cells and B cells.

For Questions 39.7–39.11, match the following answers with the phrase that best describes them. Each answer may be used once, more than once, or not at all.
a. *cytotoxic T cells*
b. *delayed sensitivity T cells*
c. *helper T cells*
d. *suppressor T cells*
e. *B cells*

e
Factual Recall

39.7 Form plasma cells that give rise to antibodies.

c
Factual Recall

39.8 Release cytokines, which activate B cells.

a
Factual Recall

39.9 Release perforin, which causes target cells to lose their cytoplasm.

c
Factual Recall

39.10 Cooperate with macrophages to enable the production of antibodies by effector cells.

b
Factual Recall

39.11 Attack and destroy intracellular pathogens such as the tuberculin bacillus.

e
Conceptual
Understanding

39.12 Which of the following participates in both the specific and nonspecific defense systems of the body?
a. pyrogens
b. complement
c. macrophages
d. B cells
e. Both b and c are correct.

c
Conceptual
Understanding

39.13 Which of the following is an example of positive feedback in the immune system?
  a. Memory cells proliferate and produce antibodies.
  b. Cytotoxic cells release substances that attract macrophages.
  c. Cells release a cytokine that stimulates these cells to divide and release more of the cytokine.
  d. Large antigens with many repeating antigenic determinant sites can stimulate B cells without the aid of T cells.
  e. Both a and b are examples of positive feedback in the immune system.

d
Application

39.14 A doctor discovers that her patient can produce antibodies against some bacterial pathogens, but he is unable to protect himself against viral infections. The doctor suspects a disorder in her patient's
  a. B cells.
  b. plasma cells.
  c. cytotoxic cells.
  d. T cells.
  e. macrophages.

e
Conceptual
Understanding

39.15 The MHC (major histocompatibility complex) is important in
  a. distinguishing self from nonself.
  b. recognizing parasitic pathogens.
  c. identifying bacterial pathogens.
  d. identifying abnormal cells.
  e. Both a and d are correct.

b
Application

39.16 Which of the following would be most beneficial in treating an individual who has been bitten by a poisonous snake which has fast-acting toxin?
  a. vaccination with a weakened form of the toxin
  b. injection of antibodies to the toxin
  c. injection of interleukin-1
  d. injection of interleukin-2
  e. injection of interferon

e
Conceptual
Understanding

39.17 Which of the following are necessary to produce monoclonal antibodies?
  a. fibroblasts
  b. plasma cells
  c. myeloma cells
  d. macrophages
  e. Both b and c are necessary.

d
Conceptual
Understanding

39.18 Which of the following is true of both T cells and B cells?
  a. They produce effector cells against specific pathogens.
  b. They are produced from stem cells of the bone marrow.
  c. They can attack and destroy invading pathogens.
  d. Both a and b are true.
  e. a, b, and c are all true.

a
Application

39.19 Which of the following could prevent the appearance of the symptoms of an allergy attack?
    a. blocking the attachment of the IgE antibodies to the mast cells
    b. blocking the antigenic determinants of the IgM antibodies
    c. reducing the number of T helper cells in the body
    d. Only a and b are correct.
    e. Only b and c are correct.

c
Application

39.20 When a physician discovers a strep infection, he or she may begin immediate treatment with antibiotics. The primary reason for this immediate use of antibiotics is to
    a. boost the formation of antibodies to the strep bacteria.
    b. control the strep bacteria while antibodies are being produced.
    c. destroy the strep bacteria before antibodies can be formed.
    d. prevent the proliferation of the strep bacteria.
    e. Both b and c are correct.

b
Application

39.21 A person suffering from AIDS would be unlikely to suffer from which of the following diseases?
    a. cancer
    b. rheumatoid arthritis
    c. hepatitis
    d. tuberculosis
    e. influenza

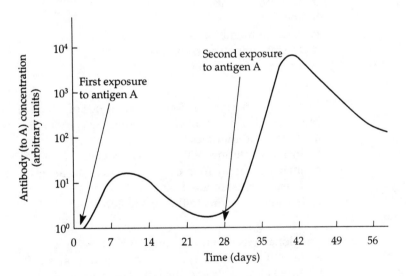

Figure 39.1

Use the graph in Figure 39.1 to answer Questions 39.22–39.24.

a
Application

39.22 When would B cells produce effector cells?
    a. between 0 and 7 days
    b. between 7 and 14 days
    c. between 28 and 35 days
    d. Both a and b are correct.
    e. Both a and c are correct.

e
Application

39.23 When would memory cells proliferate?
   a. between 0 and 7 days
   b. between 7 and 14 days
   c. between 28 and 35 days
   d. between 35 and 42 days
   e. Both a and c are correct.

e
Application

39.24 When would you find antibodies being produced?
   a. between 3 and 7 days
   b. between 14 and 21 days
   c. between 28 and 35 days
   d. Both b and c are correct.
   e. Both a and c are correct.

b
Conceptual
Understanding

39.25 Antibodies of the different classes IgM, IgG, IgA, IgD, and IgE differ from each other in
   a. the way they are produced.
   b. the way they interact with the antigen.
   c. the type of cell that produces them.
   d. the antigenic determinants that they recognize.
   e. the number of carbohydrate subunits they have.

a
Factual Recall

39.26 In mammalian defenses against invading pathogens, all of these are considered nonspecific defense mechanisms EXCEPT
   a. the immune system.
   b. the skin.
   c. mucous membranes.
   d. the inflammatory response.
   e. antimicrobial proteins.

d
Conceptual
Understanding

39.27 In the inflammatory response, the absence of which of the following would prevent all the others from happening?
   a. dilation of arterioles
   b. increased permeability of blood vessels
   c. increased population of phagocytes in the area
   d. release of histamine
   e. leakage of plasma to the affected area

b
Conceptual
Understanding

39.28 The clonal selection theory implies that
   a. related people have similar immune responses.
   b. antigens activate specific lymphocytes.
   c. only certain cells can produce interferon.
   d. memory cells are present at birth.
   e. the body selects which antigens it will respond to.

d
Factual Recall

39.29 An alarm substance that triggers an inflammatory reaction is
   a. thyroxine.
   b. adrenaline.
   c. immunoglobulin.
   d. histamine.
   e. pyrogen.

a
Factual Recall

39.30 Cell-mediated immunity is mostly the function of
   a. T cells.
   b. B cells.
   c. erythrocytes.
   d. complement cells.
   e. cytotoxic cells.

b
Factual Recall

39.31 Prevention of a disease by the injection of an antiserum containing gamma globulins is an example of
   a. active immunity.
   b. passive immunity.
   c. cell-mediated immunity.
   d. clonal selection.
   e. autoimmunity.

c
Conceptual
Understanding

39.32 All of the following are correct statements about nonspecific defenses EXCEPT:
   a. They include inflammatory responses.
   b. They include physical and chemical barriers.
   c. They must be primed by the presence of antigen.
   d. They may involve the formation of membrane attack complexes.
   e. Macrophages and natural killer cells are participants in the process.

d
Conceptual
Understanding

39.33 Which of the following statements about humoral immunity is correct?
   a. It primarily defends against fungi and protozoa.
   b. It is responsible for transplant tissue rejection.
   c. It protects the body against cells that become cancerous.
   d. It is mounted by lymphocytes that have matured in the bone marrow.
   e. It primarily defends against bacteria and viruses that have already infected cells.

a
Factual Recall

39.34 Which of the following cell types is responsible for initiating a secondary immune response?
   a. memory cells
   b. macrophages
   c. stem cells
   d. B cells
   e. T cells

c
Factual Recall

39.35 In the production of monoclonal antibodies, B lymphocytes are fused with
   a. T lymphocytes.
   b. hybridoma cells.
   c. myeloma cells.
   d. mast cells.
   e. memory cells.

c
Conceptual
Understanding

39.36 Inflammatory responses may include all of the following EXCEPT
   a. clotting proteins sealing off a localized area.
   b. increased activity of phagocytes in an inflamed area.
   c. reduced permeability of blood vessels to conserve plasma.
   d. release of substances to increase the blood supply to an inflamed area.
   e. release of substances to stimulate the release of white blood cells from bone marrow.

e
Conceptual
Understanding

39.37 All of the following are usually considered to be disorders of the immune
system EXCEPT
a. AIDS.
b. SCID.
c. lupus erythematosus.
d. allergic anaphylaxis.
e. MHC-induced transplant rejection.

c
Factual Recall

39.38 Plasma cells are
a. immature forms of T cells.
b. cells that produce few antibodies.
c. the effector cells of humoral immunity.
d. responsible for immunological memory.
e. responsible for the phagocytosis of foreign organisms.

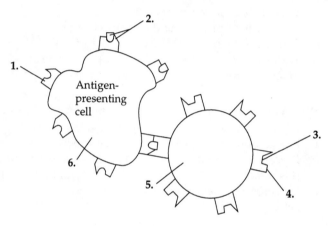

Figure 39.2

Questions 39.39–39.42 refer to the diagram in Figure 39.2.

b
Factual Recall

39.39 The cell represented by number 5 is a
a. B cell.
b. T cell.
c. mast cell.
d. macrophage.
e. plasma cell.

d
Factual Recall

39.40 The MHC binding site is represented by
a. 1.
b. 2.
c. 3.
d. 4.
e. 6.

a
Factual Recall

39.41 The MHC protein is represented by
   a. 1.
   b. 2.
   c. 3.
   d. 4.
   e. 6.

b
Conceptual
Understanding

39.42 The "self and nonself" complex is represented by
   a. 1.
   b. 2.
   c. 3.
   d. 4.
   e. 6.

Questions 39.43–39.46 refer to the following list. Match the term that best fits each of the following descriptions. Each term may be used once, more than once, or not at all.
   a. *capping*
   b. *opsonization*
   c. *complement system*
   d. *passive immunity*
   e. *vaccination*

b
Factual Recall

39.43 Coating of foreign cells by proteins to attract macrophages.

d
Factual Recall

39.44 Protection by antibodies that cross the placenta from mother to fetus.

a
Factual Recall

39.45 Mechanism by which certain antigens stimulate B cells.

e
Factual Recall

39.46 Process in which an attenuated pathogen is used to stimulate the body to produce antibodies against the pathogen.

Questions 39.47–39.49 refer to the following data.

	Case 1	Case 2	Case 3
Mother	Rh^-	Rh^-	Rh^+
Fetus	Rh^+	Rh^-	Rh^-

a
Application

39.47 In which of the following cases could the mother exhibit an anti-Rh-factor reaction to the developing fetus?
   a. case 1 only
   b. case 3 only
   c. cases 1 and 2 only
   d. cases 1, 2, and 3
   e. It cannot be determined from the data given.

c
Application

39.48 In which of the following cases would the mother NOT exhibit an anti-Rh-factor reaction to the developing fetus?
  a. case 1 only
  b. case 3 only
  c. cases 2 and 3 only
  d. cases 1, 2, and 3
  e. It cannot be determined from the data given.

a
Application

39.49 In which of the following cases would the precaution likely be taken to give the mother anti-Rh antibodies before delivering her baby?
  a. case 1 only
  b. case 3 only
  c. cases 1 and 2 only
  d. cases 1, 2, and 3
  e. It cannot be determined from the data given.

b
Conceptual
Understanding

39.50 All of the following statements about antibodies are true EXCEPT:
  a. Antibodies are immunoglobulin proteins.
  b. Antibodies bind with foreign cells and destroy them.
  c. The structure of antibodies includes both a constant and a variable region.
  d. Antibodies act as signals to blood complement proteins or phagocytes.
  e. Plasma B cells are responsible for the production of antibodies.

e
Factual Recall

39.51 The lymphatic system involves which of the following organs?
  a. spleen and lymph nodes
  b. adenoids and tonsils
  c. appendix and special portions of small intestine
  d. Only a and b are correct.
  e. a, b, and c are correct.

d
Conceptual
Understanding

39.52 A major difference between active and passive immunity is that active immunity requires
  a. acquisition and activation of antibodies.
  b. proliferation of lymphocytes in bone marrow.
  c. transfer of antibodies from the mother across the placenta.
  d. direct exposure to a living or simulated disease organism.
  e. secretion of interleukins from macrophages.

d
Conceptual
Understanding

39.53 Why can normal immune responses be described as polyclonal?
  a. Blood contains many different antibodies to many different antigens.
  b. Construction of a hybridoma requires multiple types of cells.
  c. Multiple immunoglobulins are produced from descendants of a single B cell.
  d. Diverse antibodies are produced for different epitopes of a specific antigen.
  e. Macrophages, T cells, and B cells all are involved in normal immune response.

b
Factual Recall

39.54 Which of the following are types of T cells that participate in the immune response system?
    a. CD4, CD8, and helper cells
    b. cytotoxic, suppressor, and helper cells
    c. plasma, antigen-presenting, and memory cells
    d. lymphocytes, macrophages, and coelomocytes
    e. interleukin-1, interleukin-2, and interferon

d
Application

39.55 In order to investigate the immune system of an invertebrate animal, a scientist grafts a section of epidermis from one earthworm to another. What would you expect to happen?
    a. Invertebrates do not have immune responses, so the graft will be accepted.
    b. The graft will be recognized as nonself and rejected.
    c. This graft will be accepted, but a second graft would be rejected.
    d. Amoeboid coelomocytes may be activated against the foreign tissue.
    e. Both a and d would happen.

a
Factual Recall

39.56 All of the following statements about lymphocytes are true EXCEPT:
    a. They are contained only within the lymphatic vessels and nodes.
    b. They may mature in either bone marrow or the thymus gland.
    c. All cells differentiate only from pluripotent stem cells.
    d. Many cells populate organs, such as the spleen or the tonsils, after maturity.
    e. All cells are produced in bone marrow.

c
Factual Recall

39.57 The function of CD4 and CD8 is to assist T cells in
    a. enhancing secretion of proteins such as interferon.
    b. activating B cells and other T cells.
    c. binding of the MHC–antigen complex.
    d. recognition of self cells.
    e. secretion of antibodies specific for each antigen.

Questions 39.58–39.62 refer to the following substances:
    1. *blood complement proteins*
    2. *interleukin-1*
    3. *interleukin-2*
    4. *antibodies*
    5. *perforin*

b
Factual Recall

39.58 Which substance is secreted by antigen-presenting cells?
    a. 1
    b. 2
    c. 3
    d. 4
    e. 5

a
Factual Recall

39.59 Which substances are used in the immune system to kill targeted, infected cells?
    a. 1 and 5
    b. 2 and 3
    c. 4 and 5
    d. 1 and 4
    e. 1, 4, and 5

d
Factual Recall

39.60 Which substance(s) may act to stimulate the proliferation of T cells?
   a. 1 and 4
   b. 1 only
   c. 2 and 3
   d. 3 only
   e. 5 only

e
Factual Recall

39.61 Which substance(s) is (are) used by cytotoxic T cells to lyse and kill infected cells?
   a. 1 and 4
   b. 1 only
   c. 2 and 3
   d. 3 only
   e. 5 only

c
Factual Recall

39.62 Which substance(s) activate(s) T cells?
   a. 1 and 4
   b. 1 only
   c. 2 and 3
   d. 3 only
   e. 5 only

Chapter 40

c
Factual Recall

40.1 Which of the following mechanisms would account for increased urine production as a result of drinking alcoholic beverages?
a. increased aldosterone production
b. increased blood pressure
c. decreased amount of antidiuretic hormone
d. the proximal tubule reabsorbing more water
e. the osmoregulator cells of the brain increasing their activity

c
Factual Recall

40.2 Contractile vacuoles most likely would be found in protists
a. in a marine environment.
b. that are internal parasites.
c. that are osmoregulators.
d. that are osmoconformers.
e. that are hypotonic to their environment.

e
Factual Recall

40.3 Most terrestrial animals dissipate excess heat by
a. countercurrent exchange.
b. acclimation.
c. vasoconstriction.
d. hibernation.
e. evaporation.

Refer to Figure 40.1, a diagram of a renal tubule, to answer Questions 40.4–40.6.

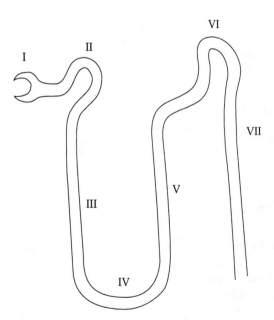

Figure 40.1

a
Factual Recall

40.4 In which region would filtration occur?
   a. I
   b. III
   c. IV
   d. V
   e. VII

e
Factual Recall

40.5 In which region would urine become more concentrated?
   a. I
   b. III
   c. IV
   d. V
   e. VII

c
Factual Recall

40.6 In which regions would K^+ and Na^+ be reabsorbed by the blood?
   a. I and VII
   b. I and V
   c. II and VI
   d. III and V
   e. I and IV

b
Conceptual
Understanding

40.7 All of the following represent adaptations by terrestrial animals to drying conditions EXCEPT
   a. anhydrobiosis.
   b. salt glands.
   c. efficient kidneys.
   d. impervious surfaces.
   e. increased thirst.

b
Application

40.8 Which organism is most completely ectothermic?
   a. lizard
   b. sea star
   c. tuna fish
   d. hummingbird
   e. winter moth

a
Conceptual
Understanding

40.9 The digestion and utilization of which nutrient creates the greatest need for osmoregulation by the kidneys?
   a. protein
   b. starch
   c. fat
   d. oil
   e. cellulose

d
Factual Recall

40.10 Which of the following mechanisms for osmoregulation/nitrogen removal is INCORRECTLY paired with its animal?
   a. nephridium—annelid
   b. Malpighian tubule—insect
   c. kidney—frog
   d. flame cell—roundworm
   e. direct cellular exchange—cnidarian

c
Factual Recall

40.11 What is the process called by which materials are returned to the blood from the nephron fluid?
   a. filtration
   b. ultrafiltration
   c. reabsorption
   d. secretion
   e. active transport

a
Factual Recall

40.12 The advantage of excreting wastes as urea rather than ammonia is that
   a. Both b and c are advantages.
   b. urea is less toxic than ammonia.
   c. urea requires less water for excretion than ammonia.
   d. urea does not affect the osmolar gradient.
   e. urea can be exchanged for Na^+.

c
Conceptual
Understanding

40.13 Compared to the sea water around them, most marine invertebrates are correctly described as which of the following?
   I. hypertonic
   II. hypotonic
   III. isotonic

   a. I only
   b. II only
   c. III only
   d. I and III only
   e. II and III only

c
Factual Recall

40.14 Urea is
   a. insoluble in water.
   b. more toxic to human cells than ammonia.
   c. the nitrogenous waste product of humans.
   d. the nitrogenous waste product of most birds.
   e. the nitrogenous waste product of most aquatic invertebrates.

c
Conceptual
Understanding

40.15 All of the following are mechanisms of thermoregulation in terrestrial mammals EXCEPT
   a. changing the rate of evaporative loss of heat.
   b. changing the rate of metabolic heat production.
   c. changing the rate of heat exchange by anhydrobiosis.
   d. changing the rate of heat loss by vasodilation and vasoconstriction.
   e. relocating to cool areas when too hot or to warm areas when too cold.

a
Factual Recall

40.16 Protonephridia are excretory structures found in
   a. flatworms.
   b. earthworms.
   c. insects.
   d. jellyfish.
   e. vertebrates.

d
Factual Recall

40.17 Terrestrial animals exchange heat with the environment by all of the following physical processes EXCEPT
    a. conduction.
    b. convection.
    c. evaporation.
    d. illumination.
    e. radiation.

b
Conceptual
Understanding

40.18 In addition to their role in gas exchange, fish gills are also directly involved in
    a. digestion.
    b. osmoregulation.
    c. thermoregulation.
    d. the excretion of uric acid.
    e. the release of atrial natriuretic proteins.

e
Factual Recall

40.19 All of the following are functions of the mammalian kidney EXCEPT
    a. water retention.
    b. filtration of blood.
    c. excretion of nitrogenous waste.
    d. regulation of salt balance in the blood.
    e. production of urea as a waste product of protein catabolism.

d
Conceptual
Understanding

40.20 Which of the following statements about the transfer of fluid from the glomerulus to Bowman's capsule is CORRECT?
    a. It results from active transport.
    b. It transfers large molecules as easily as small ones.
    c. It is very selective as to which small molecules are transferred.
    d. It is mainly a consequence of blood pressure force-filtering the fluid.
    e. It usually includes the transfer of red blood cells to the nephron tubule.

d
Application

40.21 If the concentration of glucose in the blood flowing through the kidneys is within the normal range, then which of the following statements is correct?
    a. The nephron pumps glucose into the collecting duct.
    b. Bowman's capsule secrets glucose back into the blood.
    c. No glucose passes from the glomerulus into Bowman's capsule.
    d. Most glucose within the nephron tubule is reabsorbed by the blood.
    e. Most glucose that passes from blood into the nephron tubule remains in the urine and is subsequently excreted.

d
Conceptual
Understanding

40.22 Which of the following processes of osmoregulation by the kidney is the LEAST selective?
    a. salt pumping to control osmolarity
    b. H^+ ion pumping to control pH
    c. reabsorption
    d. filtration
    e. secretion

e
Conceptual
Understanding

40.23 All of the following are aspects of temperature acclimation EXCEPT:
 a. Cells may increase the production of certain enzymes.
 b. Cells may produce enzymes with different temperature optima.
 c. Organisms may adjust some of the mechanisms that control internal temperature.
 d. Cell membranes may change their proportions of saturated and unsaturated fats.
 e. Organisms with countercurrent circulation adaptations have difficulty with thermoregulation.

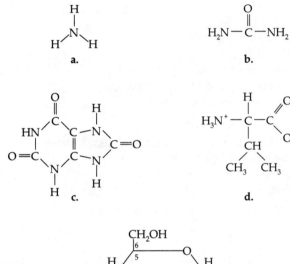

Figure 40.2

b
Conceptual
Understanding

40.24 Which of the molecules shown in Figure 40.2 represents urea?

a
Conceptual
Understanding

40.25 Which part of the vertebrate nephron consists of capillaries?
 a. glomerulus
 b. loop of Henle
 c. distal tubule
 d. Bowman's capsule
 e. collecting tubule

e
Conceptual
Understanding

40.26 Humans can produce urine that is correctly described as being which of the following?
I.   hypertonic to body fluids
II. isotonic to body fluids
III. hypotonic to body fluids

a. I only
b. II only
c. III only
d. I and II only
e. I, II, and II

a
Factual Recall

40.27 Urea is produced in the
a. liver from NH_3 and CO_2.
b. liver from glycogen.
c. kidneys from glucose.
d. kidneys from glycerol and fatty acids.
e. bladder from uric acid and H_2O.

e
Factual Recall

40.28 Which of these vertebrates are generally endotherms?
a. cartilaginous fishes
b. bony fishes
c. amphibians
d. reptiles
e. birds

b
Conceptual
Understanding

40.29 Compared to the sea water around them, most marine bony fishes are correctly described by which of the following?
I.   hypertonic
II. hypotonic
III. isotonic

a. I only
b. II only
c. III only
d. I and III only
e. II and III only

c
Factual Recall

40.30 Malpighian tubules are excretory organs found in
a. earthworms.
b. flatworms.
c. insects.
d. jellyfish.
e. vertebrates.

e
Factual Recall

40.31 The main nitrogenous waste excreted by birds is
a. ammonia.
b. nitrate.
c. nitrite.
d. urea.
e. uric acid.

Match the following terms to Questions 40.32–40.35. Each term may be used once, more than once, or not at all.

    a. ectothermy
    b. micturition
    c. evaporation
    d. torpor
    e. thermogenesis

d
Factual Recall

40.32 Diurnation.

d
Factual Recall

40.33 Aestivation.

a
Factual Recall

40.34 Absorbance of heat from the surroundings.

e
Factual Recall

40.35 Process that occurs in the brown fat of some mammals.

Questions 40.36–40.40 refer to the following structures. Each structure name may be used once, more than once, or not at all.

    a. loop of Henle
    b. collecting duct
    c. Bowman's capsule
    d. proximal convoluted tubule
    e. distal convoluted tubule

b
Factual Recall

40.36 Passes urine to the renal pelvis.

e
Factual Recall

40.37 Increases the reabsorption of Na^+ when stimulated by aldosterone.

c
Factual Recall

40.38 Possesses specialized cells called podocytes.

d
Factual Recall

40.39 Possesses a specialized epithelial lining called a brush-border.

a
Factual Recall

40.40 Descends into the renal medulla only in juxtamedullary nephrons.

a
Application

40.41 In a laboratory experiment, one group of people consumes an amount of pure water, a second group an equal amount of beer, and a third group an equal amount of concentrated salt solution that is hypertonic to their blood. Their urine production is monitored for several hours. At the end of the measurement period, which group will have produced the greatest volume of urine and which group the least?
a. beer the most, salt solution the least
b. salt solution the most, water the least
c. water the most, beer the least
d. beer the most, water the least
e. There will be no significant difference.

c
Conceptual
Understanding

40.42 Proper functioning of the human kidney requires considerable active transport of sodium in the kidney tubules. If these active transport mechanisms were to stop completely, how would urine production be affected?
a. No urine would be produced.
b. A less-than-normal volume of hypotonic urine would be produced.
c. A greater-than-normal volume of isotonic urine would be produced.
d. A greater-than-normal volume of hypertonic urine would be produced.
e. A less-than-normal volume of isotonic urine would be produced.

b
Factual Recall

40.43 Organisms categorized as osmoconformers are most likely
a. terrestrial.
b. marine.
c. amphibious.
d. found in freshwater streams.
e. found in freshwater lakes.

e
Factual Recall

40.44 Which of the following excretory systems is partly based on the filtration of fluid under high hydrostatic pressure?
a. flame cell system of flatworms
b. protonephridia of rotifers
c. metanephridia of annelids
d. Malpighian tubules of insects
e. nephrons of vertebrates

d
Application

40.45 A biologist discovers a new species of organism adapted to living in a deep underground cavern that provides no source of free water. The organism is eyeless and covered by fur, and it has a four-chambered heart with a closed circulatory system. What excretory system modifications might the biologist expect to find?
a. very long Malpighian tubules
b. very short Malpighian tubules
c. metanephridia with a large number of nephridiopores
d. kidneys with long juxtamedullary nephrons
e. kidneys with only cortical nephrons

e
Conceptual
Understanding

40.46 Which feature of osmoregulation do both marine and freshwater bony fish have?
   a. loss of water through gills
   b. gain of salt through gills
   c. large volume of urine
   d. no drinking
   e. gain of water through food

b
Application

40.47 Which of the following activities would initiate an osmoregulatory adjustment brought about primarily through the renin-angiotensin-aldosterone system?
   a. not drinking any fluids for a day or two
   b. spending several hours mowing the lawn on a hot day
   c. eating a bag of potato chips
   d. eating a pizza with green olives and pepperoni
   e. drinking several glasses of water

c
Factual Recall

40.48 The thermostat of vertebrates is located in the
   a. medulla oblongata.
   b. thyroid gland.
   c. hypothalamus.
   d. subcutaneous layer of the skin.
   d. liver.

a
Factual Recall

40.49 In vertebrates, urea and uric acid are made in
   a. the liver.
   b. the kidneys.
   c. the bladder.
   d. the pancreas.
   e. all of the organism's cells.

d
Conceptual
Understanding

40.50 Of the mechanisms by which organisms exchange heat with their surroundings, which one results in only loss of heat from the organism?
   a. conduction
   b. convection
   c. radiation
   d. evaporation
   e. metabolism

a
Factual Recall

40.51 Which of the following organisms controls its body temperature by behavior ONLY?
   a. green frog
   b. bumblebee
   c. bluefin tuna
   d. house sparrow
   e. grey wolf

c
Factual Recall

40.52 Which of the following is a nitrogenous waste that requires hardly any water
for its excretion?
a. amino acid
b. urea
c. uric acid
d. ammonia
e. nitrogen gas

d
Conceptual
Understanding

40.53 Injection of which of the following substances into the blood would NOT
produce a change in the osmoregulatory activity of the human kidney?
a. aldosterone
b. antidiuretic hormone
c. angiotensin II
d. angiotensinogen
e. renin

Chapter 41

c
Application

41.1 Which of the following examples is INCORRECTLY paired with its class?
    a. histamine—local regulator
    b. estrogen—steroid hormone
    c. prostaglandin—peptide hormone
    d. ecdysone—steroid hormone
    e. neurotransmitter—local regulator

b
Factual Recall

41.2 Which of the following hormone sequences results in the secretion of estrogens in mammals?
    a. LH → FSH → estrogen
    b. GNRH → FSH → estrogen
    c. CRH → ACTH → FSH → estrogen
    d. GNRH → LH → estrogen
    e. GNRH → FSH → LH → estrogen

b
Factual Recall

41.3 Hormones are able to control homeostasis because
    a. they are not produced by exocrine glands.
    b. they are subject to negative feedback.
    c. they may be found in the lymphatic system.
    d. they are present at low concentrations.
    e. they are steroids.

c
Factual Recall

41.4 An enzyme cascade
    a. activates a second messenger system.
    b. ends in steroid hormone release.
    c. increases the response to a hormone.
    d. activates the parathyroid gland.
    e. initiates regulation of the nervous system.

a
Factual Recall

41.5 Which of the following is an endocrine gland?
    a. parathyroid gland
    b. salivary gland
    c. sweat gland
    d. hypothalamus
    e. gall bladder

Question 41.6 refers to the following information. In an experiment, rats' ovaries were removed immediately after impregnation and then the rats were divided into two groups. Treatments and results are summarized in the table below.

	Group 1	Group 2
Daily injections of progesterone (milligrams)	0.25	2.0
Percent of rats that carried fetuses to birth	0	100

c
Application

41.6 The results most likely occurred because progesterone exerts an effect on the
    a. general health of the rat.
    b. size of the fetus.
    c. maintenance of the uterus.
    d. gestation period of rats.
    e. number of eggs fertilized.

d
Conceptual
Understanding

41.7 Which hormone exerts antagonistic action to PTH (parathyroid hormone)?
    a. thyroxin
    b. epinephrine
    c. growth hormone
    d. calcitonin
    e. glucagon

e
Application

41.8 Blood samples taken from an individual who had been fasting for 24 hours
would have which of the following?
    a. high levels of insulin
    b. high levels of glucagon
    c. low levels of insulin
    d. low levels of glucagon
    e. Both b and c would be present.

a
Application

41.9 If an individual were unable to produce or obtain tyrosine, which of the
following would you expect to occur?
    a. no "fight-or-flight" reaction
    b. loss of the ability to control blood sugar
    c. immature and nonfunctioning gonads
    d. very low metabolic rate
    e. wide fluctuations in Ca^{2+} levels in the blood

c
Factual Recall

41.10 Prolactin stimulates mammary gland growth and development in mammals
and regulates salt and water balance in freshwater fish. Many scientists
believe that this wide range of functions indicates that
    a. prolactin is a nonspecific hormone.
    b. prolactin has a unique mechanism for eliciting its effects.
    c. prolactin is an ancient hormone.
    d. prolactin is derived from two separate sources.
    e. prolactin interacts with many different receptor molecules.

a
Factual Recall

41.11 Which of the following would you expect to occur if you found an extremely
large thymus gland in an adult human?
    a. an unusual immune system
    b. a high metabolic rate
    c. decreased size of the gonads
    d. no "fight-or-flight" reaction
    e. high levels of Ca^{2+} in the blood

c
Conceptual
Understanding

41.12 A varying response to a common chemical messenger is possible because
    a. various target cells have different genes.
    b. each cell knows how it fits into the body's master plan.
    c. various target cells differ in their receptors to the same hormone.
    d. the circulatory system regulates responses to hormones by routing the hormones to specific targets.
    e. the hormone is chemically altered in different ways as it travels through different branches of the circulatory system.

b
Conceptual
Understanding

41.13 A second messenger in the response of target cells to certain hormones is
    a. ATP.
    b. cyclic AMP.
    c. adenylyl cyclase.
    d. calmodulin.
    e. protein.

d
Conceptual
Understanding

41.14 All of the following statements about hormones are correct EXCEPT:
    a. They are produced by endocrine glands.
    b. They travel to different areas of the body.
    c. They are carried by the circulatory system.
    d. They are used to communicate between different individuals.
    e. They elicit specific biological responses from target cells.

e
Factual Recall

41.15 The main target organ of ADH is the
    a. anterior pituitary.
    b. posterior pituitary.
    c. adrenal gland.
    d. bladder.
    e. kidney.

a
Factual Recall

41.16 The hypothalamus controls the anterior pituitary by means of
    a. releasing factors.
    b. second messengers.
    c. third messengers.
    d. antibodies.
    e. pyrogens.

e
Conceptual
Understanding

41.17 Functions in which prolactin participates include which of the following?
    I. delays metamorphosis in amphibians
    II. stimulates mammary gland growth in mammals
    III. regulates fat metabolism and reproduction in birds

    a. I only
    b. II only
    c. III only
    d. II and III only
    e. I, II, and III

d
Factual Recall

41.18 Which hormone is NOT a steroid?
  a. androgen
  b. cortisone
  c. estrogen
  d. insulin
  e. testosterone

b
Conceptual
Understanding

41.19 All of the following statements about adenylyl cyclase are correct EXCEPT:
  a. ATP is its substrate.
  b. It is the second messenger within cells.
  c. It mediates cellular responses to some hormones.
  d. It is found in the plasma membrane.
  e. It is an enzyme.

a
Factual Recall

41.20 Oxytocin and ADH are produced by the
  a. hypothalamus and stored in the neurohypophysis.
  b. adenohypophysis and stored in the kidneys.
  c. thymus and stored in the thyroid.
  d. adrenal cortex.
  e. gonads.

d
Application

41.21 Two kinds of cells may respond differently to the same steroid hormone because
  a. they have different receptor proteins within the cell.
  b. they have different acceptor proteins on the chromatin.
  c. steroid hormones usually transmit signals that are antagonistic.
  d. the acceptor proteins are associated with different genes in the two kinds of cells.
  e. the hormone-receptor complex is transcribed and processed differently in the two kinds of cells.

c
Conceptual
Understanding

41.22 Which of the following statements about the adrenal gland is correct?
  a. During stress, TSH stimulates the adrenal cortex and medulla to secrete acetylcholine.
  b. During stress, the alpha cells of islets secrete insulin and simultaneously the beta cells of the islets secrete glucagon.
  c. During stress, ACTH stimulates the adrenal cortex, and neurons to the sympathetic nervous system stimulate the adrenal medulla.
  d. At all times, the anterior portion secretes ACTH while the posterior portion secretes oxytocin.
  e. At all times, the adrenal gland monitors calcium levels in the blood and regulates calcium by secreting the two antagonistic hormones, epinephrine and norepinephrine.

a
Factual Recall

41.23 All of the following statements about endocrine glands are correct EXCEPT:
  a. The parathyroids regulate metabolic rate.
  b. The thyroids participate in blood calcium regulation.
  c. The pituitary participates in the regulation of the gonads.
  d. The adrenal medulla produces "fight-or-flight" responses.
  e. The pancreas helps to regulate blood sugar concentration.

d
Conceptual
Understanding

41.24 All of the following statements about the hypothalamus are correct EXCEPT:
   a. It functions as an endocrine gland.
   b. It is part of the central nervous system.
   c. It is subject to feedback inhibition by certain hormones.
   d. It secretes tropic hormones that act directly on the gonads.
   e. Its neurosecretory cells terminate in the posterior pituitary.

e
Factual Recall

41.25 Which of these hormones is a protein?
   a. epinephrine
   b. cortisone
   c. estrogen
   d. androgen
   e. insulin

b
Factual Recall

41.26 Which of the following glands is controlled directly by the hypothalamus and not the anterior pituitary?
   a. ovary
   b. adrenal medulla
   c. adrenal cortex
   d. testis
   e. thyroid

c
Factual Recall

41.27 All of the following endocrine disorders are correctly matched with the malfunctioning gland EXCEPT
   a. diabetes and pancreas.
   b. giantism and pituitary.
   c. cretinism and adrenal medulla.
   d. tetany and parathyroid.
   e. dwarfism and pituitary.

c
Conceptual
Understanding

41.28 Which of the following statements about hormones is CORRECT?
   a. Steroid and peptide hormones produce different effects but use the same biochemical mechanisms.
   b. Steroid and peptide hormones produce the same effects but differ in the mechanisms that produce the effects.
   c. Steroid hormones affect the synthesis of proteins, whereas peptide hormones affect the activity of proteins already present in the cell.
   d. Steroid hormones affect the activity of certain proteins within the cell, whereas peptide hormones directly affect the processing of mRNA.
   e. Steroid hormones affect the synthesis of proteins to be exported from the cell, whereas peptide hormones affect the synthesis of proteins that remain in the cell.

b
Factual Recall

41.29 Which of the following hormones are secreted by the adrenal gland in response to stress and promote the synthesis of glucose from noncarbohydrate substrates?
   a. glucagon
   b. glucocorticoids
   c. epinephrine
   d. thyroxine
   e. ACTH

a
Factual Recall

41.30 The hormones secreted by the adrenal cortex are
  a. steroids.
  b. polypeptides.
  c. inorganic ions.
  d. amino acids.
  e. proteins.

c
Conceptual
Understanding

41.31 Frequently, very few molecules of a hormone are required to effect changes in a target cell. This is because
  a. hormones are lipid soluble and readily penetrate the membranes of the target cell.
  b. hormones are large molecules that persist for years and can repeatedly stimulate the same cell.
  c. the mechanism of hormonal action involves an enzyme cascade that amplifies the response to a hormone.
  d. the mechanism of hormonal action involves the rapid replication of the hormone within the target cell to quickly magnify the hormone's effect.
  e. the mechanism of hormonal action involves memory cells that have had prior contact with the hormone and immediately respond to its presence.

Questions 41.32–41.34 refer to Figure 41.1, which shows the action of a hormone and inositol triphosphate.

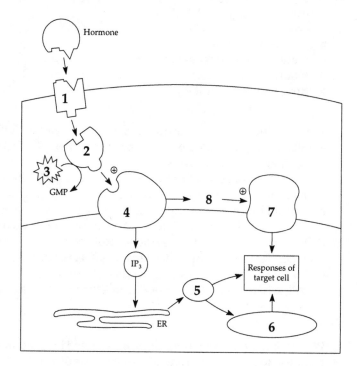

Figure 41.1

c
Factual Recall

41.32 Guanosine triphosphate is represented by number

    a. 1.
    b. 2.
    c. 3.
    d. 4.
    e. 7.

d
Factual Recall

41.33 Calcium, the third messenger, is represented by number

    a. 2.
    b. 3.
    c. 4.
    d. 5.
    e. 8.

c
Factual Recall

41.34 The substance represented by number 7 is

    a. a G protein.
    b. phospholipase C.
    c. protein kinase.
    d. diacylglycerol.
    e. calmodulin.

Questions 41.35–41.40 refer to Figure 41.2, which shows the action of two antagonistic hormones and cyclic AMP.

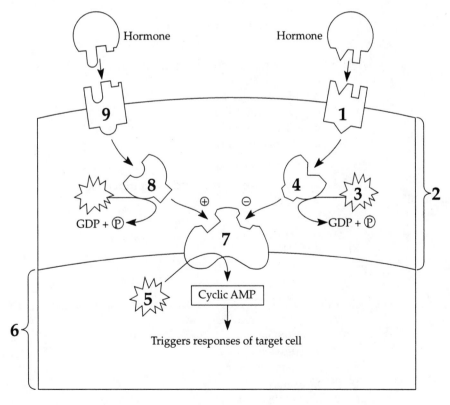

Figure 41.2

b
Conceptual
Understanding

41.35 The type of mechanism shown in Figure 41.2 is typical of many
    a. steroid hormones.
    b. peptide hormones.
    c. prostaglandins.
    d. polypeptide enzymes.
    e. ribozymes.

a
Factual Recall

41.36 The substance represented by number 7 is
    a. adenylyl cyclase.
    b. glucose-1-phosphate.
    c. glycogen phosphorylase.
    d. phosphorylase kinase.
    e. protein kinase.

e
Factual Recall

41.37 The area represented by number 2 is the
    a. cytoplasm.
    b. mitochondria.
    c. kidney cortex.
    d. adrenal cortex.
    e. plasma membrane.

d
Factual Recall

41.38 G proteins are represented by numbers
    a. 1 and 9.
    b. 1 and 4.
    c. 3 and 5.
    d. 4 and 8.
    e. 8 and 9.

a
Factual Recall

41.39 The area represented by number 6 is the
    a. cytoplasm.
    b. mitochondrion.
    c. kidney medulla.
    d. adrenal medulla.
    e. plasma membrane.

b
Factual Recall

41.40 The substance represented by number 5 is
    a. ADH.
    b. ATP.
    c. PTH.
    d. GTP.
    e. MSH.

Questions 41.41–41.45 refer to the following list of hormones. Each hormone may be used once, more than once, or not at all.
    a. ecdysone
    b. glucagon
    c. thymosin
    d. oxytocin
    e. growth hormone

b
Factual Recall

41.41 Secreted by the pancreas.

e
Factual Recall

41.42 Signals the liver to produce somatomedins.

d
Factual Recall

41.43 Stimulates the contraction of uterine muscle.

e
Factual Recall

41.44 Secreted by the anterior lobe of the pituitary.

a
Factual Recall

41.45 Steroid hormone that triggers molting in arthropods.

Questions 41.46–41.49 refer to the following list of hormones. Each hormone may be used once, more than once, or not at all.
   a. androgens
   b. estrogens
   c. progestins
   d. thymosin
   e. melatonin

a
Factual Recall

41.46 Testosterone.

b
Factual Recall

41.47 Estradiol.

e
Factual Recall

41.48 Secreted by the pineal gland.

d
Factual Recall

41.49 Stimulates the development and differentiation of T lymphocytes.

b
Conceptual
Understanding

41.50 Only certain cells in the body are target cells for the steroid hormone aldosterone. Which of the following is the best explanation for why these are the only cells that produce a response to this hormone?
   a. Only target cells are exposed to aldosterone.
   b. Only target cells contain receptors for aldosterone.
   c. Aldosterone is unable to enter nontarget cells.
   d. Nontarget cells destroy aldosterone before it can produce its effect.
   e. Nontarget cells convert aldosterone to a hormone to which they do respond.

a
Application

41.51 Hormone X produces its effect in its target cells via the cAMP second messenger system. If you expose a target cell to only a single molecule, which of the following will produce the greatest effect?
a. a molecule of hormone X applied to the extracellular fluid surrounding the cell
b. a molecule of hormone X injected into the cytoplasm of the cell
c. a molecule of cAMP applied to the extracellular fluid surrounding the cell
d. a molecule of cAMP injected into the cytoplasm of the cell
e. a molecule of activated, cAMP-dependent protein kinase injected into the cytoplasm of the cell

c
Conceptual
Understanding

41.52 The endocrine system and the nervous system are structurally related. Which of the following cells best illustrates this relationship?
a. a neuron in the spinal cord
b. a steroid-producing cell in the adrenal cortex
c. a neurosecretory cell in the hypothalamus
d. a brain cell in the cerebral cortex
e. a cell in the pancreas that produces digestive enzymes

c
Conceptual
Understanding

41.53 The endocrine system and the nervous system are chemically related. Which of the following substances best illustrates this relationship?
a. estrogen
b. calcitonin
c. norepinephrine
d. calcium
e. inositol triphosphate

d
Conceptual
Understanding

41.54 Insect brain hormone is most analogous to which of the following in humans?
a. insulin from the pancreas
b. parathyroid hormone from the parathyroid gland
c. ADH from the posterior pituitary
d. releasing hormones from the hypothalamus
e. androgens from the adrenal cortex

b
Factual Recall

41.55 Which of the following hormonal changes would result in the molt of an insect from an immature stage to an adult?
a. decrease in ecdysone, increase in juvenile hormone
b. increase in ecdysone, decrease in juvenile hormone
c. increase in both ecdysone and juvenile hormone
d. decrease in both ecdysone and juvenile hormone
e. None of these choices is correct because molting is controlled strictly by the nervous system.

a
Factual Recall

41.56 Which of the following endocrine structures is (are) NOT controlled by a tropic hormone from the anterior pituitary?
a. pancreatic islet cells
b. thyroid gland
c. adrenal cortex
d. ovaries
e. testes

d
Factual Recall

41.57 Which of the following pairs of hormones do NOT have antagonistic effects?
  a. insulin and glucagon
  b. growth hormone releasing hormone and growth hormone inhibiting hormone
  c. parathyroid hormone and calcitonin
  d. follicle stimulating hormone and luteinizing hormone
  e. aldosterone and atrial natriuretic factor

e
Factual Recall

41.58 Endocrine structures derived from nervous tissue include
  a. the thymus and thyroid glands.
  b. the ovaries and the testes.
  c. the liver and the pancreas.
  d. the anterior pituitary and the adrenal cortex.
  e. the posterior pituitary and the adrenal medulla.

c
Conceptual
Understanding

41.59 Which of the following signal transduction molecules is NOT bound to the plasma membrane?
  a. receptors for peptide hormones
  b. G proteins
  c. second messengers
  d. adenylyl cyclase
  e. phospholipase C

b
Factual Recall

41.60 The liberation of inositol triphosphate from the cell membrane is coupled with the simultaneous liberation of
  a. cAMP.
  b. diacylglycerol.
  c. prostaglandins.
  d. G protein.
  e. protein kinase.

a
Factual Recall

41.61 The function of protein kinases is to
  a. add a phosphate to other proteins.
  b. produce second messengers.
  c. activate G proteins.
  d. transport first messengers into the cell.
  e. initiate gene transcription.

Chapter 42

For Questions 42.1–42.4, choose the term from the list below that best fits each of the following descriptions. Each term may be used once, more than once, or not at all.

    a. luteinizing hormone (LH)
    b. follicle-stimulating hormone (FSH)
    c. progesterone
    d. human chorionic gonadotropin (HCG)
    e. gonadotropin-releasing hormone (GnRH)

d
Factual Recall

42.1 Embryonic hormone which maintains progesterone and estrogen secretion by the corpus luteum through the first trimester of pregnancy.

a
Factual Recall

42.2 Triggers ovulation of the secondary oocyte.

c
Factual Recall

42.3 Hormone produced by the corpus luteum when stimulated by LH.

e
Factual Recall

42.4 Hypothalamic hormone which triggers the secretion of FSH.

e
Factual Recall

42.5 During the human sexual response, vasocongestion
    a. occurs in the clitoris, vagina, and penis.
    b. occurs in the testes.
    c. can cause vaginal lubrication.
    d. Only a and b are correct.
    e. a, b, and c are correct.

e
Factual Recall

42.6 Which of the following hormones is INCORRECTLY paired with its action?
    a. GnRH—controls release of FSH and LH
    b. estrogen—responsible for primary and secondary female sex characteristics
    c. human chorionic gonadotropin—maintains secretions from the corpus luteum
    d. luteinizing hormone—stimulates ovulation
    e. progesterone—stimulates follicles to develop

e
Factual Recall

42.7 Which of the following is NOT required for internal fertilization?
    a. All of the below are necessary for internal fertilization.
    b. copulatory organ
    c. sperm receptacle
    d. behavioral interaction
    e. internal development of the embryo

d
Conceptual
Understanding

42.8 Which animal would be LEAST likely to be hermaphroditic?
   a. barnacle
   b. earthworm
   c. tapeworm
   d. lobster
   e. liver fluke

b
Factual Recall

42.9 Human fertility drugs increase the chance of multiple births, probably because they
   a. enhance implantation.
   b. stimulate follicles.
   c. mimic progesterone.
   d. stimulate sperm.
   e. prevent parturition.

c
Conceptual
Understanding

42.10 Organisms that produce amniote eggs, in general
   a. All of the below are correct.
   b. have a higher embryo mortality rate than do those with unprotected embryos.
   c. invest most of their reproductive energy in the embryonic and early postnatal development of their offspring.
   d. invest more parenting energy than do placental animals.
   e. produce more gametes than do those animals with external fertilization and development.

b
Factual Recall

42.11 Which of the following structures is INCORRECTLY paired with its function?
   a. epididymis—maturation and storage of sperm
   b. fallopian tube—site of normal embryonic implantation
   c. seminal vesicles—add sugar and mucus to semen
   d. placenta—maternal/fetal exchange organ; progesterone producing
   e. prostate gland—adds alkaline substances to semen

b
Application

42.12 The diploid chromosome number for humans is 46. How many chromatids will there be in a secondary spermatocyte?
   a. 23
   b. 46
   c. 69
   d. 92
   e. 184

a
Factual Recall

42.13 Which of the following structures is INCORRECTLY paired with its function?
   a. seminiferous tubules—add fluid containing mucus, fructose, and prostaglandin to semen
   b. scrotum—encases testes and holds them below the abdominal cavity
   c. epididymis—stores sperm
   d. prostate gland—adds alkaline secretions to semen
   e. ovary—secretes estrogen and progesterone

c
Factual Recall

42.14 One function of the corpus luteum is to
   a. nourish and protect the egg cell.
   b. produce prolactin in the alveoli.
   c. produce progesterone and estrogen.
   d. convert into a hormone-producing follicle after ovulation.
   e. stimulate ovulation.

d
Factual Recall

42.15 Which of the following conception control methods is most effective?
   a. diaphragm
   b. condom
   c. coitus interruptus
   d. vasectomy
   e. rhythm method

c
Factual Recall

42.16 The hormone progesterone is produced
   a. by the pituitary and acts directly on the ovary.
   b. in the ovary and acts directly on the testes.
   c. in the ovary and acts directly on the uterus.
   d. in the pituitary and acts directly on the uterus.
   e. in the uterus and acts directly on the pituitary.

d
Factual Recall

42.17 Which substance can be detected in the urine of females and is a positive test for pregnancy?
   a. progesterone
   b. estrogen
   c. follicle-stimulating hormone
   d. human chorionic gonadotropin
   e. hypothalamus releasing factors

c
Conceptual
Understanding

42.18 The secretory phase of the menstrual cycle
   a. is associated with dropping levels of estrogen and progesterone.
   b. is when the endometrium begins to degenerate and menstrual flow occurs.
   c. corresponds with the luteal phase of the ovarian cycle.
   d. corresponds with the follicular phase of the ovarian cycle.
   e. is the beginning of the menstrual flow.

d
Factual Recall

42.19 Which of the following is a form of asexual reproduction?
   a. protandry
   b. protogyny
   c. hermaphroditism
   d. parthenogenesis
   e. anadromesis

c
Factual Recall

42.20 Which of the following cells produce testosterone?
   a. sperm cells
   b. hypothalamus
   c. interstitial cells
   d. anterior pituitary
   e. seminiferous tubules

c
Factual Recall

42.21 After sperm cells are produced, they are mainly stored in the
    a. urethra.
    b. prostate.
    c. epididymis.
    d. seminal vesicles.
    e. bulbourethral gland.

d
Factual Recall

42.22 In men, the excretory and reproductive systems share which structure?
    a. vas deferens
    b. urinary bladder
    c. seminal vesicle
    d. urethra
    e. ureter

d
Conceptual
Understanding

42.23 All of the following statements about the human reproductive system are correct EXCEPT:
    a. The most effective means of birth control are abstinence, sterilization, and the pill.
    b. Males produce sperm continuously, whereas females ovulate according to a cycle.
    c. The external genitalia of both sexes arise from common primordia.
    d. The period of sexual activity in human females is called estrus.
    e. Mammary glands occur in both females and males.

b
Conceptual
Understanding

42.24 Diploid cells include which of the following?
    I. spermatids
    II. spermatogonia
    III. mature sperm cells

    a. I only
    b. II only
    c. I and II only
    d. II and III only
    e. I, II, and III

a
Factual Recall

42.25 Which part of the genitalia of a human female develops from the same embryonic structure as the male's scrotum?
    a. labia majora
    b. clitoris
    c. urethra
    d. hymen
    e. ovary

c
Factual Recall

42.26 During the menstrual cycle, what is the main source of progesterone in females?
    a. adrenal cortex
    b. anterior pituitary
    c. corpus luteum
    d. developing follicle
    e. placenta

d
Factual Recall

42.27 Fertilization of human eggs usually takes place in the
    a. ovary.
    b. uterus.
    c. vagina.
    d. oviduct.
    e. labia minora.

d
Factual Recall

42.28 Intrauterine devices (IUDs) probably prevent pregnancy by preventing
    a. ejaculation.
    b. fertilization.
    c. the release of gonadotropins.
    d. implantation.
    e. ovulation.

e
Factual Recall

42.29 All of the following are correct statements about reproduction in invertebrates EXCEPT:
    a. Many invertebrates have separate sexes.
    b. Many invertebrates utilize external fertilization.
    c. A few species split open to release gametes to the environment.
    d. Some invertebrates have structures that store sperm.
    e. Invertebrates do not engage in copulation.

a
Factual Recall

42.30 Pregnancy tests are based on the detection of which of the following hormones?
    a. HCG
    b. FSH
    c. GnRH
    d. estrogen
    e. progesterone

d
Conceptual
Understanding

42.31 All of the following statements about human reproduction are correct EXCEPT:
    a. The ability of a pregnant woman not to reject her "foreign" fetus may be due to the suppression of the immune response in her uterus.
    b. By the eighth week, organogenesis is complete and the embryo is referred to as a fetus.
    c. Lactation is the production and release of milk from the mammary glands.
    d. Parturition begins with conception and ends with gestation.
    e. Puberty is the onset of reproductive ability.

d
Factual Recall

42.32 What are the three phases of the ovarian cycle?
    a. embryo, fetus, and newborn
    b. first, second, and third trimesters
    c. menstrual, proliferative, and secretory
    d. follicular, ovulatory, and luteal
    e. zygote, cleavage, and blastocyst

For Questions 42.33–42.37, choose the term from the list below that best fits each description. Each term may be used once, more than once, or not at all.
    a. LH
    b. FSH
    c. ICSH
    d. GnRH
    e. estrogen

a
Factual Recall

42.33 Hormone that triggers ovulation.

e
Factual Recall

42.34 Hormone secreted by the growing follicle.

a
Factual Recall

42.35 Stimulates the corpus luteum in females and interstitial cells in males.

d
Factual Recall

42.36 Hypothalamic hormone that stimulates the secretion of gonadotropins by the anterior pituitary.

b
Factual Recall

42.37 The anterior pituitary hormone that stimulates the maturation of the follicle in the ovary during the beginning of the menstrual cycle.

For Questions 42.38–42.42, choose the term from the list below that best fits each description.
    a. HCG
    b. androgen
    c. oxytocin
    d. prolactin
    e. progesterone

b
Factual Recall

42.38 Secreted by the interstitial cells of the testes.

c
Factual Recall

42.39 Participates in the regulation of labor contractions.

e
Factual Recall

42.40 Initiates the growth of the placenta and enlargement of the uterus.

e
Factual Recall

42.41 The secretion of LH from the pituitary is inhibited by a high concentration of this hormone.

a
Factual Recall

42.42 Required so that the corpus luteum can function through the first and part of the second trimesters of pregnancy.

b
Conceptual
Understanding

42.43 The drug RU-486 functions by
    a. inhibiting release of gonadotropins from the pituitary.
    b. blocking progesterone receptors in the uterus.
    c. preventing release of the secondary oocyte from the ovary.
    d. Only a and b are correct.
    e. a, b, and c are correct.

For Questions 42.44–42.48, choose the description from the list below that best fits each term. Each description may be used once, more than once, or not at all.
    a. prevents release of mature eggs and/or sperm from the gonads
    b. prevents fertilization by keeping sperm and egg physically separated by a barrier
    c. prevents implantation of an embryo
    d. prevents sperm from entering the urethra
    e. prevents oocytes from traveling into the uterus

a
Factual Recall

42.44 Birth control pill.

c
Factual Recall

42.45 Intrauterine device.

e
Factual Recall

42.46 Tubal ligation.

d
Factual Recall

42.47 Vasectomy.

b
Factual Recall

42.48 Diaphragm.

d
Factual Recall

42.49 Which of the following is the least reliable method of birth (conception) control?
    a. birth control pills
    b. intrauterine device
    c. tubal ligation
    d. coitus interruptus
    e. diaphragm

c
Conceptual
Understanding

42.50 Inhibition of the release of GnRH from the hypothalamus will
    a. stimulate production of estrogen and progesterone.
    b. initiate ovulation.
    c. inhibit secretion of gonadotropins from the pituitary.
    d. stimulate secretion of LH and FSH.
    e. initiate the flow phase of the menstrual cycle.

d
Factual Recall

42.51 Which of the following are modes of asexual reproduction found in animals?
    a. fission and budding
    b. fragmentation and gemmule production
    c. regeneration
    d. Only a and b are correct.
    e. a, b, and c are correct.

c
Conceptual
Understanding

42.52 Which of the following are possible advantages of asexual reproduction?
    a. It allows the species to endure periods of fluctuating or unstable environmental conditions.
    b. It enhances genetic variability in the species.
    c. It enables the species to rapidly colonize new regions.
    d. Both a and b are advantages.
    e. a, b, and c are all advantages.

a
Application

42.53 In which of the following environments would you most likely find many invertebrate species that utilize asexual reproduction?
    a. tropical forests
    b. temperate mountain woodlands
    c. deserts
    d. ponds and streams
    e. temperate prairies

b
Factual Recall

42.54 A cloaca is an anatomical structure found in most vertebrates, which functions as
    a. a specialized sperm-transfer device produced by males.
    b. a common exit for the digestive, excretory, and reproductive systems.
    c. a region bordered by the labia minora and clitoris in females.
    d. a source of nutrients for developing sperm in the testes.
    e. a gland that secretes mucus to lubricate the vaginal opening.

b
Application

42.55 You observe an organism with the following characteristics: parthenogenetic reproduction, internal development of embryos, presence of an amnion, lack of parental care of young. Of the following, the organism is probably a (an)
    a. earthworm.
    b. lizard.
    c. bird.
    d. frog.
    e. mammal.

b
Factual Recall

42.56 Suspending the testes in a scrotum outside the abdominal cavity in male mammals is functional because
    a. this arrangement is more attractive to females.
    b. lower temperatures promote normal maturation of sperm in seminiferous tubules.
    c. the scrotum serves as a shifting counterweight during running locomotion.
    d. lower temperatures enhance the release of secretions from the prostate gland.
    e. this arrangement assists sperm and semen to flow into the urethra.

c
Conceptual
Understanding

42.57 In vertebrate animals, spermatogenesis and oogenesis differ, in that
  a. oogenesis begins at the onset of sexual maturity.
  b. oogenesis produces four haploid cells, whereas spermatogenesis produces only one functional spermatozoa.
  c. oogenesis produces one functional gamete, whereas spermatogenesis produces four functional spermatozoa.
  d. spermatogenesis begins before birth.
  e. spermatogenesis is not complete until fertilization occurs.

b
Factual Recall

42.58 Estrous and menstrual cycles in mammals differ, in that females that exhibit estrus
  a. are always sexually receptive to males.
  b. resorb rather than shed endometrial tissues.
  c. have cycles that are always less than 28 days.
  d. experience a lower body temperature during cycles.
  e. do not ovulate until after copulation.

c
Application

42.59 If the release of LH were inhibited in a human female, which of the following events would NOT occur?
  a. release of FSH from the pituitary
  b. maturation of a primary follicle and oocyte
  c. ovulation of a secondary oocyte
  d. release of GnRH from the hypothalamus
  e. production of estrogen by follicle cells

e
Application

42.60 If a man were given a vasectomy, which of the following structures would no longer contribute to the production of semen?
  a. vas deferens
  b. seminal vesicles
  c. ejaculatory duct
  d. prostate gland
  e. seminiferous tubules

e
Application

42.61 What would happen if a woman in the later stages of pregnancy were given a combination of estrogen and oxytocin?
  a. Oxytocin receptors would develop on uterine smooth muscle cells.
  b. Prostaglandins would be secreted from the placenta.
  c. Contractions of uterine muscles would begin.
  d. Only a and c are correct.
  e. a, b, and c are correct.

Chapter 43

43.1 What does the archenteron eventually become?
   a. a body cavity
   b. the anus in deuterostomes
   c. the blastocoel
   d. the digestive tract
   e. the neural tube

43.2 In a frog embryo, gastrulation
   a. produces a blastocoel displaced into the animal hemisphere.
   b. occurs along the primitive streak in the animal hemisphere.
   c. is impossible because of the large amount of yolk in the ovum.
   d. proceeds by involution as cells roll over the dorsal lip of the blastopore.
   e. occurs within the inner cell mass that is embedded in the large amount of yolk.

43.3 Gastrulation and subsequent development in a chick are very different from these processes in amphioxus because
   a. amphioxus is more primitive.
   b. chicken eggs have more yolk.
   c. the chick has paired appendages.
   d. amphioxus is aquatic.
   e. the chick has more germ layers.

Figure 43.1

43.4 The drawing in Figure 43.1 is from what stage of amphibian development?
   a. blastula
   b. neurula
   c. early gastrula
   d. late gastrula
   e. gray crescent stage

c
Factual Recall

43.5 Which of the following is mismatched?
    a. mesoderm—notochord
    b. endoderm—lungs
    c. ectoderm—liver
    d. mesoderm—somites
    e. ectoderm—eye

a
Factual Recall

43.6 The cortical reaction functions directly in the
    a. formation of a fertilization membrane.
    b. production of a fast block to polyspermy.
    c. release of hydrolytic enzymes from the sperm cell.
    d. generation of a nervelike impulse by the egg cell.
    e. the fusion of egg and sperm nuclei.

d
Factual Recall

43.7 The archenteron of the developing frog eventually develops into which structure?
    a. reproductive organs
    b. the blastocoel
    c. heart and lungs
    d. digestive tract
    e. brain and spinal cord

b
Conceptual
Understanding

43.8 What is the difference between differentiated and determined cells?
    a. A determined cell induces the differentiation of neighboring cells.
    b. A determined cell has its developmental fate set, even though it may not yet be differentiated.
    c. A differentiated cell has its developmental fate set, even though it may not yet be determined.
    d. A differentiated cell is found in blastula cells, whereas a determined cell is found only in adult structures.
    e. Cytoplasmic determinants have regulated some genes in a determined cell; differentiated cells are not regulated.

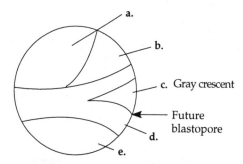

Figure 43.2

c
Factual Recall

43.9 Which labeled area in the frog blastula diagram in Figure 43.2 forms opposite the point of sperm penetration?

c
Factual Recall

43.10 Which of the following are TRUE concerning the vitelline layer of the sea urchin egg?
   a. It is outside the fertilization membrane.
   b. It releases calcium, which initiates the cortical reaction.
   c. It has receptor molecules that are specific for binding acrosomal proteins.
   d. Only a and b are correct.
   e. a, b, and c are correct.

e
Conceptual
Understanding

43.11 Arrange the following stages of fertilization and early development into a proper sequence.
   I.   onset of new DNA synthesis
   II.   cortical reaction
   III. first cell division
   IV. acrosomal reaction; plasma membrane depolarization
   V.   fusion of egg and sperm nuclei complete

   a. III, V, I, IV, II
   b. V, I, IV, II, III
   c. I, III, II, IV, V
   d. V, III, I, II, IV
   e. IV, II, V, I, III

d
Factual Recall

43.12 In the development of an amphibian embryo, what is the "organizer"?
   a. neural tube
   b. notochord
   c. archenteron
   d. dorsal lip of blastopore
   e. dorsal ectoderm

Use the following information to answer Questions 43.13–43.15. In a study of the development of frog embryos, several early gastrula were stained with vital dyes. The location of the dyes after gastrulation was noted. The results are shown in the following table.

Tissue	Stain
Brain	red
Notochord	yellow
Liver	green
Lens of the eye	blue
Lining of the digestive tract	purple

d
Conceptual
Understanding

43.13 Ectoderm would give rise to tissues containing which of the following colors?
   a. yellow and purple
   b. purple and green
   c. green and red
   d. red and blue
   e. red and yellow

b
Conceptual
Understanding

43.14 The mesoderm was probably stained with which color?
 a. blue
 b. yellow
 c. red
 d. purple
 e. green

c
Conceptual
Understanding

43.15 The endoderm was probably which color?
 a. red and yellow
 b. yellow and green
 c. green and purple
 d. blue and yellow
 e. purple and red

d
Factual Recall

43.16 The "slow block" to polyspermy is largely due to
 a. a transient voltage change across the membrane.
 b. consumption of yolk protein.
 c. the jelly coat.
 d. the fertilization membrane.
 e. inactivation of the sperm.

a
Conceptual
Understanding

43.17 All of the following are TRUE concerning homeotic genes EXCEPT:
 a. They are the primary inducer of frog morphogenesis.
 b. A DNA sequence of 180 nucleotides is common to all of the genes.
 c. They are translated into peptide sequences called homeodomains.
 d. The peptide gene product is a regulatory protein that controls transcription.
 e. A mutation may cause misplacement of body segments.

c
Factual Recall

43.18 Which developmental sequence is correct?
 a. cleavage, blastula, gastrula, and morula
 b. cleavage, gastrula, morula, and blastula
 c. cleavage, morula, blastula, and gastrula
 d. gastrula, morula, blastula, and cleavage
 e. morula, cleavage, gastrula, and blastula

d
Application

43.19 Which of the following does NOT occur during early cleavage of an animal zygote?
 a. The cell undergoes mitosis.
 b. The nuclear-to-cytoplasmic ratio of the cells increases.
 c. The ratio of surface area to volume of the cells increases.
 d. The embryo grows significantly in mass.
 e. The cell undergoes cytokinesis.

d
Factual Recall

43.20 A primitive streak forms during the early embryonic development of which of the following?
 I. birds
 II. frogs
 III. humans

 a. I only
 b. II only
 c. I and II only
 d. I and III only
 e. II and III only

c
Conceptual
Understanding

43.21 Which of the following is a function of the acrosome during fertilization?
   a. to block polyspermy
   b. to help propel the sperm
   c. to digest the exterior coats of the egg
   d. to nourish the mitochondria of the sperm
   e. to trigger the completion of meiosis by the sperm cell

e
Conceptual
Understanding

43.22 Which of the following is correct about the yolk of the frog egg?
   a. It prevents gastrulation.
   b. It is concentrated at the animal pole.
   c. It is homogeneously arranged in the egg.
   d. It impedes the formation of a primitive streak.
   e. It leads to unequal rates of cleavage for the animal pole compared to the vegetal pole.

a
Conceptual
Understanding

43.23 The sudden burst in protein synthesis when an egg is fertilized is due to an activation of
   a. translation of mRNA.
   b. transcription of DNA.
   c. DNA polymerase.
   d. RNA polymerase.
   e. polycistronic genes.

d
Conceptual
Understanding

43.24 All of the following statements about developmental mechanisms are correct EXCEPT:
   a. Cell differentiation is contolled by the serial process of determination.
   b. Some groups of cells can influence the development of nearby cells by induction.
   c. Morphogenetic movements rely on the extensor, contractile, and adhesive properties of cells.
   d. In mosaic development, cells remain totipotent longer than do the cells in regulative development.
   e. Unequal concentrations of cytoplasmic determinates in specific blastomeres can fix the developmental fates of different regions of the embryo very early.

b
Factual Recall

43.25 Meroblastic cleavage occurs in which of the following?
   I.   sea urchins
   II. humans
   III. birds

   a. I only
   b. III only
   c. I and II only
   d. I and III only
   e. II and III only

a
Conceptual
Understanding

43.26 Which structure in bird and mammalian embryos functions like the blastopore of frog embryos?
   a. primitive streak
   b. neural plate
   c. archenteron
   d. notochord
   e. somites

e
Conceptual
Understanding

43.27 Extraembryonic membranes develop in which of the following?
    I. mammals
    II. birds
    III. lizards

    a. I only
    b. II only
    c. I and II only
    d. II and III only
    e. I, II, and III

d
Factual Recall

43.28 The least amount of yolk would be found in the egg of a
    a. bird.
    b. fish.
    c. frog.
    d. placental mammal.
    e. terrestrial reptile.

e
Conceptual
Understanding

43.29 The embryonic portion of the mammalian placenta comes from which of the following?
    I. inner cell mass
    II. trophoblast
    III. chorion

    a. I only
    b. II only
    c. III only
    d. I and II only
    e. II and III only

b
Factual Recall

43.30 Which extraembryonic membrane of a chick embryo is a receptacle for uric acid wastes?
    a. amnion
    b. allantois
    c. chorion
    d. trophoblast
    e. yolk sac

b
Factual Recall

43.31 If the cells of a young embryo can be separated and each cell develops into a complete embryo, the development is said to be
    a. mosaic.
    b. regulative.
    c. determinate.
    d. meroblastic.
    e. holoblastic.

e
Factual Recall

43.32 Hans Spemann has referred to which of the following structures as the primary organizer in the early development of amphibian embryos?
    a. optic cup
    b. notochord
    c. neural tube
    d. dorsal ectoderm
    e. dorsal lip of the blastopore

a
Factual Recall

43.33 At the time of implantation, the human embryo is called a(n)
  a. blastocyst.
  b. embryo.
  c. fetus.
  d. somite.
  e. zygote.

b
Factual Recall

43.34 All of the following statements about fertilization are correct EXCEPT:
  a. It reinstates diploidy.
  b. It invaginates the blastula to form the gastrula.
  c. Egg cell depolarization initiates the cortical reaction.
  d. Gamete fusion depolarizes the egg cell membrane and sets up a fast block to polyspermy.
  e. A slow block to polyspermy occurs when cortical granules erect a fertilization membrane.

Questions 43.35–43.39 refer to the diagram of the embryo in Figure 43.3.

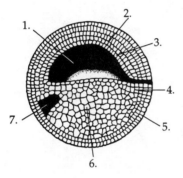

Figure 43.3

b
Factual Recall

43.35 The embryo shown probably belongs to a
  a. dog.
  b. frog.
  c. chick.
  d. snake.
  e. human.

a
Conceptual
Understanding

43.36 Which of the following was the most recent developmental process that directly affected this embryo?
  a. gastrulation
  b. fertilization
  c. blastula formation
  d. morula formation
  e. cleavage

c
Application

43.37 Ectoderm, mesoderm, and endoderm are, respectively,
    a. 1, 7, and 6.
    b. 3, 4, and 5.
    c. 5, 2, and 3.
    d. 5, 6, and 3.
    e. 7, 2, and 6.

a
Factual Recall

43.38 Which of the following will form the lumen of the digestive tract?
    a. 1
    b. 4
    c. 5
    d. 6
    e. 7

a
Factual Recall

43.39 The blastopore in this organism will become the
    a. anus.
    b. ears.
    c. eyes.
    d. nose.
    e. mouth.

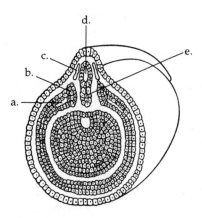

Figure 43.4

Questions 43.40–43.44 refer to the diagram of an embryo in Figure 43.4. Match the word or statement to the lettered structures.

a
Factual Recall

43.40 Coelom.

b
Factual Recall

43.41 Somites.

e
Factual Recall

43.42 Notochord.

b
Factual Recall

43.43 Gives rise to the muscles.

d
Factual Recall

43.44 Gives rise to the brain and spinal cord.

d
Conceptual
Understanding

43.45 Which of the following statements concerning homeotic genes is correct?
   a. There is a sequence of 180 nucleotides common to all the genes.
   b. They are translated into homeodomains that function as transcription factors.
   c. They are egg-polarity genes that code for morphogens.
   d. Only a and b are correct.
   e. a, b, and c are correct.

b
Application

43.46 Suppose a mutation occurred in *Drosophilia* in the region of DNA that codes for the protein called bicoid. What is most likely to happen during development?
   a. Two sets of limbs will form in a mirror-image arrangement.
   b. The polarity of the fertilized egg will be disrupted.
   c. The transcription of developmental genes will stop.
   d. The embryos will express their father's genotype.
   e. The fertilized egg will be bipolar.

a
Factual Recall

43.47 The shaping of an animal and its individual parts into a body form with specialized organs and tissues is called
   a. pattern formation.
   b. induction.
   c. differentiation.
   d. determination.
   e. organogenesis.

Questions 43.48–43.51 refer to the following list of terms. Each term may be used once, more than once, or not at all.
   a. *homeodomains*
   b. *morphogen*
   c. *egg-polarity genes*
   d. *zone of polarizing activity*
   e. *cell adhesion molecules*

e
Conceptual
Understanding

43.48 Contribute to selective reaggregation of dissociated gastrula cells.

d
Conceptual
Understanding

43.49 Assigns positional information along the anterior-posterior axis to developing avian limb buds.

c
Conceptual
Understanding

43.50 When translated, provide positional information for development of the body plan on the coarsest level.

b
Conceptual
Understanding

43.51 A substance such as retinoic acid that varies in concentration and therefore enables cells to resolve their position along a gradient.

e
Factual Recall

43.52 What is the process called that involves the movement of cells into new relative positions in an embryo and results in the establishment of three tissue layers?
    a. determination
    b. cleavage
    c. fertilization
    d. induction
    e. gastrulation

d
Application

43.53 You observe an embryo in which the initial cleavage divisions are radial and meroblastic, extraembryonic membranes develop, and a primitive streak is formed. How would you identify this organism, based on the information given?
    a. invertebrate
    b. fish or amphibian
    c. reptile or bird
    d. bird or mammal
    e. mammal

c
Factual Recall

43.54 After gastrulation, the outer-to-inner sequence of tissue layers in a vertebrate is
    a. endoderm, ectoderm, mesoderm.
    b. mesoderm, endoderm, ectoderm.
    c. ectoderm, mesoderm, endoderm.
    d. ectoderm, endoderm, mesoderm.
    e. endoderm, mesoderm, ectoderm.

c
Factual Recall

43.55 In humans, the increase in Ca^{2+} concentrations in the cytoplasm of the oocyte causes
    a. an increase in vaginal acidity, deactivation of the developing oocyte, and the inhibition of fertilization.
    b. the acrosome to discharge hydrolytic enzymes, resulting in the deactivation of the sperm flagellum and release of the sperm nucleus.
    c. cortical granules to fuse with the oocyte plasma membrane, ultimately causing the elevation of the vitelline membrane.
    d. the development of the vitelline membrane around the oocyte, causing the sperm to lose its flagellum.
    e. depolarization of the sperm plasma membrane that releases acrosomal enzymes.

d
Factual Recall

43.56 Depending upon the species, the initial polarity of an animal embryo might be determined by
    a. the gray crescent.
    b. concentrations of mRNA.
    c. the primitive streak.
    d. Only a and c are correct.
    e. a, b, and c are correct.

c
Conceptual
Understanding

43.57 If an amphibian zygote is manipulated so that the first cleavage plane does NOT divide the gray crescent, what is the expected fate of the two daughter cells?
    a. The daughter cell with the entire gray crescent will die.
    b. Both daughter cells will develop normally because amphibians are totipotent at this stage.
    c. Only the daughter cell with the gray crescent will develop normally.
    d. Both daughter cells will develop abnormally.
    e. Both daughter cells will die immediately.

d
Factual Recall

43.58 Of the following proteins, which is (are) NOT useful to cells during morphogenetic movements?
    a. All of the below are useful.
    b. bindin
    c. cell adhesion molecules
    d. laminin
    e. fibronectins

c
Factual Recall

43.59 Which of the following is NOT an extraembryonic membrane that develops from the embryos of reptiles, birds, and mammals?
    a. chorion
    b. yolk sac
    c. egg shell
    d. amnion
    e. allantois

d
Factual Recall

43.60 In placental mammals, what is the major function of the yolk sac during development?
    a. It transfers nutrients from the yolk to the embryo.
    b. It differentiates into the placenta.
    c. It becomes a fluid-filled sac that surrounds and protects the embryo.
    d. It produces blood cells which then migrate into the embryo.
    e. It stores waste products from the embryo until the placenta develops.

d
Factual Recall

43.61 The term *homeobox* refers to
    a. a group of genes that determine polarity during development.
    b. peptide sequences of 60 amino acids that turn other genes on or off.
    c. zones of polarizing activity commonly present during limb formation.
    d. a specific nucleotide sequence of some genes that regulate development.
    e. glycoproteins that assist cells during morphogenetic movements.

b
Application

43.62 A group of cells, destined to form cartilage, begins producing a protein, type II collagen. At this point, these cells have just completed which developmental process?
    a. determination
    b. differentiation
    c. organogenesis
    d. morphogenesis
    e. pattern formation

d
Application

43.63 Thalidomide was a chemical prescribed as a sedative in the early 1960s. If taken by women in their first trimester of pregnancy, the children born had deformities of the arms and legs. What developmental process was disturbed by this drug?
a. early cleavage divisions
b. determination of the polarity of the zygote
c. differentiation of bone tissue
d. morphogenesis
e. organogenesis

Chapter 44

a
Factual Recall

44.1 Which of the following activities would be associated with the parasympathetic division of the nervous system?
  a. resting and digesting
  b. release of both acetylcholine and epinephrine
  c. increased metabolic rate
  d. "fight-or-flight" response
  e. release of epinephrine only

c
Conceptual
Understanding

44.2 The postsynaptic membrane of a nerve may be stimulated by certain neurotransmitters to permit the influx of negative chloride ions into the cell. This process will result in
  a. membrane depolarization.
  b. an action potential.
  c. the production of an IPSP.
  d. the production of an EPSP.
  e. the membrane becoming more positive.

b
Factual Recall

44.3 After an action potential, how is the resting potential restored?
  a. the opening of sodium activation gates
  b. the opening of voltage-sensitive potassium channels and the closing of sodium activation gates
  c. an increase in the membrane's permeability to potassium and chloride ions
  d. the delay in the action of the sodium-potassium pump
  e. the refractory period in which the membrane is hyperpolarized

d
Factual Recall

44.4 The threshold potential of a membrane
  a. is equal to about 35 mV.
  b. is equal to about 70 mV.
  c. opens voltage-sensitive gates that result in the rapid outflow of sodium ions.
  d. is the depolarization that is needed to generate an action potential.
  e. is a graded potential that is proportional to the strength of a stimulus.

b
Factual Recall

44.5 Which of the following statements is TRUE regarding temporal summation?
  a. The sum of simultaneously arriving neurotransmitters from different presynaptic nerve cells determines whether the postsynaptic cell fires.
  b. Several action potentials arrive in fast succession without allowing the postsynaptic cell to return to its resting potential.
  c. Several IPSPs arrive concurrently, bringing the presynaptic cell closer to its threshold.
  d. Several postsynaptic cells fire at the same time when neurotransmitters are released from several synaptic terminals simultaneously.
  e. The voltage spike of the action potential that is initiated is higher than normal.

a
Factual Recall

44.6 Which part of the vertebrate nervous system is most involved in preparation for "fight or flight"?
  a. sympathetic
  b. somatic
  c. central
  d. visceral
  e. parasympathetic

e
Conceptual
Understanding

44.7 Membrane proteins are critical components in neuron function. Which neuronal process is NOT mediated by one or more membrane proteins?
  a. active transport of sodium ions
  b. active transport of potassium ions
  c. reception of acetylcholine
  d. propagation of an action potential
  e. release of acetylcholine

For Questions 44.8–44.12, choose the best answer from the following list. Each answer may be used once, more than once, or not at all.
  a. cerebrum
  b. cerebellum
  c. thalamus
  d. hypothalmus
  e. medulla oblongata

d
Factual Recall

44.8 Produces hormones which are secreted by the pituitary gland.

b
Factual Recall

44.9 Coordinates muscle actions.

d
Factual Recall

44.10 Regulates body temperature.

e
Factual Recall

44.11 Contains regulatory centers for the respiratory and circulatory systems.

d
Factual Recall

44.12 Contains regions that help regulate hunger and thirst.

For Questions 44.13–44.15, refer to the graph of an action potential in Figure 44.1 and use the lettered line to indicate your answer.

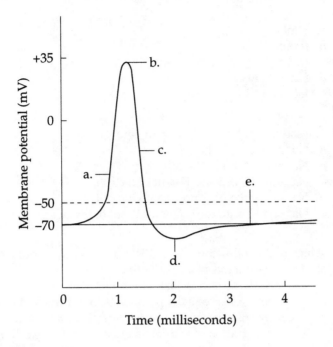

Figure 44.1

d
Conceptual
Understanding

44.13 The membrane is unable to respond to any further stimulation, regardless of intensity.

a
Conceptual
Understanding

44.14 The sodium gates open.

a
Conceptual
Understanding

44.15 The threshold potential is reached.

c
Factual Recall

44.16 Which of the following offers the best description of neural transmission across a mammalian synaptic gap?
   a. Neural impulses involve the flow of K^+ and Na^+ across the gap.
   b. Neural impulses travel across the gap as electrical currents.
   c. Neural impulses cause the release of chemicals that diffuse across the gap.
   d. Neural impulses travel across the gap in both directions.
   e. The calcium within the axons and dendrites of nerves adjacent to a synapse acts as the neurotransmitter.

e
Conceptual
Understanding

44.17 The general functions of the nervous system include which of the following?
    I. integration
    II. motor output
    III. sensory input

    a. I only
    b. II only
    c. III only
    d. I and II only
    e. I, II, and III

d
Factual Recall

44.18 The sodium-potassium pump of neurons pumps
    a. Na^+ and K^+ into the cell.
    b. Na^+ and K^+ out of the cell.
    c. Na^+ into the cell and K^+ out of the cell.
    d. Na^+ out of the cell and K^+ into the cell.
    e. Na^+ and K^+ into the cell and H^+ out of the cell through an antiport.

d
Factual Recall

44.19 Which of the following is a CORRECT statement about a resting neuron?
    a. It is releasing lots of acetylcholine.
    b. The membrane is very leaky to sodium.
    c. The membrane is equally permeable to sodium and potassium.
    d. The membrane potential is more negative than the threshold potential.
    e. The concentration of sodium is greater inside the cell than outside.

a
Factual Recall

44.20 In the sequence of permeability changes that depolarizes and then repolarizes the membrane of a neuron during an action potential, which of the following changes occurs first?
    a. Sodium gates open.
    b. The Na^+–K^+ pump shuts down.
    c. The Na^+–K^+ pump is activated.
    d. Potassium gates close.
    e. Potassium gates open.

e
Factual Recall

44.21 Synaptic vesicles discharge their contents by exocytosis at the
    a. dendrite.
    b. axon hillock.
    c. nodes of Ranvier.
    d. postsynaptic membrane.
    e. presynaptic membrane.

Question 44.22 refers to the following information:
    1. *neurotransmitter binds with receptor*
    2. *sodium ions rush into neuron's cytoplasm*
    3. *action potential depolarizes the presynaptic membrane*
    4. *ion gate opens to allow particular ion to enter cell*
    5. *synaptic vesicles release neurotransmitter into the synaptic cleft*

c
Conceptual
Understanding

44.22 Given the steps shown above, which of the following is the correct sequence for transmission at a chemical synapse?
a. 1, 2, 3, 4, 5
b. 2, 3, 5, 4, 1
c. 3, 2, 5, 1, 4
d. 4, 3, 1, 2, 5
e. 5, 1, 2, 4, 3

e
Conceptual
Understanding

44.23 Repolarization of the membrane of a neuron after an action potential is a consequence of which of the following?
I. Ca^{2+} gates opening
II. Na^+ gates inactivating
III. K^+ gates opening

a. I only
b. II only
c. III only
d. I and II only
e. II and III only

a
Factual Recall

44.24 During an IPSP, the postsynaptic membrane becomes more permeable to
a. K^+.
b. Na^+.
c. Ca^{2+}.
d. GABA.
e. serotonin.

c
Conceptual
Understanding

44.25 Wernicke's and Broca's regions of the brain affect different aspects of
a. olfaction.
b. vision.
c. speech.
d. memory.
e. hearing.

b
Conceptual
Understanding

44.26 All of the following statements about transmission along neurons are correct EXCEPT:
a. The rate of transmission of a nerve impulse is directly related to the diameter of the axon.
b. The intensity of a nerve impulse is directly related to the size of the voltage change.
c. The resting potential is maintained by differential ion permeabilities and the sodium-potassium pump.
d. Once initiated, local depolarizations stimulate a propagation of serial action potentials down the axon.
e. A stimulus that affects the membrane's permeability to ions can either depolarize or hyperpolarize the membrane.

d
Conceptual
Understanding

44.27 A drug might act as a stimulant of the somatic nervous system if it
- a. makes the membrane permanently impermeable to sodium.
- b. stimulates the activity of acetylcholinesterase in the synaptic cleft.
- c. increases the release of substances that cause the hyperpolarization of the neurons.
- d. increases the sensitivity of the postsynaptic membrane to acetylcholine.
- e. increases the sensitivity of the presynaptic membrane to acetylcholine.

a
Factual Recall

44.28 An EPSP facilitates depolarization of the postsynaptic membrane by
- a. increasing the permeability of the membrane to Na^+.
- b. increasing the permeability of the membrane to K^+.
- c. insulating the hillock region of the axon.
- d. allowing Cl^- to enter the cell.
- e. stimulating the $Na^+–K^+$ pump.

b
Factual Recall

44.29 A change at a synapse that increases the magnitude of the potential of the postsynaptic membrane by making the inside of the cell more negative is termed
- a. an EPSP.
- b. an IPSP.
- c. an action potential.
- d. a threshold potential.
- e. a depolarized potential.

a
Factual Recall

44.30 The main neurotransmitter of the parasympathetic system is
- a. acetylcholine.
- b. cholinesterase.
- c. epinephrine.
- d. adrenaline.
- e. dopamine.

a
Conceptual
Understanding

44.31 All of the following statements about the nervous system are correct EXCEPT:
- a. The three evolutionary changes in the vertebrate brain include increases in relative size, increases in compartmentalization of function, and decreases in cephalization.
- b. The size of the primary motor and sensory areas of the cortex devoted to controlling each part of the body is proportional to the importance of that part of the body.
- c. Human emotions are believed to originate from interactions between the cerebral cortex and the limbic system.
- d. Human memory may be stored by chemical and structural changes in the neurons of the sensory cortex.
- e. The autonomic nervous system is subdivided into the parasympathetic and sympathetic nervous systems.

b
Factual Recall

44.32 Heart rate is controlled by the
- a. neocortex.
- b. medulla.
- c. thalamus.
- d. pituitary.
- e. cerebellum.

a
Factual Recall

44.33 The motor cortex is part of the
    a. cerebrum.
    b. cerebellum.
    c. spinal cord.
    d. midbrain.
    e. medulla.

a
Factual Recall

44.34 Integration of simple responses to certain stimuli, such as the patellar reflex, is accomplished by the
    a. spinal cord.
    b. hypothalamus.
    c. corpus callosum.
    d. cerebellum.
    e. medulla.

Questions 44.35 and 44.36 refer to the following choices of neurotransmitters. Each choice may be used once, more than once, or not at all.
    a. acetylcholine
    b. epinephrine
    c. endorphin
    d. serotonin
    e. GABA

c
Factual Recall

44.35 A neuropeptide that functions as a natural analgesic.

e
Factual Recall

44.36 An amino acid that operates at inhibitory synapses in the brain.

Questions 44.37 and 44.38 refer to the following terms. Each term may be used once, more than once, or not at all.
    a. meninges
    b. telodendria
    c. axon hillocks
    d. myelin sheaths
    e. postsynaptic membranes

e
Factual Recall

44.37 Possess neurotransmitter receptors.

c
Factual Recall

44.38 Usually the sites of the initial action potential in neurons.

e
Factual Recall

44.39 Neurotransmitters are released from presynaptic axon terminals into the synaptic cleft by which of the following mechanisms?
    a. osmosis
    b. active transport
    c. diffusion
    d. endocytosis
    e. exocytosis

a

Application

44.40 A single inhibitory postsynaptic potential has a magnitude of 0.5 mV at the axon hillock and a single excitatory postsynaptic potential has a magnitude of 0.5 mV. What will be the membrane potential at the hillock after the spatial summation of 6 IPSPs and 2 EPSPs, if the initial membrane potential is –70 mV?
   a. –72 mV
   b. –71 mV
   c. –70 mV
   d. –69 mV
   e. –68 mV

b

Factual Recall

44.41 A ganglion is a group of nerve cell bodies
   a. in the central nervous system.
   b. in the peripheral nervous system.
   c. anywhere in the nervous system.
   d. within the brain.
   e. within the spinal cord.

b

Factual Recall

44.42 Neurons at rest are not at the equilibrium potential for K^+ because the cell membrane is
   a. only permeable to K^+.
   b. slightly permeable to Na^+.
   c. not permeable to Na^+.
   d. not permeable to K^+.
   e. only permeable to Na^+.

e

Factual Recall

44.43 Saltatory conduction is a term applied to conduction of impulses
   a. across electrical synapses.
   b. along the postsynaptic membrane from dendrite to axon hillock.
   c. in two directions at the same time.
   d. from one neuron to another.
   e. along myelinated nerve fibers.

c

Application

44.44 If the concentration of potassium in the cytoplasm of a nerve cell with a resting membrane potential of –70 mV were elevated above normal, the new resting potential would
   a. still be –70 mV.
   b. be –69 mV or higher.
   c. be –71 mV or lower.
   d. be 0 mV.
   e. reverse polarity.

d
Application

44.45 Action potentials are normally carried in one direction from the axon hillock to the axon terminals. By using an electronic probe, you experimentally depolarize the middle of the axon to threshold. What do you expect?
a. No action potential will be initiated.
b. An action potential will be initiated and proceed in the normal direction toward the axon terminal.
c. An action potential will be initiated and proceed back toward the axon hillock.
d. Two action potentials will be initiated, one going toward the axon terminal and one going back toward the hillock.
e. An action potential will be initiated, but it will die out before it reaches the axon terminal.

e
Application

44.46 You have discovered an unknown organism. It contains long cells with excitable membranes that you suspect are used for rapid information transfer. The membrane of the cell is permeable only to ion X, which carries a negative charge. Active transport pumps in the membrane move X into the cell while simultaneously moving ion Y, also carrying a negative charge, out of the cell. Which of the following is true about the establishment of the resting membrane potential in this cell?
a. The resting potential of this cell will be zero.
b. The resting potential of this cell will be negative.
c. A negative resting potential is directly produced by the pump moving negative charge into the cell.
d. A negative resting potential is directly produced by the diffusion of Y⁻ into the cell.
e. A positive resting potential is directly produced by the diffusion of X⁻ out of the cell.

a
Application

44.47 When neurotransmitter Z is released into the extracellular fluid in contact with a portion of the cell from Question 44.46, membrane channels open that allow both X⁻ and Y⁻ through the membrane. Which of the following is INCORRECT?
a. The magnitude of the potential will immediately increase.
b. Y⁻ will diffuse into the cell.
c. X⁻ will diffuse out of the cell.
d. The membrane will depolarize.
e. The channels are chemically gated.

b
Factual Recall

44.48 Voltage-gated ion channels are NOT found in which of the following locations?
a. the membranes of axon terminals
b. the postsynaptic membranes of dendrites
c. the membrane of the axon hillock
d. the axon membrane of unmyelinated neurons
e. the cell membrane at the nodes of Ranvier of myelinated neurons

c
Factual Recall

44.49 The area of the brain most intimately associated with the unconscious control of respiration and circulation is the
  a. thalamus.
  b. cerebellum.
  c. medulla.
  d. corpus callosum.
  e. cerebrum.

d
Factual Recall

44.50 The motor division of the PNS can be divided into
  a. the brain and spinal cord.
  b. the sympathetic and parasympathetic system.
  c. the central nervous and sensory systems.
  d. the somatic and autonomic systems.
  e. muscles and glands.

b
Factual Recall

44.51 Centralization of the nervous system seems to be associated with the evolution of
  a. a complete gut.
  b. bilateral symmetry.
  c. radial symmetry.
  d. a closed circulatory system.
  e. excitable membranes.

e
Factual Recall

44.52 The major inhibitory neurotransmitter of the brain is
  a. acetylcholine.
  b. cholinesterase.
  c. norepinephrine.
  d. dopamine.
  e. GABA.

c
Conceptual
Understanding

44.53 Neurotransmitters categorized as inhibitory would NOT be expected to
  a. bind to receptors.
  b. open K^+ channels.
  c. open Na^+ channels.
  d. open Cl^- channels.
  e. hyperpolarize the membrane.

Chapter 45

45.1 Hydrostatic skeletons are normally used for movement by all of the following EXCEPT
  a. annelids.
  b. cnidarians.
  c. crustaceans.
  d. nematodes.
  e. flatworms.

45.2 Animals perceive their world in a variety of ways, many of them very different from the perception of human animals. What is the primary reason for these differences?
  a. Animals differ in morphological complexity.
  b. Animals differ in size.
  c. Animals that are adapted perceive what they need for survival.
  d. Animals that are adapted perceive the strongest signals from around them.
  e. Animals have different reproductive methods.

45.3 Which type(s) of skeleton(s) does an elephant have?
  a. cartilaginous endoskeleton
  b. bony endoskeleton
  c. cartilaginous and bony endoskeletons
  d. cartilaginous endoskeleton and hydrostatic skeleton
  e. cartilaginous and bony endoskeletons and hydrostatic skeleton

45.4 The ability of a receptor to absorb the energy of a stimulus is called
  a. integration.
  b. transmission.
  c. transduction.
  d. reception.
  e. amplification.

45.5 A muscle cell is properly referred to as a
  a. muscle fiber.
  b. myofibril.
  c. myofilament.
  d. belly of the muscle.
  e. sarcomere.

45.6 Which of the following receptors is INCORRECTLY paired to a type of energy it transduces?
  a. mechanoreceptors—sound
  b. electromagnetic receptors—magnetism
  c. chemoreceptors—solute concentrations
  d. thermoreceptors—heat
  e. pain receptors—electricity

e
Factual Recall

45.7 Which of the following receptors is INCORRECTLY paired with its category?
    a. hair cell—mechanoreceptor
    b. muscle spindle—mechanoreceptor
    c. gustatory receptor—chemoreceptor
    d. rod—photoreceptor
    e. cone—deep-pressure receptor

Question 45.8 refers to the following information.
    1. Tropomyosin shifts and unblocks the cross-bridge binding sites.
    2. Calcium is released and binds to troponin.
    3. Transverse tubules depolarize the sarcoplasmic reticulum.
    4. The thin filaments are ratcheted across the thick filaments by the heads of the myosin molecules and ATP.
    5. An action potential in a motor neuron causes the axon to release acetylcholine, which depolarizes the muscle cell membrane.

e
Application

45.8 For the events listed above, which of the following is the correct sequence for their occurrence during the excitation and contraction of a muscle cell?
    a. 1, 2, 3, 4, 5
    b. 2, 1, 3, 5, 4
    c. 2, 3, 4, 1, 5
    d. 5, 3, 1, 2, 4
    e. 5, 3, 2, 1, 4

a
Conceptual
Understanding

45.9 All of the following are CORRECT statements about the vertebrate eye EXCEPT:
    a. The vitreous humor regulates the amount of light entering the pupil.
    b. The transparent cornea is an extension of the sclera.
    c. The fovea is the center of the visual field and contains only cones.
    d. The ciliary muscle functions in accommodation.
    e. The retina lies just inside the choroid and contains the photoreceptor cells.

b
Factual Recall

45.10 What is the role of calcium in muscle contractions?
    a. to break the cross-bridges as a cofactor in the hydrolysis of ATP
    b. to bind with troponin, changing its shape so that the actin filament is exposed
    c. to transmit the action potential across the neuromuscular junction
    d. to spread the action potential through the T-tubules
    e. to reestablish the polarization of the plasma's membrane following an action potential

a
Factual Recall

45.11 People are not constantly aware of the rings, watches, or clothes that they may be wearing because of a phenomenon called
    a. sensory adaptation.
    b. fovea accommodation.
    c. rhodopsin bleaching.
    d. motor unit recruitment.
    e. receptor amplification.

a
Factual Recall

45.12 Which of the following is a CORRECT statement about the cells of the human retina?
   a. Cone cells can detect color and rod cells cannot.
   b. Cone cells are more sensitive to light than rod cells are.
   c. Cone cells, but not rod cells, have a visual pigment.
   d. Rod cells are most highly concentrated in the center of the retina.
   e. Rod cells require higher illumination for stimulation than do cone cells.

b
Factual Recall

45.13 A sustained contraction of muscle due to a succession of stimuli with no time between the stimuli for relaxation is called
   a. tonus.
   b. tetanus.
   c. an all-or-none response.
   d. fatigue.
   e. a spasm.

a
Conceptual
Understanding

45.14 All of the following are CORRECT statements about sensory receptors EXCEPT:
   a. A sensory receptor is usually a modified neuron that can collect and transmit information from the central nervous system.
   b. Transduction is the conversion of the stimulus energy into the electrochemical energy of changes in membrane potential and action potentials.
   c. Bending or stretching of the sensory receptor cell membrane in response to stimuli affects ion permeabilities and creates a receptor potential.
   d. Sensations are action potentials traveling along sensory neurons that are interpreted by different parts of the brain as perception.
   e. Changes in the spontaneous firing rate of sensory cells provide information about the presence or absence of the stimulus and its intensity.

b
Factual Recall

45.15 Which type of muscle is responsible for peristalsis along the digestive tract?
   a. cardiac
   b. smooth
   c. skeletal
   d. striated
   e. voluntary

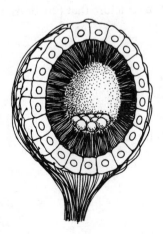

Figure 45.1

b
Factual Recall

45.16 The structure diagrammed in Figure 45.1 is
  a. a neuromast.
  b. a statocyst.
  c. a taste bud.
  d. an ommatidium.
  e. an olfactory bulb.

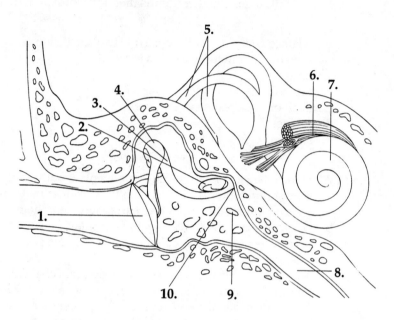

Figure 45.2

Questions 45.17–45.21 refer to the diagram of the ear (Figure 45.2).

c
Factual Recall

45.17 The number that represents the structure equalizing the pressure between the ear and the atmosphere is
  a. 1.
  b. 7.
  c. 8.
  d. 9.
  e. 10.

d
Factual Recall

45.18 The number that represents a structure that functions in balance and equilibrium is
  a. 2.
  b. 3.
  c. 4.
  d. 5.
  e. 8.

b
Factual Recall

45.19 The number that represents the stapes is
  a. 1.
  b. 2.
  c. 3.
  d. 4.
  e. 5.

e
Factual Recall

45.20 The organ of Corti is located in the structure represented by number
    a. 3.
    b. 4.
    c. 5.
    d. 6.
    e. 7.

c
Factual Recall

45.21 Hair cells are found in structures represented by numbers
    a. 1 and 2.
    b. 3 and 4.
    c. 5 and 7.
    d. 6 and 8.
    e. 9 and 10.

e
Factual Recall

45.22 Focusing the eye by changing the shape of the lens is called
    a. zooming.
    b. refraction.
    c. conditioning.
    d. habituation.
    e. accommodation.

c
Conceptual
Understanding

45.23 The perceived pitch of a sound depends partly on
    a. the amplitude of the sound waves.
    b. which bones of the middle ear move.
    c. which hair cells of the cochlea are stimulated.
    d. where particles settle in the semicircular canals.
    e. whether it is the round window or the oval window that vibrates.

d
Conceptual
Understanding

45.24 All of the following statements about vision are correct EXCEPT:
    a. Perception of visual information takes place in the brain.
    b. Rods contain the light-absorbing molecule called rhodopsin.
    c. Rods are more light sensitive than cones and are responsible for night vision.
    d. All information from the left eye goes to the right visual cortex and all information from the right eye goes to the left visual cortex.
    e. Color vision results from the presence of three subclasses of cones in the retina, each with its own type of opsin associated with retinal.

b
Conceptual
Understanding

45.25 Muscle cells are stimulated by neurotransmitters released from the tips of
    a. T-tubules.
    b. motor cell axons.
    c. sensory cell axons.
    d. motor cell dendrites.
    e. sensory cell dendrites.

Question 45.26 refers to the following information.
1. *Receptor cell releases less neurotransmitter.*
2. *Fewer signals cross synapses to bipolar cells.*
3. *Fewer signals cross synapses to ganglion cells.*
4. *The membranes of receptor cells are hyperpolarized by decreased permeability to sodium ions.*

e
Conceptual
Understanding

45.26 For the events listed above, which of the following is the correct sequence for their occurrence, starting with the excitation of cone cells by light and ending with the transmission of action potentials along the optic nerve?
   a. 1, 2, 3, 4
   b. 2, 3, 4, 1
   c. 2, 4, 1, 3
   d. 3, 4, 1, 2
   e. 4, 1, 2, 3

d
Factual Recall

45.27 During muscle contraction, the ion that leaks out of the sarcoplasmic reticulum and induces myofibrils to contract is
   a. K^+.
   b. Cl^-.
   c. Na^+.
   d. Ca^{2+}.
   e. Mg^{2+}.

c
Factual Recall

45.28 All of the following are correct statements about hearing and balance EXCEPT:
   a. The semicircular canals respond to rotation of the head.
   b. Fish have inner ears that sense vibrations in the water.
   c. The volume of sound is a function of the size of the action potentials that reach the brain.
   d. In mammals, the tympanic membrane transmits sound to the three middle ear bones.
   e. In mammals, the middle ear bones transmit sound through the oval window to the coiled cochlea of the inner ear.

Questions 45.29–45.31 refer to Figure 45.3. Match each of the following phrases with the letter of its structure in the figure.

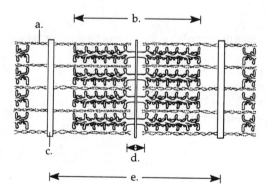

Figure 45.3

e
Factual Recall

45.29 A sarcomere.

d
Factual Recall

45.30 Consists only of myosin filaments.

b
Factual Recall

45.31 Consists of both actin and myosin filaments.

a
Application

45.32 Which of the following is a good example of sensory adaptation?
a. olfactory receptors ceasing after several minutes to produce receptor potentials triggered by perfume molecules
b. hair cells in the organ of Corti not responding to high-pitched sounds after prolonged exposure to high levels of sound at a concert
c. cones in the human eye failing to respond to light in the infrared range
d. hair cells in the utricle and saccule responding to a change in orientation of the head
e. rods in the human eye responding to mechanical stimulation from a blow to the eye so that a flash of light is perceived

b
Factual Recall

45.33 In the vertebrate eye, the material known as transducin is best categorized as a(n)
a. visual pigment.
b. G protein.
c. second messenger.
d. ion channel.
e. receptor protein.

c
Factual Recall

45.34 Hair cells of the organ of Corti are deflected when they brush against the
a. basilar membrane.
b. tympanic membrane.
c. tectorial membrane.
d. plasma membrane.
e. ciliary body.

c
Factual Recall

45.35 Which of the following structures is LAST encountered by sensory information during visual processing?
a. ganglion cells
b. bipolar cells
c. primary visual cortex
d. optic chiasma
e. lateral geniculate nuclei

a
Conceptual
Understanding

45.36 Information carried by your optic nerve is perceived as "sight," whereas information carried by your auditory nerve is perceived as "sound." Which of the following statements best explains this?
a. The information is carried to different areas of your brain.
b. The structure of neurons in the optic nerve differs from those in the auditory nerve.
c. Light energy and sound waves are different from each other.
d. Different ions enter and leave the axons of the two different nerves.
e. Action potentials that carry visual information are of a different amplitude and frequency than those carrying sound information.

c
Factual Recall

45.37 Hair cells can be found in all of the following locations EXCEPT
    a. the statocysts of invertebrates.
    b. the lateral line system of fish.
    c. the retina of mammals.
    d. the organ of Corti in humans.
    e. the semicircular canals in mammals.

b
Application

45.38 When an organism dies, its muscles remain in a contracted state termed *rigor mortis* for a brief period of time. Which of the following most directly contributes to this phenomenon?
    a. no ATP to move cross-bridges
    b. no ATP to break bonds between thick and thin filaments
    c. no calcium to bind to troponin
    d. no oxygen supplied to muscle
    e. no energy for the synthesis of actin and myosin

e
Conceptual
Understanding

45.39 Which of the following are shared by skeletal, cardiac, and smooth muscle?
    a. A bands and I bands
    b. transverse tubules
    c. gap junctions
    d. motor units
    e. thick and thin filaments

c
Application

45.40 The breast muscle of turkeys and chickens is usually referred to as light meat, whereas that of wild ducks and geese is described as dark meat. Which of the following is consistent with this observation?
    a. Turkeys and chickens are not closely related to ducks and geese.
    b. Turkeys and chickens do not use their breast muscle, whereas ducks and geese do.
    c. Turkeys and chickens do not fly for sustained periods; ducks and geese do.
    d. The muscles of these two groups of birds contain different filamentous proteins.
    e. The darker body color of ducks and geese provides protective camouflage against predators.

d
Factual Recall

45.41 Which of the following does NOT form part of the thin filaments of a muscle cell?
    a. actin
    b. troponin
    c. tropomyosin
    d. myosin
    e. calcium-binding site

a
Conceptual
Understanding

45.42 Which of the following could you find in the lumen of a transverse tubule?
    a. extracellular fluid
    b. cytoplasm
    c. actin
    d. myosin
    e. sarcomeres

Chapter 46

b
Factual Recall

46.1 Two plant species live in the same biome but on different continents. Although these two are not at all closely related, they may appear quite similar as a result of
a. parallel evolution.
b. convergent evolution.
c. allopatric speciation.
d. introgression.
e. gene flow.

e
Conceptual
Understanding

46.2 The pine forest floor would be a relatively fine-grained environment for which of the following?
a. a nematode
b. a toad
c. an herbaceous plant
d. an insect
e. a bear

d
Conceptual
Understanding

46.3 Organisms respond to environmental changes (such as global warming) in several ways. Which response is the slowest, and thus least likely in the event of rapid environmental change?
a. physiological adaptation
b. morphological adaptation
c. migration
d. evolutionary adaptation
e. behavioral adaptation

e
Factual Recall

46.4 Which biome is able to support many large animals despite receiving moderate amounts of rainfall?
a. tropical forest
b. temperate forest
c. chaparral
d. taiga
e. savanna

b
Factual Recall

46.5 Phytoplankton is most frequently found in which of the following zones?
a. tidal
b. photic
c. benthic
d. abyssal
e. intertidal

d
Factual Recall

46.6 Probably the most important factors affecting the distribution of biomes are
a. wind and water current patterns.
b. species diversity and abundance.
c. community succession and climate.
d. climate and topography.
e. day length and rainfall.

e
Conceptual
Understanding

46.7 "How does the foraging of animals on tree seeds affect the distribution and abundance of the trees?" This question
   a. is a valid ecological question.
   b. is difficult to answer because a large experimental area would be required.
   c. is difficult to answer because a long-term experiment would be required.
   d. Both a and b are correct.
   e. a, b, and c are correct.

a
Factual Recall

46.8 All of the following are important factors in the development of terrestrial biomes EXCEPT
   a. the species of colonizing animals.
   b. prevailing temperature.
   c. prevailing rainfall.
   d. mineral nutrient availability.
   e. soil structure.

e
Factual Recall

46.9 Which of the following terrestrial biomes is (are) adapted to frequent fires?
   a. savanna
   b. chaparral
   c. temperate grasslands
   d. Only a and b are correct.
   e. a, b, and c are correct.

c
Factual Recall

46.10 Where would an ecologist find the most phytoplankton in a lake?
   a. profundal zone
   b. benthic zone
   c. limnetic zone
   d. oligotrophic zone
   e. aphotic zone

a
Application

46.11 If a meteor impact or volcanic eruption injected a lot of dust into the atmosphere and reduced sunlight reaching the Earth's surface by seventy percent for one year, all of the following marine communities would be greatly affected EXCEPT a
   a. deep-sea vent community.
   b. coral reef community.
   c. benthic community.
   d. pelagic community.
   e. estuary community.

d
Conceptual
Understanding

46.12 In which community would organisms most likely have evolved to respond to different photoperiods?
   a. tropical forest
   b. coral reef
   c. savanna
   d. temperate forest
   e. abyssal

b
Factual Recall

46.13 All of the following would have a direct effect on the amount of precipitation in an area EXCEPT
   a. air circulation cells.
   b. continental drift.
   c. ocean currents.
   d. mountain ranges.
   e. evaporation from vegetation.

a
Factual Recall

46.14 When the environment of an animal changes, the animal may respond in several ways. Which of the following represents a correct sequence (from most rapid to slowest) of potential animal responses?
   a. migration, acclimation, morphological change, evolution
   b. acclimation, migration, evolution, morphological change
   c. migration, evolution, acclimation, morphological change
   d. migration, evolution, morphological change, acclimation
   e. acclimation, morphological change, migration, evolution

e
Application

46.15 Because of dune erosion, cold salt water begins to mix with the warm fresh water in a pond. Which of the following best describes the responses of organisms to this change?
   a. Conforming invertebrates would survive.
   b. Organisms with cell walls would survive.
   c. Ectotherms would generate more energy for osmoregulation.
   d. Fish would evolve higher metabolic rates.
   e. Most organisms would be unable to survive the change.

a
Application

46.16 There are a few vertebrates that seem to lack respiratory pigments. Which of the following is most likely to be an example of such an animal?
   a. a fish in the Antarctic Ocean
   b. a bird in north Alaska
   c. a mammal in an Amazon jungle
   d. a fish in a shallow pool in a sunny meadow in Texas
   e. a frog in a pond in North Dakota

c
Factual Recall

46.17 All of the following statements about ecology are correct EXCEPT:
   a. Ecologists may study populations and communities of organisms.
   b. Ecological studies may involve the use of models and computers.
   c. Ecology is a discipline that is independent from natural selection and evolutionary history.
   d. Ecology spans increasingly comprehensive levels of organization, from individuals to ecosystems.
   e. Ecology is the study of the interactions between biotic and abiotic aspects of the environment.

b
Conceptual
Understanding

46.18 Which ecological unit incorporates abiotic factors?
   a. community
   b. ecosystem
   c. population
   d. species
   e. symbiosis

e
Conceptual
Understanding

46.19 Important abiotic factors in ecosystems include which of the following?
  I. temperature
  II. water
  III. wind

  a. I only
  b. II only
  c. III only
  d. I and II only
  e. I, II, and III

e
Application

46.20 Imagine some cosmic catastrophe that jolts Earth so that its axis is perpendicular to the line between the Earth and the Sun. The most obvious effect of this change would be
  a. the elimination of tides.
  b. an increase in the length of night.
  c. an increase in the length of a year.
  d. the elimination of the greenhouse effect and a cooling of the equator.
  e. the elimination of seasonal variation.

d
Factual Recall

46.21 Biomes that are maintained by periodic fires include which of the following?
  I. desert
  II. chaparral
  III. temperate grassland

  a. I only
  b. II only
  c. III only
  d. II and III only
  e. I, II, and III

e
Conceptual
Understanding

46.22 All of the following statements about biomes are correct EXCEPT:
  a. Biomes are major terrestrial communities.
  b. Within biomes there may be extensive patchiness.
  c. Climographs are often used to demonstrate climatic differences between biomes.
  d. Temperature and precipitation account for most of the variation between biomes.
  e. Biomes can be recognized as separate entities because they have sharp, well-defined boundaries.

c
Factual Recall

46.23 Tropical grasslands are also known as
  a. taiga.
  b. tundra.
  c. savanna.
  d. chaparral.
  e. temperate plains.

c
Conceptual
Understanding

46.24 Which of the following are CORRECT statements about light in aquatic environments?
  I. Water selectively reflects and absorbs certain wavelengths of light.
  II. Photosynthetic organisms that live in deep water probably utilize red light.
  III. Light intensity is an important abiotic factor in limiting the distribution of photosynthetic organisms.

  a. I only
  b. II only
  c. I and III only
  d. II and III only
  e. I, II, and III

b
Factual Recall

46.25 Which of the following levels of organization is arranged in the correct sequence from most to least inclusive?
  a. community, ecosystem, individual, population
  b. ecosystem, community, population, individual
  c. population, ecosystem, individual, community
  d. individual, population, community, ecosystem
  e. individual, community, population, ecosystem

e
Factual Recall

46.26 The growing season would generally be shortest in which of the following biomes?
  a. savanna
  b. deciduous forest
  c. temperate grassland
  d. tropical rain forest
  e. taiga

c
Conceptual
Understanding

46.27 Which of the following statements follows from the principle of allocation?
  a. The allocation of an organism's resources is independent of environmental fluctuations.
  b. The physiological responses to changing environmental conditions can shift the tolerance ranges of organisms.
  c. Natural selection has resulted in different budgeting strategies for each organism's limited amount of energy.
  d. The classification of regions in marine environments is based on long-term energy fluxes.
  e. An ecosystem's carrying capacity is based on the total energy available to the ecosystem's biota.

d
Factual Recall

46.28 Generally speaking, deserts are located in places where air masses are usually
  a. cold.
  b. humid.
  c. rising.
  d. falling.
  e. expanding.

c
Factual Recall

46.29 Biologists term an organism's physiological adjustments to a change in an environmental factor as
a. evolution.
b. adaptation.
c. acclimation.
d. habituation.
e. transformation.

d
Factual Recall

46.30 Desert animal adaptations to low amounts of water and high temperatures include which of the following?
I.   living in burrows
II. possessing dark-colored fur
III. being active only during the night

a. I only
b. II only
c. III only
d. I and III only
e. II and III only

a
Factual Recall

46.31 In temperate lakes, the surface water is replenished with minerals during turnovers that occur in the
a. fall and spring.
b. fall and winter.
c. spring and summer.
d. summer and winter.
e. summer and fall.

c
Factual Recall

46.32 Which of the following abiotic factors has the greatest influence on the metabolic rates of plants and animals?
a. water
b. wind
c. temperature
d. rocks and soil
e. disturbances

c
Conceptual
Understanding

46.33 Which of the following statements best describes the effect of climate on biome distribution?
a. Knowledge of annual temperature and precipitation is sufficient to predict which biome will inhabit an area.
b. Fluctuation of environmental variables is not important if areas have the same annual temperature and precipitation means.
c. The distribution of biomes depends, in part, on the mean annual temperature and precipitation of an area.
d. Temperate forests, coniferous forests, and grasslands all have the same mean annual temperatures and precipitation.
e. Correlation of climate with biome distribution is sufficient to determine the cause of biome patterns.

b
Conceptual
Understanding

46.34 Which of the following statements best describes the interaction between fire and ecosystems?
   a. The chance of fire in a given ecosystem is highly predictable over the short term.
   b. Many kinds of plants and plant communities have adapted to frequent fires.
   c. The prevention of forest fires has allowed more productive and stable plant communities to develop.
   d. Chaparral communities have evolved to the extent that they rarely burn.
   e. Fire is unnatural in ecosystems and should be prevented.

e
Conceptual
Understanding

46.35 Which of the following causes the Earth's seasons?
   a. global air circulation
   b. global wind patterns
   c. ocean currents
   d. changes in the Earth's distance from the sun
   e. the tilt of the Earth's axis

b
Conceptual
Understanding

46.36 Polar regions are cooler than the equator because
   a. there is more ice at the poles.
   b. sunlight strikes the poles at an oblique angle.
   c. the poles are farther from the sun.
   d. the poles have a thicker atmosphere.
   e. the poles are permanently tilted away from the sun.

a
Conceptual
Understanding

46.37 Which of the following is responsible for the summer and winter stratification of lakes?
   a. Water is densest at 4° C.
   b. Oxygen is most abundant in deeper waters.
   c. Winter ice sinks in the summer.
   d. Stratification is caused by a thermocline.
   e. Stratification always follows the fall and spring turnovers.

e
Conceptual
Understanding

46.38 The principle of allocation implies that
   a. organisms are regulators or conformers.
   b. organisms tolerate a relatively narrow range of environmental variables.
   c. organisms can respond to environmental changes.
   d. organisms have mechanisms to maintain homeostasis.
   e. organisms have mechanisms that effectively distribute their resources.

d
Factual Recall

46.39 Which type of biome would most likely occur in a climate with mild, rainy winters and hot, dry summers?
   a. desert
   b. taiga
   c. temperate grassland
   d. chaparral
   e. savanna

b
Conceptual
Understanding

46.40 Which marine zone would have the lowest rates of primary productivity (photosynthesis)?
  a. pelagic
  b. abyssal
  c. neritic
  d. estuary
  e. intertidal

e
Conceptual
Understanding

46.41 Plants typically can respond to changes in their environment in all of the following ways EXCEPT by
  a. acclimation.
  b. maintaining homeostasis.
  c. evolutionary adaptation.
  d. morphological responses.
  e. behavioral responses.

e
Conceptual
Understanding

46.42 Which of the following is a fine-grained environment?
  a. a sandy beach
  b. a field of wildflowers
  c. a small island
  d. an inner-city park
  e. Before this question can be answered, more information is needed about the organisms that inhabit these environments.

a
Conceptual
Understanding

46.43 Which of the following environmental features might influence microclimates?
  a. All of the below might influence microclimates.
  b. a forest tree
  c. a fallen log
  d. a large stone
  e. a discarded soft drink can

d
Conceptual
Understanding

46.44 When moving along altitudinal and latitudinal gradients, you generally find the same sequence of biomes. In both cases, there are equivalent changes along the gradient in all of the following EXCEPT
  a. temperature.
  b. humidity.
  c. vegetation.
  d. day length.
  e. communities.

Chapter 47

47.1 The most common kind of dispersion is
   a. clumped.
   b. random.
   c. uniform.
   d. indeterminate.
   e. dispersive.

47.2 A table listing such items as age, observed number of organisms alive each year, and life expectancy is known as a
   a. life table.
   b. mortality table.
   c. survivorship table.
   d. rate table.
   e. insurance table.

Use the following choices to answer Questions 47.3–47.5. Each choice may be used once, more than once, or not at all.

   a. $\dfrac{rN}{K}$

   b. rN

   c. $rN(K+N)$

   d. $rN\dfrac{(K-N)}{K}$

   e. $rN\dfrac{(N-K)}{K}$

47.3 Logistic growth of a population is represented by $dN/dt =$

47.4 Exponential growth of a population is represented by $dN/dt =$

47.5 Which of the following is the best example of K-selected species?
   a. desert annual flowers
   b. humans
   c. salmon
   d. spiders
   e. mosquitos

b
Conceptual
Understanding

47.6 A type I survival curve (level at first, with a rapid increase in mortality in old age) is most likely to be found in
   a. an opportunistic species.
   b. an equilibrial species.
   c. a species that undergoes periodic molting.
   d. a species that is territorial.
   e. an invertebrate species.

a
Factual Recall

47.7 Which of the following is characteristic of an opportunistic species?
   a. usually one reproductive episode per lifetime with many offspring
   b. extensive homeostatic capability to deal with environmental fluctuations'
   c. long maturation time and long generation time
   d. large body size and large offspring or eggs
   e. several reproductive episodes with parental care provided to offspring

Use the survivorship curves in Figure 47.1 to answer Questions 47.8–47.12.

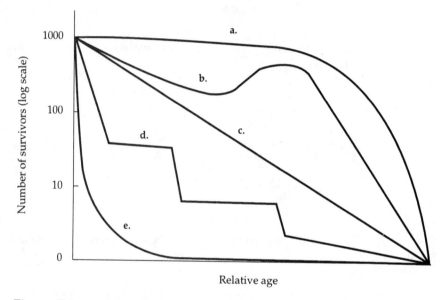

Figure 47.1

e
Conceptual
Understanding

47.8 Which curve best describes survivorship in barnacles?

e
Conceptual
Understanding

47.9 Which curve best describes survivorship in oak trees?

d
Conceptual
Understanding

47.10 Which curve best describes survivorship in a marine crustacean that molts?

a
Conceptual
Understanding

47.11 Which curve best describes survivorship in humans who live in developed nations?

b
Conceptual
Understanding

47.12 Which curve is impossible?

c
Factual Recall

47.13 A species that is relatively *r*-selected might have all of the following characteristics EXCEPT
    a. a disturbed habitat.
    b. small offspring.
    c. parental care of offspring.
    d. numerous offspring.
    e. little homeostatic capability.

b
Factual Recall

47.14 How would the dispersion of humans in the United States best be described?
    a. dense
    b. clumped
    c. random
    d. intrinsic
    e. uniform

d
Factual Recall

47.15 All of the following phrases could characterize a population EXCEPT
    a. fluctuating numbers.
    b. variable dispersion.
    c. measurable numbers.
    d. several species.
    e. geographical boundaries.

e
Factual Recall

47.16 Natural selection involves energetic trade-offs between or among life history traits such as
    a. clutch size.
    b. number of reproductive episodes per lifetime.
    c. age at first reproduction.
    d. Only a and c are correct.
    e. a, b, and c are correct.

a
Factual Recall

47.17 Which of the following characterizes *K*-selected populations?
    a. offspring with good chances of survival
    b. large clutch sizes
    c. small offspring
    d. a high intrinsic rate of increase
    e. early parental reproduction

c
Application

47.18 In which of the following habitats would you expect to find the largest number of *K*-selected individuals?
    a. an abandoned field in Ohio
    b. the sand dunes south of Lake Michigan
    c. the rain forests of Brazil
    d. South Florida after a hurricane
    e. a newly emergent volcanic island

d
Factual Recall

47.19 Natural selection has led to the evolution of diverse natural history strategies which have in common
    a. large clutch sizes.
    b. limitation by density-dependent limiting factors.
    c. adaptation to stable environments.
    d. maximum lifetime reproductive success.
    e. relatively large offspring.

e
Conceptual
Understanding

47.20 Which of the following would be most helpful in solving the world's environmental problems?
    a. increased agricultural productivity
    b. new energy sources
    c. increased life expectancy
    d. more food from the oceans
    e. decreased human birth rates

a
Factual Recall

47.21 All of the following have contributed to the growth of the human population EXCEPT
    a. environmental degradation.
    b. improved nutrition.
    c. vaccines.
    d. pesticides.
    e. improved sanitation.

b
Application

47.22 Which of the following is a density-independent factor limiting human population growth?
    a. social pressure for birth control
    b. earthquakes
    c. plagues
    d. famines
    e. military conflicts

d
Conceptual
Understanding

47.23 Which of the following is an important variable contributing to the rapid growth of human populations?
    a. high percentage of young people
    b. average age to first give birth
    c. carrying capacity of the environment
    d. Only a and b are correct.
    e. a, b, and c are correct.

b
Conceptual
Understanding

47.24 The pattern of dispersion for a certain species of kelp is clumped. The pattern of dispersion for a certain species of snail that only lives on this kelp would be
a. absolute.
b. clumped.
c. homogeneous.
d. random.
e. uniform.

c
Conceptual
Understanding

47.25 A population is correctly defined as having which of the following characteristics?
I.   inhabiting a specific geographic range
II. individuals belonging to the same species
III. possessing a constant and uniform density and dispersion

a. I only
b. III only
c. I and II only
d. II and III only
e. I, II, and III

d
Conceptual
Understanding

47.26 All of the following statements about the logistic model of population growth are correct EXCEPT:
a. It fits an S-shaped curve.
b. It incorporates the concept of carrying capacity.
c. It describes population density shifts over time.
d. It exactly predicts the population growth of most populations.
e. It predicts an eventual state in which birth rate equals death rate.

a
Conceptual
Understanding

47.27 A biologist reported that a sample of ocean water had 5 million diatoms of the species *Coscinodiscus centralis* per cubic meter. What was the biologist measuring?
a. density
b. dispersion
c. carrying capacity
d. quadrats
e. range

b
Application

47.28 To measure the population density of monarch butterflies occupying a particular park, 100 butterflies are captured, marked with a small dot on a wing, and then released. The next day, another 100 butterflies are captured, including the recapture of 20 marked butterflies. One would correctly estimate the population to be
a. 200.
b. 500.
c. 1,000.
d. 10,000.
e. 900,000.

d
Conceptual
Understanding

47.29 Which of the following statements about human birth and death rates is CORRECT?
a. Both death rates and birth rates are highest in 30-year-olds.
b. Both death rates and birth rates are highest in teenagers.
c. Death rates are highest in middle-aged adults, whereas the birth rates are highest in teenagers.
d. Death rates are highest in newborns and in the elderly, whereas birth rates are highest in 20-year-olds.
e. Death rates are highest in the elderly, whereas birth rates are highest in newborns.

b
Conceptual
Understanding

47.30 All of the following are correct statements about the regulation of populations EXCEPT:
a. The logistic equation reflects the effect of density-dependent factors, which can ultimately stabilize populations around the carrying capacity.
b. Density-independent factors have a greater effect as a population's density increases.
c. High densities in a population may cause physiological changes that inhibit reproduction.
d. Because of the overlapping nature of population-regulating factors, it is often difficult to precisely determine their cause-and-effect relationships.
e. The occurrence of population cycles in some populations may be the result of crowding or lag times in the response to density-dependent factors.

e
Conceptual
Understanding

47.31 Life tables are useful in determining which of the following?
I.   carrying capacity
II. mortality rates
III. the fate of a cohort of newborn organisms throughout their lives

a. I only
b. II only
c. III only
d. I and III only
e. II and III only

d
Conceptual
Understanding

47.32 Uniform spacing patterns in plants such as the creosote bush are most often associated with which of the following?
a. chance
b. patterns of high humidity
c. the random distribution of seeds
d. antagonistic interactions of individuals in the population
e. the concentration of resources within the population's range

b
Conceptual
Understanding

47.33 As N approaches K for a certain population, which of the following is predicted by the logistic equation?
a. The growth rate will not change.
b. The growth rate will approach zero.
c. The population will show an Allee effect.
d. The population will increase exponentially.
e. The carrying capacity of the environment will increase.

c
Conceptual
Understanding

47.34 Which of the following statements about the evolution of life histories is correct?
a. Stable environments with limited resources favor *r*-selected populations.
b. *K*-selected populations are most often found in environments where density-independent factors are important regulators of population size.
c. Most populations have both *r*- and *K*-selected characteristics that vary under different environmental conditions.
d. The reproductive efforts of *r*-selected populations are directed at producing just a few offspring with good competitive abilities.
e. *K*-selected populations rarely approach carrying capacity.

e
Application

47.35 Consider several human populations of equal size and net reproductive rate, but different in age structure. The population that is likely to grow the most during the next 30 years is the one with the greatest fraction of people in which age range?
a. 50 to 60 years
b. 40 to 50 years
c. 30 to 40 years
d. 20 to 30 years
e. 10 to 20 years

d
Factual Recall

47.36 All of the following characteristics are typical of an *r*-selected population EXCEPT
a. high mortality rates.
b. high intrinsic rate of growth.
c. onset of reproduction at an early age.
d. extensive parental care of offspring.
e. population control by density-independent factors.

Questions 47.37–47.41 refer to the terms below. Each term may be used once, more than once, or not at all.
a. *cohort*
b. *dispersion*
c. *Allee effect*
d. *iteroparous*
e. *semelparous*

c
Factual Recall

47.37 A density-dependent factor.

e
Factual Recall

47.38 Pacific salmon or annual plants.

d
Factual Recall

47.39 Reproduce more than once in a lifetime.

b
Factual Recall

47.40 Pattern of spacing for individuals within the boundaries of the population.

c
Application

47.41 A predator might be more likely to be spotted if a large number of prey are all together than it would be by a single prey animal.

d
Conceptual
Understanding

47.42 Life history strategies result from
   a. environmental pressures.
   b. natural selection.
   c. conscious choice.
   d. Both a and b are correct.
   e. a, b, and c are correct.

d
Conceptual
Understanding

47.43 In the logistic equation $dN/dt = rN \dfrac{(K-N)}{K}$, r is a measure of the population's intrinsic rate of increase. It is determined by which of the following?
   a. birth rate
   b. death rate
   c. density
   d. Both a and b are correct.
   e. a, b, and c are correct.

Questions 47.44 and 47.45 refer to Figure 47.2, which depicts the age structure of three populations.

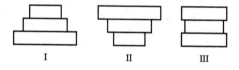

I   II   III

Figure 47.2

b
Application

47.44 Which population is in the process of decreasing?
   a. I only
   b. II only
   c. III only
   d. I and II
   e. II and III

c
Application

47.45 Which population appears to be stable?
   a. I only
   b. II only
   c. III only
   d. I and II
   e. II and III

d
Conceptual
Understanding

47.46 Often the growth cycle of one population has an effect on the cycle of another—as moose populations increase, wolf populations follow. Thus, if we are considering the logistic equation for the wolf population,

$$dN/dt = rN \frac{(K - N)}{K}$$

which of the factors accounts for the effect of the moose population?
a. r
b. N
c. rN
d. K
e. dt

e
Conceptual
Understanding

47.47 In a mature forest of oak, maple, and hickory trees, a disease causes a reduction in the number of acorns produced by oak trees. Which of the following would LEAST likely be a result of this?
a. There might be fewer squirrels because they feed on acorns.
b. There might be fewer mice and seed-eating birds because squirrels would eat more seeds and compete with the mice and birds.
c. There might be an increase in the number of hickory trees because the competition between hickory nuts and acorns for germination sites would be reduced or eliminated.
d. There might be fewer owls because they feed on baby squirrels, mice, and young seed-eating birds, whose populations would be reduced.
e. There might be a decrease in the number of maple seeds as the disease spreads to other trees in the forest.

d
Conceptual
Understanding

47.48 Your friend Forrest comes to you with a problem. It seems his shrimp boats aren't catching nearly as much shrimp as they used to. He can't understand it because originally he caught all the shrimp he could handle. Each year, he added a new boat, and for a long time, each boat caught tons of shrimp. As he added more boats, there came a time when each boat caught a little less shrimp, and now, each boat is catching a lot less shrimp. Which of the following topics might help Forrest understand the source of his problem?
a. density-dependent population regulation
b. logistic growth and intrinsic characteristics of population growth
c. density-independent population regulation
d. Both a and b are correct.
e. a, b, and c are correct.

Chapter 48

e
Conceptual
Understanding

48.1 Which of the following could cause a realized niche to differ from a fundamental niche?
    a. suitable habitat
    b. food size and availability
    c. temperature limitations
    d. water availability
    e. competition from other species

e
Conceptual
Understanding

48.2 The following factors were found to limit the distribution of fish populations: temperature, oxygen content of the water, and free protein in the water. This is a
    a. dimensional profile.
    b. mutualistic niche.
    c. realized niche.
    d. resource profile.
    e. fundamental niche.

b
Application

48.3 In a tide pool, 15 species of invertebrates were reduced to eight after one species was removed. The species removed was likely a(n)
    a. community facilitator.
    b. keystone predator.
    c. herbivore.
    d. resource partitioner.
    e. mutualistic organism.

c
Factual Recall

48.4 What is it called when organisms in an ecosystem make their environment more favorable for competitors than for themselves?
    a. coevolution
    b. competitive exclusion
    c. facilitation
    d. inhibition
    e. competitive replacement

d
Factual Recall

48.5 What does the species equitability of a community refer to?
    a. the species diversity
    b. the number of different species
    c. both the species diversity and the number of different species
    d. the relative numbers of individuals in each species
    e. the feeding relationships or trophic structure

b
Conceptual
Understanding

48.6 Resource partitioning is best described by which of the following statements?
    a. Competitive exclusion results in the success of the superior species.
    b. Slight variations in niche allow closely related species to coexist.
    c. Two species can coevolve and share the same realized niche.
    d. Species diversity is maintained by switching between prey species.
    e. A climax community is reached when no new niches are available.

e
Conceptual
Understanding

48.7 The realized niche of species often differs from their fundamental niche. Why this happens could be explained by any of the following EXCEPT
a. competitive exclusion.
b. predation when densities are high.
c. parasitism when densities are high.
d. interspecific competition.   -
e. vegetation changes.

a
Conceptual
Understanding

48.8 The traditional concept of succession includes the idea of an equilibrium state called a climax community. Ecologists now think there may be no such thing as a climax community because
a. disturbance is ongoing in ecosystems.
b. all organisms eventually die.
c. species diversity generally increases.
d. extinction increases in stable environments.
e. each succession is different from others.

c
Factual Recall

48.9 Which of the following types of interactions is INCORRECTLY paired to its effects on the two interacting species?
a. predation—one benefits, one loses
b. parasitism—one benefits, one loses
c. commensalism—both benefit
d. mutualism—both benefit
e. competition—both lose

c
Factual Recall

48.10 An insect that has evolved to closely resemble a twig will probably be able to avoid
a. parasitism.
b. symbiosis.
c. predation.
d. competition.
e. commensalism.

d
Conceptual
Understanding

48.11 Mimicry systems depend on all of the following EXCEPT
a. the models being noxious or disagreeable.
b. the mimics being less common than their models.
c. the ability of predators to "learn" characteristics of their prey.
d. the models being cryptically colored.
e. the models being easily recognized.

d
Conceptual
Understanding

48.12 To be certain that two species had coevolved, one would ideally have to establish that
a. the two species are adapted to a common set of environmental conditions.
b. each species has an impact on the population density of the other species.
c. the local extinction of one species results in the extinction of the other species.
d. one species has adaptations that specifically followed evolutionary change in the other species.
e. the two species originated at about the same time.

c
Factual Recall

48.13 An example of Batesian mimicry is
   a. an insect that resembles a twig.
   b. a butterfly that resembles a leaf.
   c. a nonpoisonous snake that looks like a poisonous snake.
   d. a fawn with fur coloring that camouflages it in the forest environment.
   e. a snapping turtle that uses its tongue to mimic a worm, thus attracting fish.

b
Conceptual
Understanding

48.14 All of the following statements about the biogeographical aspects of diversity
   are correct EXCEPT:
   a. The patterns of continental drift are important considerations in the study
      of the past and present distributions of species.
   b. The magnitude of photosynthesis is the factor that accounts for the major
      variations in species diversity over Earth's large areas.
   c. Species richness on an island reaches an equilibrium point when immi-
      gration equals extinction.
   d. A species may be limited to a particular range because it never dispersed
      beyond that range, or it dispersed but failed to survive in other locations.
   e. Island biogeographical theory applies to the relatively short period of time
      when colonization is the important process determining species compo-
      sition; over a longer time, actual speciation affects the composition.

e
Factual Recall

48.15 The sum total of an organism's interaction with the biotic and abiotic
   resources of its environment is called its
   a. habitat.
   b. logistic growth.
   c. biotic potential.
   d. microclimax.
   e. niche.

d
Conceptual
Understanding

48.16 Communities can be structured by which of the following?
   I.   predation
   II. systematics
   III. competition

   a. I only
   b. III only
   c. I and II only
   d. I and III only
   e. I, II, and III

a
Factual Recall

48.17 Which of the following is considered by ecologists as a measure either of the
   ability of a community to resist change or of the ability of a community to
   recover to its original state after change?
   a. stability
   b. succession
   c. partitioning
   d. productivity
   e. competitive exclusion

b
Factual Recall

48.18 According to the competitive exclusion principle, two species cannot continue to occupy the same
   a. habitat.
   b. niche.
   c. territory.
   d. range.
   e. biome.

d
Conceptual
Understanding

48.19 All of the following statements about communities are correct EXCEPT:
   a. Many plant species in communities seem to be independently distributed.
   b. Some animal species distributions within a community are linked to other species.
   c. The distribution of almost all organisms is probably affected to some extent by both abiotic gradients and interactions with other species.
   d. Ecologists refer to species equitability as the number of species within a community.
   e. The trophic structure of a community describes the feeding relationships within a community.

e
Conceptual
Understanding

48.20 According to the nonequilibrium model, species diversity is likely to be greatest in a community where disturbance is
   a. mild and rare.
   b. mild and frequent.
   c. severe and rare.
   d. severe and frequent.
   e. moderate in severity and frequency.

c
Conceptual
Understanding

48.21 Which of the following statements about succession is correct?
   a. Secondary succession occurs where no soil has previously existed.
   b. Primary succession occurs in areas where soil remains after a disturbance.
   c. Disturbances can stabilize community structure or initiate or alter succession, depending on the severity and the duration of the disturbance.
   d. Some cases of succession involve facilitation, a phenomenon in which species inhibit the growth of newcomers.
   e. Through successional dynamics, most communities will eventually reach a stable climax with maximum diversity.

c
Factual Recall

48.22 The study of the past and present distribution of species is called
   a. demography.
   b. conservation biology.
   c. biogeography.
   d. geographic ecology.
   e. population ecology.

e
Factual Recall

48.23 An example of cryptic coloration is the
   a. stripes of a skunk.
   b. green color of a plant.
   c. markings of a viceroy butterfly.
   d. colors of an insect-pollinated flower.
   e. mottled coloring of peppered moths living in the unpolluted regions of England.

d
Factual Recall

48.24 In a particular case of secondary succession, three species of wild grass all invaded a field the first growing season after a farmer abandoned the field. By the second season, a single one of the wild grasses dominated the field. A possible factor in this succession was
  a. equilibrium.
  b. facilitation.
  c. immigration.
  d. inhibition.
  e. mutualism.

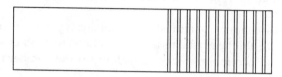

Figure 48.1

b
Conceptual
Understanding

48.25 The entire box shown in Figure 48.1 represents the fundamental niche of species A. Species A is biologically constrained from the striped area of its fundamental niche by species B. The unstriped portion most likely represents species A's
  a. cline.
  b. realized niche.
  c. achieved habitat.
  d. biome.
  e. zone.

e
Factual Recall

48.26 An example of Müllerian mimicry is
  a. an insect that resembles a twig.
  b. a butterfly that resembles a leaf.
  c. a beetle that resembles a scorpion.
  d. a moth with spots that look like large eyes.
  e. two poisonous frogs that resemble one another in coloration.

c
Conceptual
Understanding

48.27 All of the following statements about community interactions are correct EXCEPT:
  a. Closely related species may be able to coexist if there is at least one significant difference in their niches.
  b. Plants can defend themselves against herbivores by the production of compounds that are irritating or toxic.
  c. Keystone predators reduce diversity in a community by holding down or wiping out prey populations.
  d. Mutualism is an important biotic interaction that occurs in communities.
  e. Some predators use mimicry to attract prey.

a
Factual Recall

48.28 An example of aposematic coloration is the
    a. stripes of a skunk.
    b. eye color in humans.
    c. green color of a plant.
    d. colors of an insect-pollinated flower.
    e. mottled coloring of peppered moths living in the unpolluted regions of England.

Questions 48.29–48.32 refer to the biogeographical regions below. Each region may be used once, more than once, or not at all.
    a. *Ethiopian*
    b. *Nearctic*
    c. *Neotropical*
    d. *Oriental*
    e. *Palearctic*

a
Factual Recall

48.29 Most of Africa.

b
Factual Recall

48.30 Most of North America.

c
Factual Recall

48.31 South America.

e
Factual Recall

48.32 Europe.

Questions 48.33–48.36 refer to the diagram in Figure 48.2 of five islands formed at about the same time near a particular mainland.

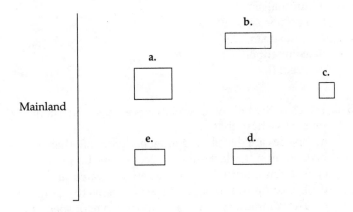

Figure 48.2

a
Application

48.33 Island with the greatest number of species.

c
Application

48.34 Island with the least number of species.

a
Application

48.35 Island with the lowest extinction rate.

c
Application

48.36 Island with the lowest immigration rate.

Questions 48.37–48.40 refer to the following list of terms. Each term may be used once, more than once, or not at all.
   a. *parasitism*
   b. *mutualism*
   c. *inhibition*
   d. *facilitation*
   e. *commensalism*

b
Factual Recall

48.37 The relationship existing between ants and acacia trees.

e
Factual Recall

48.38 The relationship existing between cattle egrets and cattle.

b
Factual Recall

48.39 The relationship existing between legumes and nitrogen-fixing bacteria.

d
Factual Recall

48.40 Successional event in which one organism makes the environment suitable for another organism

a
Application

48.41 Evidence shows that some grasses benefit from being grazed. Which of the following terms would best describe this plant–herbivore interaction?
   a. mutualism
   b. commensalism
   c. parasitism
   d. competition
   e. predation

c
Conceptual
Understanding

48.42 Which of the following statements is most consistent with F. E. Clements' interactive hypothesis?
   a. Species are distributed independently of other species.
   b. Communities lack discrete geographic boundaries.
   c. The community functions as an integrated unit.
   d. The composition of plant species seems to change on a continuum.
   e. The community is a chance assemblage of species.

e
Factual Recall

48.43 Which of the following terms best describes the interaction between passion-flower vines and *Heliconius* butterflies?
   a. pollination
   b. mutualism
   c. competitive exclusion
   d. Batesian mimicry
   e. coevolution

e
Conceptual
Understanding

48.44 Which of the following is least likely to kill the organism it feeds on?
   a. parasitoid
   b. predator
   c. seed eater
   d. carnivore
   e. parasite

b
Factual Recall

48.45 Which of the following is NOT an example of a plant defense against herbivory?
   a. nicotine
   b. cryptic coloration
   c. spines
   d. thorns
   e. microscopic crystals

d
Factual Recall

48.46 All of the following represent ways that animals defend themselves against predators EXCEPT
   a. incorporating plant toxins into their tissues.
   b. cryptic coloration.
   c. mobbing.
   d. interspecific competition.
   e. distraction displays.

d
Conceptual
Understanding

48.47 Which of the following statements is consistent with the competitive exclusion principle?
   a. Bird species generally do not compete for nesting sites.
   b. The density of one competing species will have a positive impact on the population growth of the other competing species.
   c. Two species with the same fundamental niche will exclude other competing species.
   d. Even a slight reproductive advantage will eventually lead to the elimination of inferior competitors.
   e. Evolution tends to increase competition between related species.

b
Conceptual
Understanding

48.48 All of the following describe possible results of competition EXCEPT
   a. competitive exclusion.
   b. aposematic coloration.
   c. resource partitioning.
   d. character displacement.
   e. extinction.

a
Conceptual
Understanding

48.49 All of the following act to increase species diversity EXCEPT
    a. competitive exclusion.
    b. keystone predators.
    c. patchy environments.
    d. moderate disturbances.
    e. migration of populations.

d
Conceptual
Understanding

48.50 All of the following are terms that ecologists use to describe communities EXCEPT
    a. species richness.
    b. species diversity.
    c. resilience.
    d. symbiosis.
    e. stability.

c
Conceptual
Understanding

48.51 Which of the following best describes plants that are found in late successional stages?
    a. excellent dispersal mechanisms
    b. weedy
    c. *K*-selected
    d. annual
    e. good colonizer

e
Factual Recall

48.52 Which animal has created the greatest disturbances and thus had the biggest impact on world ecosystems?
    a. goat
    b. beaver
    c. gypsy moth
    d. zebra mussel
    e. human

a
Conceptual
Understanding

48.53 According to the theory of island biogeography, all of the following contribute to greater species diversity on an island EXCEPT
    a. a relatively recent formation of the island.
    b. a shorter distance of the island from the mainland.
    c. a bigger island.
    d. a lower extinction rate on the island.
    e. higher rates of migration to and from the island.

Chapter 49

Use Figure 49.1 to answer Questions. 49.1–49.5. Examine this food web for a particular terrestrial ecosystem. Each letter is a species. The arrows represent energy flow. Answer the following questions concerning this figure.

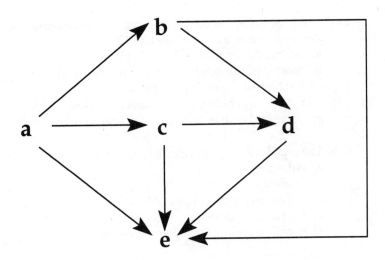

Figure 49.1

a
Conceptual
Understanding

49.1 Which species is autotrophic?

e
Conceptual
Understanding

49.2 Which species is most likely the decomposer?

d
Conceptual
Understanding

49.3 A toxic pollutant would probably reach its highest concentration in which species?

b
Application

49.4 Species c makes its predators sick. Which species is most likely to benefit from being a mimic of c?

d
Conceptual
Understanding

49.5 Excluding the decomposer, biomass would probably be smallest for which species?

c
Application

49.6 If the flow of energy in an Arctic ecosystem goes through a simple food chain from seaweeds to fish to seals to polar bears, then which of the following is true?
a. Polar bears can provide more food for Eskimos than seals can.
b. The total energy content of the seaweeds is lower than that of the seals.
c. Polar bear meat probably contains the highest concentrations of fat-soluble toxins.
d. Seals are more numerous than fish.
e. The carnivores can provide more food for the Eskimos than the herbivores can.

e
Factual Recall

49.7 All of the following are likely results of land-clearing operations such as deforestation and agriculture EXCEPT
a. destruction of plant and animal habitats.
b. erosion of soil due to increased water runoff.
c. leaching of minerals from the soil.
d. rapid eutrophication of streams and lakes.
e. decreased carbon dioxide in the atmosphere.

c
Factual Recall

49.8 The high levels of pesticides found in birds of prey is an example of
a. eutrophication.
b. predation.
c. biological magnification.
d. the Gaia hypothesis.
e. chemical cycling through an ecosystem.

Refer to Figure 49.2, the diagram of a food web, for Questions 49.9–49.11. (Arrows represent energy flow and letters represent species.)

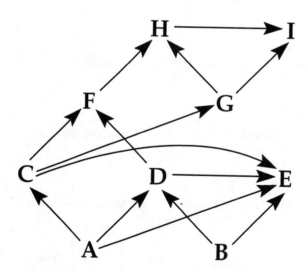

Figure 49.2

d
Application

49.9 If this is a terrestrial food web, the combined biomass of C + D is probably
a. greater than the biomass of A.
b. less than the biomass of H.
c. greater than the biomass of B.
d. less than the biomass of A + B.
e. less than the biomass of E.

a
Application

49.10 If this is a marine food web, the smallest organism could be
a. A.
b. F.
c. C.
d. I.
e. E.

e
Application

49.11 Which species could be described as an omnivore?
a. F
b. B
c. I
d. D
e. E

a
Conceptual
Understanding

49.12 In which cycle are bacteria important for processes other than decomposition?
a. nitrogen cycle
b. hydrologic cycle
c. carbon cycle
d. phosphorous cycle
e. energy cycle

e
Factual Recall

49.13 How does phosphorus normally enter the atmosphere?
a. respiration
b. photosynthesis
c. rock weathering
d. geological uplifting (subduction and vulcanism)
e. It does not enter the atmosphere in biologically significant amounts.

d
Factual Recall

49.14 Which of these ecosystems accounts for the largest amount of Earth's primary productivity?
a. tundra
b. savanna
c. salt marsh
d. open ocean
e. tropical rain forest

e
Conceptual
Understanding

49.15 The producers in ecosystems include which of the following?
I.   photosynthetic protists
II. cyanobacteria
III. plants

a. I only
b. II only
c. III only
d. I and III only
e. I, II, and III

d
Conceptual
Understanding

49.16 Which of these ecosystems has the highest primary productivity per square meter?
    a. savanna
    b. open ocean
    c. boreal forest
    d. tropical rain forest
    e. temperate forest

e
Factual Recall

49.17 Compared to the open ocean, marine life is especially abundant and diverse near the shore because
    a. the open ocean is too saline.
    b. the water is calmer near the shore.
    c. the water is warmer near the shore.
    d. there is less competition for light near the shore.
    e. inorganic nutrients are more plentiful near the shore.

d
Conceptual
Understanding

49.18 All of the following statements about energy flow are correct EXCEPT:
    a. Secondary productivity declines with each trophic level.
    b. Only net primary productivity is available to consumers.
    c. About 90% of the energy at one trophic level does not appear at the next.
    d. Eating meat is probably the most efficient way of tapping photosynthetic productivity.
    e. Only about one thousandth of the chemical energy fixed by photosynthesis actually reaches a tertiary-level consumer.

a
Factual Recall

49.19 In general, the total biomass in a terrestrial ecosystem will be greatest for which trophic level?
    a. producers
    b. herbivores
    c. primary consumers
    d. tertiary consumers
    e. secondary consumers

c
Conceptual
Understanding

49.20 Aquatic primary productivity is often limited by which of the following?
    I. light
    II. nutrients
    III. pressure

    a. II only
    b. III only
    c. I and II only
    d. II and III only
    e. I, II, and III

d
Factual Recall

49.21 The rate at which solar energy is converted to the chemical energy of organic compounds by autotrophs is termed
    a. biomass.
    b. standing crop.
    c. biomagnification.
    d. primary productivity.
    e. secondary productivity.

b
Factual Recall

49.22 Which of the following statements is CORRECT about biogeochemical cycling?
   a. The phosphorus cycle involves the rapid recycling of atmospheric phosphorus.
   b. The phosphorus cycle is a sedimentary cycle that involves the weathering of rocks.
   c. The carbon cycle is a localized cycle that primarily reflects the burning of fossil fuels.
   d. The carbon cycle has maintained the atmospheric concentration of CO_2 constant for the past million years.
   e. The equation $6 \, CO_2 + 6 \, H_2O \rightarrow C_6H_{12}O_6 + 6 \, O_2$ accounts for the tremendous CO_2 buffering capabilities of the ocean.

e
Conceptual
Understanding

49.23 Which of the following statements about human intrusions in ecosystem dynamics is CORRECT?
   a. Burning fossil fuels has decreased the rate of global eutrophication.
   b. Deforestation has few consequences that extend beyond the affected forest.
   c. Dumping nutrient wastes into aquatic habitats accelerates the greenhouse effect.
   d. The release of toxic substances is a temporary nuisance because bacterial decomposers soon break them all down.
   e. Agricultural practices result in the constant removal of nutrients from ecosystems, so that large supplements are continuously required.

d
Conceptual
Understanding

49.24 Detritivores in ecosystems include which of the following?
   I.   phytoplankton
   II. bacteria
   III. fungi

   a. I only
   b. I and II only
   c. I and III only
   d. II and III only
   e. I, II, and III

b
Factual Recall

49.25 All organisms capable of fixing nitrogen belong to the kingdom
   a. Protista.
   b. Monera.
   c. Fungi.
   d. Plantae.
   e. Animalia.

a
Factual Recall

49.26 Cows belong to which trophic level?
   a. primary consumers
   b. secondary consumers
   c. decomposers
   d. autotrophs
   e. producers

b
Factual Recall

49.27 Which of the following organisms fix nitrogen in aquatic ecosystems?
 a. *Rhizobium*
 b. cyanobacteria
 c. chemoautotrophs
 d. phytoplankton
 e. legumes

c
Factual Recall

49.28 In the nitrogen cycle, the bacteria that replenish the atmosphere with N_2 are
 a. *Rhizobium* bacteria.
 b. nitrifying bacteria.
 c. denitrifying bacteria.
 d. methanogenic protozoans.
 e. nitrogen-fixing bacteria.

Questions 49.29–49.33 refer to the organisms in a grassland ecosystem listed below. Each term may be used once, more than once, or not at all.
 a. *hawks*
 b. *snakes*
 c. *shrews*
 d. *grasshoppers*
 e. *grass*

e
Factual Recall

49.29 An autotroph.

d
Factual Recall

49.30 An herbivore.

a
Factual Recall

49.31 Smallest biomass.

e
Factual Recall

49.32 Incorporates greatest amount of energy.

a
Factual Recall

49.33 Probably the highest internal concentration of toxic pollutants.

Questions 49.34–49.38 refer to the terms below. Each term may be used once, more than once, or not at all.
 a. *Gaia hypothesis*
 b. *turnover*
 c. *biomagnification*
 d. *greenhouse effect*
 e. *cultural eutrophication*

d
Factual Recall

49.34 Caused by increasing concentrations of CO_2.

e
Factual Recall

49.35 Caused by excessive nutrient input into lakes.

c
Factual Recall

49.36 Caused excessive levels of DDT in fish-eating birds.

b
Factual Recall

49.37 Occurs at a high rate for nutrients in tropical rain forests.

b
Factual Recall

49.38 Occurs at a high rate for phytoplankton in the English Channel.

b
Conceptual
Understanding

49.39 To recycle nutrients, the minimum an ecosystem must have is
a. producers.
b. producers and decomposers.
c. producers, primary consumers, and decomposers.
d. producers, primary consumers, secondary consumers, and decomposers.
e. producers, primary consumers, secondary consumers, top carnivores, and decomposers.

c
Conceptual
Understanding

49.40 A friend is astounded by the fact that one bird called a purple martin can eat millions of mosquitoes and other small flying insects each year. Which of the following would NOT be involved in your explanation of this to your friend?
a. Insects are small primary consumers and martins are medium-sized secondary consumers.
b. Bird flight requires great amounts of metabolic energy, which must be obtained by taking in large amounts of food.
c. As a result of the intake of all those insects, martins produce tremendous amounts of CO_2, which is a contributing factor to the greenhouse effect.
d. Relying on insects as a source of energy is rather inefficient because so much of the insect's energy goes into making its exoskeleton, which is nearly impossible to digest.
e. Martins optimize their foraging by feeding in the early evening when the greatest number of small insects are flying.

a
Conceptual
Understanding

49.41 The fundamental difference between materials and energy is that
a. materials are cycled through ecosystems; energy is not.
b. energy is cycled through ecosystems; materials are not.
c. energy can be converted into materials; materials cannot be converted into energy.
d. materials can be converted into energy; energy cannot be converted into materials.
e. ecosystems are much more efficient in their transfer of energy than in their transfer of materials.

b
Conceptual
Understanding

49.42 For most terrestrial ecosystems, pyramids of numbers, biomass, and energy are essentially the same: they have a broad base and a narrow top. The primary reason for this pattern is that
  a. secondary consumers and top carnivores require less energy than producers.
  b. at each step, energy is lost from the system as a result of keeping the organisms alive.
  c. as materials pass through ecosystems, some of them are lost to the environment.
  d. biomagnification of toxic materials limits the secondary consumers and top carnivores.
  e. top carnivores and secondary consumers have a more general diet than primary producers.

c
Conceptual
Understanding

49.43 Most homeowners mow their lawns during the summer and collect the clippings, which are then hauled to the local landfill each week. Which of the following would be the ecologically best way to alter this process?
  a. Don't mow the lawn—have a goat graze on it and put the goat's feces into the landfill.
  b. Collect the clippings and burn them.
  c. Either collect the clippings and add them to a compost pile, or don't collect the clippings and let them decompose in the lawn.
  d. Collect the clippings and wash them into the nearest storm sewer that feeds into the local lake.
  e. Dig up the lawn and cover the yard with asphalt.

d
Conceptual
Understanding

49.44 When exotic species are introduced into an ecosystem, most of them fail to become established. This is a result of which of the following?
  a. physiological limitations
  b. lack of success in competing for resources against established members of the ecosystem
  c. lack of efficiency in cycling materials
  d. Both a and b are correct.
  e. a, b, and c are correct.

d
Application

49.45 When levels of CO_2 are experimentally increased, C_3 plants generally respond with a greater increase in productivity than C_4 plants. This is because
  a. C_3 plants are more efficient in their use of CO_2.
  b. C_3 plants are able to obtain the same amount of CO_2 by keeping their stomata open for shorter periods of time.
  c. C_4 plants don't use CO_2 as their source of carbon.
  d. the rate of photosynthesis is limited more by CO_2 in C_3 plants than in C_4 plants.
  e. as CO_2 levels increase, animals become less active, and C_3 plants have more predators than C_4 plants.

Chapter 50

a
Factual Recall

50.1 Which of the following is true about imprinting?
   a. It may be triggered by visual or chemical stimuli.
   b. It happens to many adult animals, but not to their young.
   c. It is a type of learning involving no innate behavior.
   d. It occurs only in birds.
   e. It causes behaviors that last for only a short time (the critical period).

c
Factual Recall

50.2 Animals that help other animals of the same species are expected to
   a. have excess energy reserves.
   b. be bigger and stronger than the other animals.
   c. be genetically related to the other animals.
   d. be male.
   e. have defective genes controlling behavior.

a
Application

50.3 Fred and Joe, two unrelated, mature male gorillas, encounter one another. Fred is courting a female. Fred grunts as Joe comes near. As Joe continues to advance, Fred begins drumming (pounding his chest) and bares his teeth. At this, Joe rolls on the ground on his back, then gets up and quickly leaves. This behavioral pattern is repeated several times during the mating season. Choose the most specific behavior described by this example.
   a. agonistic behavior
   b. territorial behavior
   c. learned behavior
   d. social behavior
   e. fixed-action patterns

c
Factual Recall

50.4 Learning in which a new sign stimulus may be used to elicit the same behavioral response as the original sign stimulus is called
   a. concept formation.
   b. trial and error.
   c. classical conditioning.
   d. training.
   e. habituation.

d
Factual Recall

50.5 Which of the following groups of scientists is closely associated with ethology?
   a. Watson, Crick, and Franklin
   b. McClintock, Goodall, and Lyon
   c. Fossey, Hershey, and Chase
   d. von Frisch, Lorenz, and Tinbergen
   e. Hardy, Weinberg, and Castle

e
Conceptual
Understanding

50.6 Which of the following statements are true of fixed-action patterns?
    a. They are highly stereotyped, instinctive behaviors.
    b. They are triggered by sign stimuli in the environment and, once begun, are continued to completion.
    c. A supernormal stimulus often produces a stronger response.
    d. Only a and b are correct.
    e. a, b, and c are correct.

b
Factual Recall

50.7 Animals tend to maximize their energy intake-to-expenditure ratio. What is this behavior called?
    a. agonistic behavior
    b. optimal foraging
    c. dominance hierarchies
    d. animal cognition
    e. territoriality

Use the following terms to answer Questions 50.8–50.13. Match the term that best fits each of the following descriptions of behavior. Each term may be used once, more than once, or not at all.
    a. fixed-action pattern releaser
    b. habituation
    c. imprinting
    d. classical conditioning
    e. operant conditioning

e
Factual Recall

50.8 Tits (chickadee-like birds) learned to peck through the paper tops of milk bottles left on doorsteps and drink the cream from the top.

b
Factual Recall

50.9 Male insects attempt to mate with orchids but eventually stop responding to them.

c
Factual Recall

50.10 A returning salmon goes back to its own home stream to spawn.

a
Factual Recall

50.11 A stickleback fish will attack a nonfish model as long as the model has red color.

a
Factual Recall

50.12 Parental protective behavior in turkeys is triggered by the cheeping sound produced by the young chicks.

c
Factual Recall

50.13 Sparrows are receptive to learning songs only during a critical period.

e
Conceptual
Understanding

50.14 Many animals increase their activity when they are hungry. Which of the following explanations for this is the most anthropomorphic?
    a. This behavior is a product of natural selection.
    b. The hunger sensation stimulates the activity level of the animal.
    c. There is a competitive advantage in displaying this behavior.
    d. Nervous activity is a function of nutrient levels in the bloodstream.
    e. The animal becomes more active because it knows that's a better way of finding food.

b
Factual Recall

50.15 The proximate causes of behavior are interactions with the environment, whereas the ultimate cause of behavior is
    a. hormones.
    b. evolution.
    c. sexuality.
    d. pheromones.
    e. the nervous system.

b
Factual Recall

50.16 In the territorial behavior of the stickleback fish, the red belly of one male elicits attack from another male by functioning as
    a. a pheromone.
    b. a releaser.
    c. a *Zeitgeber*.
    d. a search image.
    e. an imprint stimulus.

b
Factual Recall

50.17 A sign stimulus that functions as a signal that triggers a certain behavior in another member of the same species is specifically called
    a. a ritual.
    b. a releaser.
    c. a search image.
    d. an agonistic sign.
    e. a fixed-action stimulus.

d
Conceptual
Understanding

50.18 All of the following are correct statements about innate components of behavior EXCEPT:
    a. fixed-action patterns are stereotypic innate behaviors.
    b. fixed-action patterns continue to completion after initiation.
    c. Sign stimuli tend to be based on simple cues associated with the relevant activity.
    d. The concepts of innate releasing mechanisms have shed light on the actual mechanisms of behavior.
    e. Exaggeration of a relevant stimulus in the form of a supernormal stimulus often produces a stronger response.

e
Conceptual
Understanding

50.19 Learning to ignore unimportant stimuli is called
    a. adapting.
    b. spacing.
    c. conditioning.
    d. imprinting.
    e. habituation.

b
Factual Recall

50.20 A type of learning that can occur only during a brief period of early life and results in a behavior that is difficult to modify through later experiences is called
    a. insight.
    b. imprinting.
    c. habituation.
    d. operant conditioning.
    e. trial-and-error learning.

e
Factual Recall

50.21 The swimming response of the nudibranch *Tritonia*, which is caused by the presence of a predator, is brought about by
a. habituation.
b. behaviorism.
c. imprinting during the critical period.
d. an internal biological clock.
e. an innate releasing mechanism.

d
Factual Recall

50.22 The type of learning that causes specially trained dogs to salivate when they hear bells is called
a. insight.
b. imprinting.
c. habituation.
d. classical conditioning.
e. trial-and-error learning.

c
Conceptual
Understanding

50.23 All of the following statements about learning and behavior are correct EXCEPT:
a. Insight learning involves the ability to reason.
b. Associative learning involves linking one stimulus with another.
c. Operant conditioning is a type of innate behavior that involves drive.
d. Behavior can be modified by learning, but some apparent learning is due to maturation.
e. Imprinting is a learned behavior with an innate component acquired during a critical period.

c
Factual Recall

50.24 The congregation of lice in a moist location due to greater activity in dry areas is an example of
a. taxis.
b. tropism.
c. kinesis.
d. insight.
e. net reflex.

a
Conceptual
Understanding

50.25 Which of the following statements about behavioral rhythms is CORRECT?
a. Daily behaviors are regulated by an endogenous clock set by external signs.
b. Exogenous cues have little influence on biological clocks.
c. The proposed mechanism of the biological clock involves the resonance of magnetite.
d. Circannual behaviors are controlled by endogenous clocks.
e. Hibernation is controlled by an internal clock.

e
Factual Recall

50.26 Which of the following have been shown as reference frameworks used by some birds for navigation during migration?
I.   the Sun
II. constellations
III. Earth's magnetic field

a. I only
b. II only
c. III only
d. I and II only
e. I, II, and III

c
Factual Recall

50.27 A chemical produced by an animal that elicits a specific response by another animal of the same species is called
   a. a marker.
   b. an inducer.
   c. a pheromone.
   d. an imprinter.
   e. an agonistic chemical.

d
Conceptual
Understanding

50.28 The central concept of sociobiology is that
   a. human behavior is rigidly predetermined.
   b. the behavior of an individual cannot be modified.
   c. our behavior consists mainly of fixed-action patterns.
   d. most aspects of our social behavior have an evolutionary basis.
   e. the social behavior of humans is homologous to the social behavior of honeybees.

e
Factual Recall

50.29 Feeding behavior that has a high energy intake-to-expenditure ratio is called
   a. herbivory.
   b. autotrophy.
   c. heterotrophy.
   d. search scavenging.
   e. optimal foraging.

d
Conceptual
Understanding

50.30 All of the following statements about mating behavior are correct EXCEPT:
   a. Some aspects of courtship behavior may have evolved from agonistic interactions.
   b. Courtship interactions ensure that the participating individuals are nonthreatening and of the proper species, sex, and physiological condition for mating.
   c. The degree to which evolution affects mating relationships depends on the degree of prenatal and postnatal input the parents are required to make.
   d. The mating relationship in most mammals is monogamous, to ensure the reproductive success of the pair.
   e. Polygamous relationships most often involve a single male and many females, but in some species this is reversed.

b
Conceptual
Understanding

50.31 The presence of altruistic behavior in animals is most likely due to kin selection, a theory that maintains that
   a. aggression between sexes promotes the survival of the fittest individuals.
   b. genes enhance survival of copies of themselves by directing organisms to assist others who share those genes.
   c. companionship is advantageous to animals because in the future they can help each other.
   d. critical thinking abilities are normal traits for animals and they have arisen, like other traits, through natural selection.
   e. natural selection has generally favored the evolution of exaggerated aggressive and submissive behaviors to resolve conflict without grave harm to participants.

Questions 50.32–50.34 refer to the following list of scientists.
  a. *Karl von Frisch*
  b. *Niko Tinbergen*
  c. *Konrad Lorenz*
  d. *B. F. Skinner*
  e. *Ivan Pavlov*

a
Factual Recall

50.32 Studied communication in bees.

d
Factual Recall

50.33 Studied operant conditioning in rats.

c
Factual Recall

50.34 Studied imprinting of greylag geese.

Questions 50.35–50.37 refer to the following list of scientists.
  a. *E. O. Wilson*
  b. *Jane Goodall*
  c. *J. B. S. Haldane*
  d. *Donald Griffin*
  e. *Jean-Henri Fabre*

d
Factual Recall

50.35 Proponent of cognitive ethology.

a
Factual Recall

50.36 Studied social behavior of insects on an evolutionary basis.

e
Factual Recall

50.37 Studied mechanistic fixed-action patterns in digger wasps.

Use the following information for Questions 50.38 and 50.39. Sunshine is a female cat. When she comes into heat, she urinates more frequently and in a large number of places around her house. Several male cats from around the neighborhood congregate at the door of the house where Sunshine lives. They fight with each other and call to Sunshine through the door.

c
Application

50.38 Which of the following is a proximate cause of Sunshine's behavior of increased urination?
  a. It announces to the males that she is in heat.
  b. Female cats that did this in the past attracted more males.
  c. It is a result of hormonal changes associated with her reproductive cycle.
  d. She saw the neighbor cat, Tulip, doing it, and it worked for her.
  e. In the past, when she did it, more males were attracted.

c
Application

50.39 Which of the following would be an ultimate cause of the male cats' response to Sunshine's behavior?
  a. The males have learned to recognize the specific odor of the urine of a female in heat.
  b. By smelling the odor, various neurons in the males' brains are stimulated.
  c. Male cats respond to the odor because it is a means of locating females in heat.
  d. Male cats' hormones are triggered by the odor released by the female.
  e. The odor serves as a releaser for the instinctive behavior of the males.

d
Conceptual
Understanding

50.40 *Mary had a little lamb*
    Its fleece was white as snow.
And everywhere that Mary went
    The lamb was sure to go.

Which of the following statements is CORRECT regarding Mary and her lamb?
  a. Mary raised the lamb from shortly after birth.
  b. The lamb had little contact with its mother.
  c. The lamb will eventually outgrow this attachment.
  d. Both a and b are correct.
  e. a, b, and c are correct.

a
Conceptual
Understanding

50.41 You are watching some ducks. You see a male wood duck apparently courting a female mallard. Which of the following statements is most likely to be CORRECT concerning this behavior?
  a. The male wood duck was reared by a mallard female.
  b. The female mallard was reared by a female wood duck.
  c. The male will be unsuccessful and will eventually learn that he should court a female of his own species.
  d. The female will eventually give an appropriate response to the male, and the two will mate.
  e. Although the two will mate and produce offspring, the offspring will be so confused that they will never mate.

e
Conceptual
Understanding

50.42 Your friend Jim comes to you with a problem: His dog barks too much. He tells you that it is getting worse and the only way he can get his dog to stop barking is to give it a treat. Which of the following might you use to explain what has happened to your friend?
  a. The dog is displaying an instinctive fixed-action pattern of barking that is triggered by a specific releaser. Somehow, Jim is doing something that serves as the releaser.
  b. The dog is performing a social behavior and is considering Jim a potential rival dog. Jim needs to roll on his back as an appeasement display.
  c. The dog is trying to protect Jim from something that it perceives as being harmful. Jim needs to identify what it is and get rid of it.
  d. The dog has been classically conditioned, in that some inappropriate stimulus has become associated with barking.
  e. The dog's behavior is a result of operant conditioning. Every time the dog barks, Jim rewards it with a treat. Thus, the dog will bark more to get more treats.

d
Conceptual
Understanding

50.43 Which of the following is NOT associated with a species in which the female makes a greater contribution to rearing offspring than the male?
  a. dominance hierarchies in males
  b. polygyny
  c. short courtship
  d. females having multiple male partners
  e. agonistic behavior

d
Conceptual
Understanding

50.44 Which of the following would you classify as habituation?
  a. You enter a room and hear the motor of a fan. After a period of time, you are no longer aware of the sound of the fan motor.
  b. You are driving your car and you hear a horn blow. You step on the brakes, but see that it is someone on a side street. So, you resume your previous speed.
  c. You are sitting in a room and suddenly hear a steady beep—beep—beep. You look out the window and see a garbage truck backing up. A week later, you hear the same beep—beep—beep but don't bother to look out the window.
  d. Both a and c are correct.
  e. a, b, and c are correct.

d
Conceptual
Understanding

50.45 Shadow is a dog who spends most of his time sleeping in various locations around the house. When a bag of potato chips is moved in the kitchen, Shadow comes running to the person with the potato chips. He stops and sits perfectly still (usually salivating profusely), looking at the person with the potato chips. This usually results in Shadow getting a potato chip. Which of the following types of behavior are involved here?
  a. fixed-action patterns
  b. classical conditioning
  c. operant conditioning
  d. Both b and c are correct.
  e. a, b, and c are correct.

b
Conceptual
Understanding

50.46 Some male mosquitoes locate females by the sound of their wings during flight (the characteristic buzzing of a mosquito). A cage with male mosquitoes in it has a small earphone placed on top, through which the sound of a female mosquito is played. All the males immediately fly to the earphone and thrust their abdomens through the fabric of the cage. Which of the following best describes this?
  a. The males learn to associate the sound with a female and are thus attracted to it.
  b. Copulation is a fixed-action pattern and the female flight sound is its releaser.
  c. The sound from the earphone irritated the male mosquitoes, causing them to attempt to sting it.
  d. The reproductive drive is so strong that when males are deprived of females, they will attempt to mate with anything that has even the slightest female characteristic.
  e. Through classical conditioning, the male mosquitoes have associated the inappropriate stimulus from the earphone with the normal response of copulation.

a
Application

50.47 You look around the room during a test. You see that many students are rapidly tapping their feet, drumming on their desks with their pencils, frequently changing position in their chairs, playing with their hair, and chewing on various items. These activities can best be described as
a. kineses.
b. taxes.
c. fixed-action patterns.
d. Both a and b are correct.
e. a, b, and c are correct.